Catalysis of Organic Reactions

CHEMICAL INDUSTRIES

A Series of Reference Books and Textbooks

Consulting Editor
HEINZ HEINEMANN
Heinz Heinemann, Inc.,
Berkeley, California

Volume 1: Fluid Catalytic Cracking with Zeolite Catalysts, *Paul B. Venuto and E. Thomas Habib, Jr.*

Volume 2: Ethylene: Keystone to the Petrochemical Industry, *Ludwig Kniel, Olaf Winter, and Karl Stork*

Volume 3: The Chemistry and Technology of Petroleum, *James G. Speight*

Volume 4: The Desulfurization of Heavy Oils and Residua, *James G. Speight*

Volume 5: Catalysis of Organic Reactions, *edited by William R. Moser*

Volume 6: Acetylene-Based Chemicals from Coal and Other Natural Resources, *Robert J. Tedeschi*

Volume 7: Chemically Resistant Masonry, *Walter Lee Sheppard, Jr.*

Volume 8: Compressors and Expanders: Selection and Application for the Process Industry, *Heinz P. Bloch, Joseph A. Cameron, Frank M. Danowski, Jr., Ralph James, Jr., Judson S. Swearingen, and Marilyn E. Weightman*

Volume 9: Metering Pumps: Selection and Application, *James P. Poynton*

Volume 10: Hydrocarbons from Methanol, *Clarence D. Chang*

Volume 11: Foam Flotation: Theory and Applications, *Ann N. Clarke and David J. Wilson*

Volume 12: The Chemistry and Technology of Coal, *James G. Speight*

Volume 13: Pneumatic and Hydraulic Conveying of Solids, *O. A. Williams*

Volume 14: Catalyst Manufacture: Laboratory and Commercial Preparations, *Alvin B. Stiles*

Volume 15: Characterization of Heterogeneous Catalysts, *edited by Francis Delannay*

Volume 16: BASIC Programs for Chemical Engineering Design, *James H. Weber*

Volume 17: Catalyst Poisoning, *L. Louis Hegedus and Robert W. McCabe*

Volume 18: Catalysis of Organic Reactions, *edited by John R. Kosak*

Volume 19: Adsorption Technology: A Step-by-Step Approach to Process Evaluation and Application, *edited by Frank L. Slejko*

Volume 20: Deactivation and Poisoning of Catalysts, *edited by Jacques Oudar and Henry Wise*

Volume 21: Catalysis and Surface Science: Developments in Chemicals from Methanol, Hydrotreating of Hydrocarbons, Catalyst Preparation, Monomers and Polymers, Photocatalysis and Photovoltaics, *edited by Heinz Heinemann and Gabor A. Somorjai*

Volume 22: Catalysis of Organic Reactions, *edited by Robert L. Augustine*

Volume 23: Modern Control Techniques for the Processing Industries, *T. H. Tsai, J. W. Lane, and C. S. Lin*

Volume 24: Temperature-Programmed Reduction for Solid Materials Characterization, *Alan Jones and Brian McNicol*

Volume 25: Catalytic Cracking: Catalysts, Chemistry, and Kinetics, *Bohdan W. Wojciechowski and Avelino Corma*

Volume 26: Chemical Reaction and Reactor Engineering, *edited by J. J. Carberry and A. Varma*

Volume 27: Filtration: Principles and Practices, second edition, *edited by Michael J. Matteson and Clyde Orr*

Volume 28: Corrosion Mechanisms, *edited by Florian Mansfeld*

Volume 29: Catalysis and Surface Properties of Liquid Metals and Alloys, *Yoshisada Ogino*

Volume 30: Catalyst Deactivation, *edited by Eugene E. Petersen and Alexis T. Bell*

Volume 31: Hydrogen Effects in Catalysis: Fundamentals and Practical Applications, *edited by Zoltán Paál and P. G. Menon*

Volume 32: Flow Management for Engineers and Scientists, *Nicholas P. Cheremisinoff and Paul N. Cheremisinoff*

Volume 33: Catalysis of Organic Reactions, *edited by Paul N. Rylander, Harold Greenfield, and Robert L. Augustine*

Volume 34: Powder and Bulk Solids Handling Processes: Instrumentation and Control, *Koichi Iinoya, Hiroaki Masuda, and Kinnosuke Watanabe*

Volume 35: Reverse Osmosis Technology: Applications for High-Purity-Water Production, *edited by Bipin S. Parekh*

Volume 36: Shape Selective Catalysis in Industrial Applications, *N. Y. Chen William E. Garwood, and Frank G. Dwyer*

Volume 37: Alpha Olefins Applications Handbook, *edited by George R. Lappin and Joseph L. Sauer*

Volume 38: Process Modeling and Control in Chemical Industries, *edited by Kaddour Najim*

Volume 39: Clathrate Hydrates of Natural Gases, *E. Dendy Sloan, Jr.*

Volume 40: Catalysis of Organic Reactions, *edited by Dale W. Blackburn*

Additional Volumes in Preparation

Fuel Science and Technology Handbook, *edited by James G. Speight*

Oxygen in Catalysis, *A. Bielanski and Jerzy Haber*

Industrial Drying Equipment: Selection and Application, *Kees van't Land*

Catalysis of Organic Reactions

edited by

Dale W. Blackburn
The Organic Reactions Catalysis Society
Moorestown, New Jersey

CRC Press is an imprint of the
Taylor & Francis Group, an **informa** business

First published 1990 by Marcel Dekker, Inc.

Published 2019 by CRC Press
Taylor & Francis Group
6000 Broken Sound Parkway NW, Suite 300
Boca Raton, FL 33487-2742

First issued in paperback 2019

ISBN 13: 978-0-367-45091-5 (pbk)
ISBN 13: 978-0-8247-8286-3 (hbk)

Visit the Taylor & Francis Web site at
http://www.taylorandfrancis.com

and the CRC Press Web site at
http://www.crcpress.com

Library of Congress Cataloging-in-Publication Data

Catalysis of organic reactions / [edited by] Dale W. Blackburn
p. cm. -- (Chemical industries ; v.)
"Papers presented at the Twelfth Conference on the Catalysis of Organic Reactions in San Antonio, Texas, in April, 1988 ... sponsored by the Organic Reactions Catalysis Society"--Pref.
Includes bibliographical references.
ISBN 0-8247-8286-0 (alk. paper)
1. Chemistry, Organic--Synthesis--Congresses. 2. Catalysis -Congresses. I. Blackburn, Dale W. II. Conference on the Catalysis of Organic Reactions (12th : 1988 : San Antonio, Tex.) III. Organic Reactions Catalysis Society. IV. Series.
QD262.C355 1989
547.1'395--dc20 89-23722
CIP

Preface

This volume is a collection of the papers presented at the Twelfth Conference on the Catalysis of Organic Reactions in San Antonio, Texas, in April, 1988. The conference was sponsored by the Organic Reactions Catalysis Society (ORCS), an affiliate of the Catalysis Society of North America.

The ORCS had its first meeting under the auspices of the New York Academy of Sciences in 1966 as The Conference on Catalytic Hydrogenation and Analogous Pressure Reactions to foster the communication of the practical aspects of the application of catalysis to organic reactions, in particular the reactions employing or generating pressure as one of the conditions. The name of the society was changed to The Organic Reactions Catalysis Society with the fifth conference held in 1975 when we became affiliated with the Catalysis Society of North America. Through the years the catalysis or pressure laboratory scientist and organic chemist from the pharmaceutical and chemical industries have been brought together with academia and the representative from the catalyst manufacture to discuss the technology involved. Thus a forum was provided for the presentation of papers and discussion of the art and techniques involved. All phases of catalysis, including heterogeneous catalytic hydrogenation, catalytic oxidation, homogeneous catalysis, carbonylation, and amination have been discussed. Also, papers have been presented on the design, operation, and safety of facilities and equipment. The diversity of papers presented reflects the large and changing scope of the meetings.

In this conference a special session on chiral catalysis was arranged by Patrick Burk to present a review and recent progress in enantioselective catalytic syntheses. The first paper on enantioselective catalysis with transition metal compounds by Henri Brunner was very well received. Later, a session on selected special topics included the application of the acid catalyst strontium hydrogen phosphate to the selective catalytic synthesis of mixed alkyl amines and polyfunctional amines. Also presented in the session was the application of small-pore zeolites to the increased selective amination of alcohols and olefins.

The financial support of this conference from the following corporations is gratefully acknowledged: Air Products and Chemicals, Allied Signal, Degussa Corporation, E. I. du Pont de Nemours, Eastman Kodak Company, Engelhard Corporation, Ethyl Corporation, Eli Lilly and Company, Harshaw/Filtrol, Hercules Incorporated, Johnson Matthey, Parr Instrument Company, G. D. Searle & Co., Uniroyal Chemical Company, and Warner Lambert/Parke-Davis.

I wish to express my appreciation to the speakers and session leaders for their participation in the program. Their early acceptance contributed to the success of the conference. I also want to express my appreciation to the Executive Committee, in particular John Kosak, Paul Rylander, Bob Augustine, Tom Johnson, and Russ Malz, for their help in organizing this conference. And finally I wish to acknowledge the help of Dennis Morrell in editing the papers in his session.

DALE W. BLACKBURN

Contents

Contributors

Juan G. Andrade Degūssa Corporation, Ridgefield Park, New Jersey

Robert L. Augustine Professor and Head, Department of Chemistry, Seton Hall University, South Orange, New Jersey

Mihály Bartók Professor, Department of Organic Chemistry, Attila József University, Szeged, Hungary

Mark D. Bednarski Assistant Professor, Department of Chemistry, University of California at Berkeley, Berkeley, California

Henri Brunner Professor Dr., Institut für Anorganische Chemie, Universität Regensburg, Regensburg, Federal Republic of Germany

Dennis E. Butterfield Research Scientist, Chemical Engineering Research Laboratory, Eastman Kodak Company, Rochester, New York

M. Deeba Specialty Chemical Division, Engelhard Corporation, Edison, New Jersey

Klaus M. Deller Manager, Inorganic Chemical Products Division, Degūssa AG, Hanau, Federal Republic of Germany

Joseph K. Doles Supervising Technician, Organic Process Division, Eastman Kodak Company, Rochester, New York

Michael E. Ford Lead Research Chemist, Specialty Chemicals Division, Air Products and Chemicals, Inc., Allentown, Pennsylvania

Gerald L. Goe Director, Research Department, Reilly Industries, Inc., Indianapolis, Indiana

Richard F. Heck Professor, Department of Chemistry and Biochemistry, University of Delaware, Newark, Delaware

John E. Hengeveld Senior Research and Development Scientist, Process Development, Abbott Laboratories, Abbott Park, Illinois

*Michael J. Hennessy** Departmental Fellow, Department of Chemistry and Biochemistry, University of Delaware, Newark, Delaware

R. Scott Hoerrner† Research Assistant/Graduate Student, Department of Chemistry and Biochemistry, University of Delaware, Newark, Delaware

Brian R. James Professor, Department of Chemistry, University of British Columbia, Vancouver, British Columbia, Canada

Thomas A. Johnson Principal Research Associate, Specialty Chemicals Department, Air Products and Chemicals, Inc., Allentown, Pennsylvania

Ajey M. Joshi Ph.D. Student, Department of Chemistry, University of British Columbia, Vancouver, British Columbia, Canada

Makarand G. Joshi Senior Research Scientist, Chemical Engineering Research Laboratory, Eastman Kodak Company, Rochester, New York

Nicholas K. Kildahl Associate Professor, Department of Chemistry, Worcester Polytechnic Institute, Worcester, Massachusetts

John F. Knifton Senior Research Associate, Texaco Chemical Company, Austin, Texas

Harold H. Kung Professor, Department of Chemical Engineering, Northwestern University, Evanston, Illinois

Mayfair C. Kung Research Associate, Department of Chemical Engineering, Northwestern University, Evanston, Illinois

Pál Kvintovics‡ Assistant Professor, Department of Chemistry, University of British Columbia, Vancouver, British Columbia, Canada

---

**Current affiliation*: Development Chemist, Chemicals and Pigments Department, E. I. du Pont de Nemours & Co., Inc., Deepwater, New Jersey.

†*Current affiliation*: Postdoctoral Research Fellow, Department of Chemistry, Northwestern University, Evanston, Illinois.

‡*Current affiliation*: Research Group for Petrochemistry of the Hungarian Academy of Sciences, University of Chemical Engineering, Veszprém, Hungary.

J. M. Lambert, Jr. Manager, Advanced Systems Technology, Division of Autoclave Engineers, Inc., Chemical Data Systems, Erie, Pennsylvania

Charles R. Marston Senior Research Chemist, Applications Group, Reilly Industries, Inc., Indianapolis, Indiana

*Robert M. Morris** Associate Professor, Department of Chemistry, University of British Columbia, Vancouver, British Columbia, Canada

William R. Moser Professor, Department of Chemical Engineering, Worcester Polytechnic Institute, Worcester, Massachusetts

Victor L. Mylroie Director of High Pressure Laboratory, Chemical Development Division, Eastman Kodak Company, Rochester, New York

Ulrich Nagel Professor Dr., Institut für Anorganische Chemie, University of Munich, Munich, Federal Republic of Germany

Kimmai Thi Nguyen Graduate Student, Chemical Engineering Department, Northwestern University, Evanston, Illinois

Ferenc Notheisz Associate Professor, Department of Organic Chemistry, Attila József University, Szeged, Hungary

Patrick J. O'Hagan Student, Department of Chemistry, Seton Hall University, South Orange, New Jersey

Daniel J. Ostgard Ph.D. Student, Department of Chemistry and Biochemistry, Southern Illinois University, Carbondale, Illinois

István Pálinkó Assistant Professor, Department of Organic Chemistry Attila József University, Szeged, Hungary

Walt Partenheimer Senior Research Associate, Research and Development, Amoco Chemical Company, Naperville, Illinois

Kenneth M. Partyka Research Chemist, High Pressure Laboratory, Abbott Laboratories, Abbott Park, Illinois

William E. Pascoe Senior Research Scientist, Chemical Engineering Research Laboratory, Eastman Kodak Company, Rochester, New York

Deepak Patel Graduate Student, Chemical Engineering Department, Northwestern University, Evanston, Illinois

Guenter Prescher Degussa Corporation, Ridgefield Park, New Jersey

Krishna Raman† University of Delaware, Newark, Delaware

**Current affiliation*: Department of Chemistry, University of Toronto, Toronto, Ontario, Canada.

†*Current affiliation*: Himont, Inc., Wilmington, Delaware.

Arnold L. Rheingold Professor, Department of Chemistry and Biochemistry, University of Delaware, Newark, Delaware

*Robert E. Ruckle, Jr.** University of Delaware, Newark, Delaware

Adolf Schaefer Department of Organic and Biochemical Research, Degussa AG, Hanau, Federal Republic of Germany

Jonathan S. Schuchardt† University of Delaware, Newark, Delaware

Eric F. V. Scriven Section Head, Research and Development, Reilly Industries, Inc., Indianapolis, Indiana

Louis S. Seif Group Leader, High Pressure Laboratory, Abbott Laboratories, Abbott Park, Illinois

Gerard V. Smith Director, Molecular Science Program, and Professor, Department of Chemistry, Southern Illinois University at Carbondale, Carbondale, Illinois

Edward E. Sowers Senior Section Head, Research Department, Reilly Industries, Inc., Indianapolis, Indiana

Douglass F. Taber Associate Professor, Department of Chemistry and Biochemistry, University of Delaware, Newark, Delaware

Weijing Tao Graduate Student, Department of Chemistry and Biochemistry, University of Delaware, Newark, Delaware

Ian S. Thorburn Student, Univeristy of British Columbia, Vancouver, British Columbia, Canada

Andrew W. Wang Department of Chemical Engineering, Worcester Polytechnic Institute, Worcester, Massachusetts

Guangzhong Wu Ph.D. Student, Department of Chemistry and Biochemistry, University of Delaware, Newark, Delaware

Ágnes G. Zsigmond Assistant Professor, Department of Organic Chemistry, Attila József University, Szeged, Hungary

---

**Current affiliation*: Department of Chemistry, The Pennsylvania State University, State College, Pennsylvania

†*Current affiliation*: ARCO Chemical Co., Newton Square, Pennsylvania

Catalysis of Organic Reactions

Part I

Chiral Catalysis

1

Enantioselective Catalysis with Transition Metal Compounds

HENRI BRUNNER

Institut für Anorganische Chemie, Universität Regensburg, Regensburg, Federal Republic of Germany

ABSTRACT

Prochiral olefins, imines, and ketones can be reduced enantioselectively using optically active transition metal catalysts. With this concept, a small amount of an optically active catalyst is sufficient to produce a large amount of an optically active product, corresponding to a multiplication of the chiral information contained in the catalyst. The different reduction variants, hydrogenation with gaseous H_2, transfer hydrogenation, hydrosilylation, etc., are compared. The scope of the substrates as well as the enantioselective catalysts is discussed. An example is the enantioselective hydrosilylation of acetophenone with diphenylsilane which gives α-phenylethanol in close to 100% ee after hydrolysis. The development of new enantioselective catalysts and the extension of the concept of enantioselective catalysis to new organic reactions will be presented.

Optically active compounds are of increasing importance in industry and science. Therefore, efficient and economic chiral syntheses are required. The most elegant approach is enantioselective catalysis. The application of an enantioselective catalyst results in a multiplication of its chiral information because it

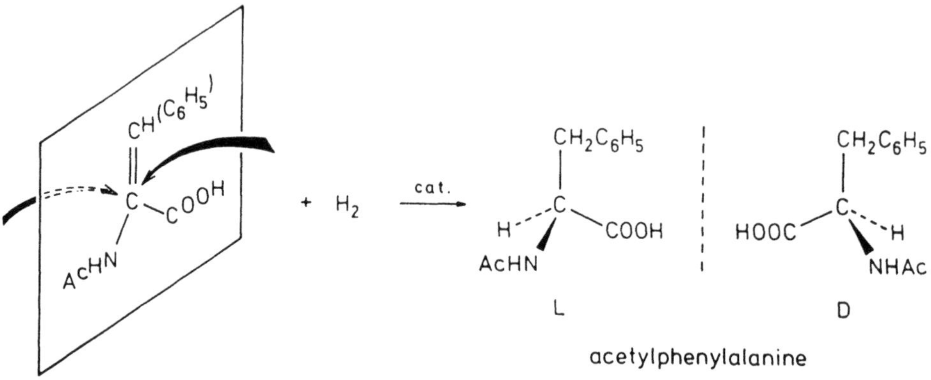

Scheme 1

enters anew in each catalytic cycle, so that with a small amount of an enantioselective catalyst, large amounts of an optically active product can be obtained. The most well-known enantiospecific catalysts are the enzymes. A relatively new approach is the use of optically active transition metal compounds, some aspects of which will be dealt with in the present chapter.

Outstanding success in enantioselective catalysis with transition metal compounds has been achieved in the hydrogenation of dehydroamino acids to give optically active amino acids. An example is the conversion of (Z)-α-acetamidocinnamic acid into *N*-acetylphenylalanine, shown in Scheme 1, which is frequently used as a standard reaction in asymmetric hydrogenation. The enantioselective catalysts control the formation of the natural L and the unnatural D isomer by differentiating the addition of the hydrogen atom to the prochiral carbon atom according to the arrows in Scheme 1.

Rhodium complexes of the type of the Wilkinson catalyst, containing optically active phosphines, are the usual catalysts in such hydrogenation reactions. One of the many optically active phosphines that have been reported is the celebrated Diop[1] derived from tartaric acid (Scheme 2).

Another example is Norphos (Scheme 2), the phosphine oxide of which is resolved according to Scheme 3 using (*R*,*R*)-dibenzoyltartaric acid, abbreviated (-)-DBT.[2,3] A racemic mixture of NorphosO treated

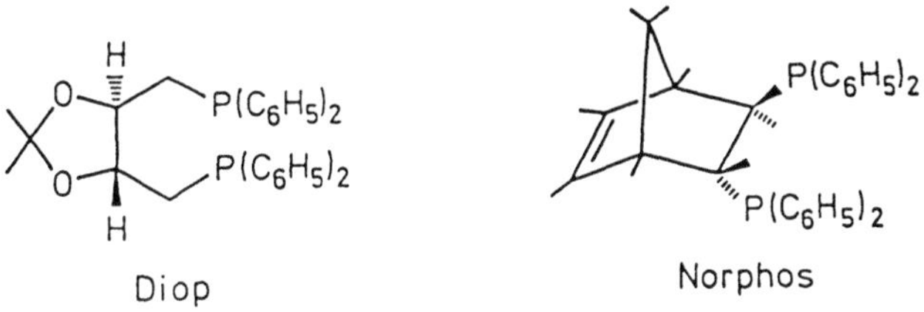
H
O
O
H
$P(C_6H_5)_2$
$P(C_6H_5)_2$
Diop
$P(C_6H_5)_2$
$P(C_6H_5)_2$
Norphos

Scheme 2

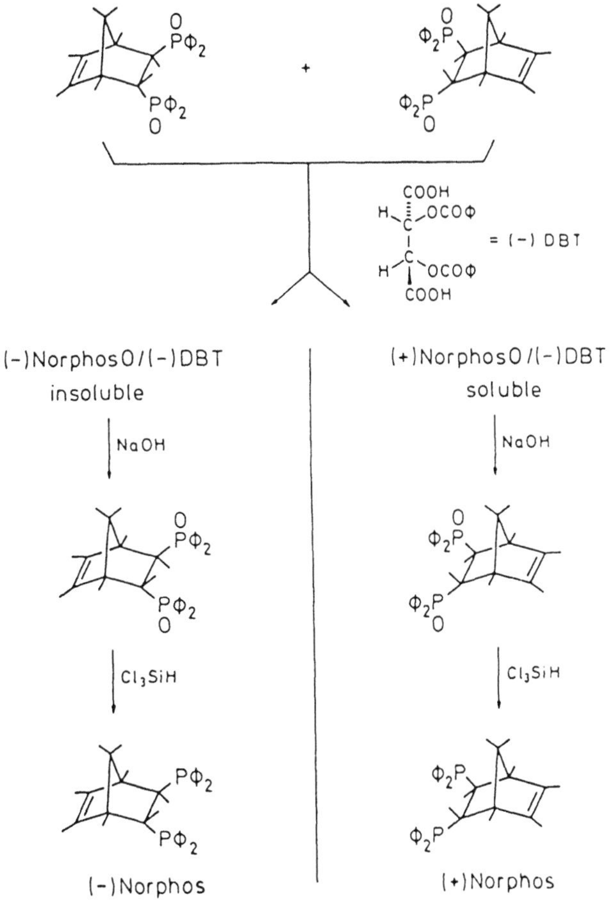
O
PΦ2
PΦ2
O
+
COOH
H
OCOΦ
C
C
H
OCOΦ
COOH
= (-) DBT
(-)NorphosO/(-)DBT
insoluble
(+)NorphosO/(-)DBT
soluble
NaOH
NaOH
Cl3SiH
Cl3SiH
(-)Norphos
(+)Norphos

Scheme 3

with (-)-DBT precipitates (-)-NorphosO/(-)-DBT from chloroform, whereas (+)-NorphosO/(-)-DBT stays in solution. After recrystallization the less soluble diastereomer (-)-NorphosO/(-)-DBT is optically pure. The optically active auxiliary (-)-DBT can be removed with base and subsequently NorphosO is reduced with trichlorosilane to give the optically pure Norphos (Scheme 3).

There are some variants concerning this resolution. (+)-NorphosO can be obtained in a similar way as (-)-NorphosO described above by using (+)-DBT. Also, with half of the equivalent amount of (-)-DBT first (-)-NorphosO/(-)-DBT and then with half of the equivalent amount of (+)-DBT (+)-NorphosO/(+)-DBT can be precipitated. The resolution of phosphine oxides with DBT has successfully been applied to the oxides of Binap[4] and DPCB.[5] The two ligands Diop and Norphos shown in Scheme 2 have been used in the enantioselective hydrogenation of Scheme 1. The in situ catalysts $[Rh(cod)Cl]_2$/Diop (cod = 1.5-cyclooctadiene) and $[Rh(cod)Cl]_2$/Norphos gave 91 and 97% ee, respectively.

Most hydrogenation reactions are carried out with gaseous hydrogen. In transfer hydrogenations, however, the hydrogen is derived from a hydrogen donor, e.g., isopropanol. These transfer hydrogenations play a role in the enantioselective reduction of prochiral ketones. Also formic acid is a frequently used hydrogen donor which on hydrogen transfer liberates CO_2. In a new approach, the commercially available azeotrope formic acid/triethylamine 5:2 was used for a transfer hydrogenation of dehydroamino acids, e.g., (Z)-α-acetamidocinnamic acid, shown in Scheme 4.

50% ee was obtained in the solvent DMSO using the in situ catalyst $[Rh(cod)Cl]_2$/Diop.[6] With the same ee, itaconic acid could be

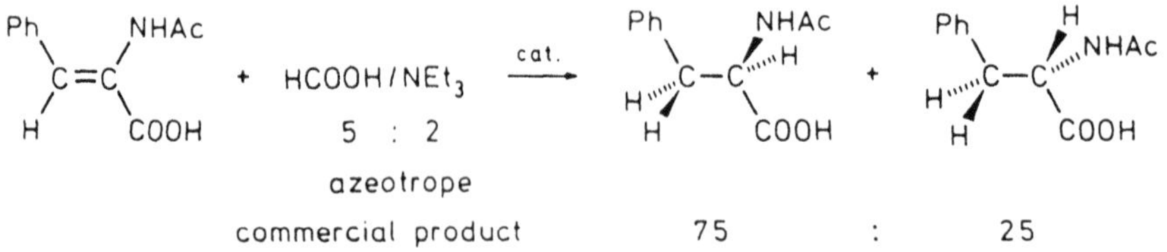

Ph(H)C=C(NHAc)COOH + $HCOOH/NEt_3$ (5 : 2, azeotrope, commercial product) —cat.→ 75 : 25 (two enantiomers of $PhCH_2$–CH(NHAc)COOH)

Scheme 4

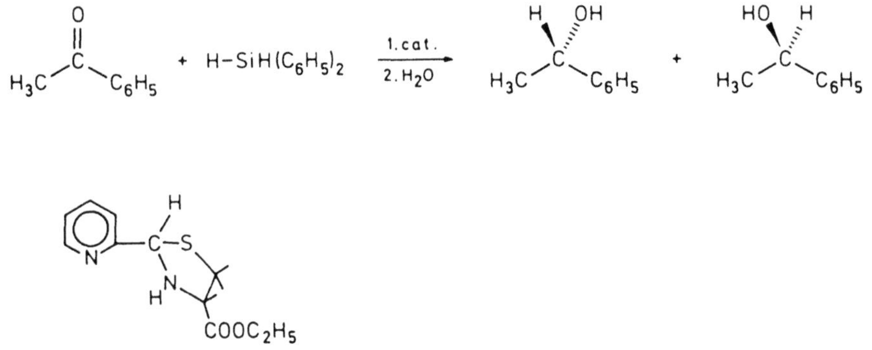

$$H_3C{-}C({=}O){-}C_6H_5 + H{-}SiH(C_6H_5)_2 \xrightarrow[2.\ H_2O]{1.\ cat.} H_3C{-}CH(OH){-}C_6H_5 + H_3C{-}CH(OH){-}C_6H_5$$

Scheme 5

reduced to methylsuccinic acid. The new methodology is a room temperature procedure which avoids the risks of gaseous hydrogen. It represents an improvement of the transfer hydrogenation with boiling 80% aqueous formic acid developed previously.[7]

The reduction of prochiral ketones to optically active secondary alcohols is a reaction of fundamental importance. In addition to enzymatic procedures, there are chemical reduction systems applied in stoichiometric amounts. The catalytic hydrogenation of ketones using rhodium/phosphine catalysts is reluctant, alkylated phosphine ligands being superior to arylated phosphine ligands. Another approach to the enantioselective reduction of prochiral ketones is hydrosilylation followed by hydrolysis. So, as shown in Scheme 5, acetophenone can be catalytically hydrosilylated with diphenylsilane. After hydrolysis of the silylether α-phenylethanol is the product.

In situ catalysts composed of $[Rh(cod)Cl]_2$ and optically active phosphines turned out to give poor enantioselectivities. However, the advent of optically active nitrogen ligands distinctly improved the situation. $[Rh(cod)Cl]_2$ together with Pythia, the pyridinethiazolidine shown in Scheme 5, achieved optical inductions close to 100%.[8] In a recent study, 58 prochiral ketones were reduced enantioselectively with the system $[Rh(cod)Cl]_2$/Pythia, most of them giving high enantiomeric excess.[9] It should be added that nitrogen ligands of the Pythia type can be prepared in one step from readily available

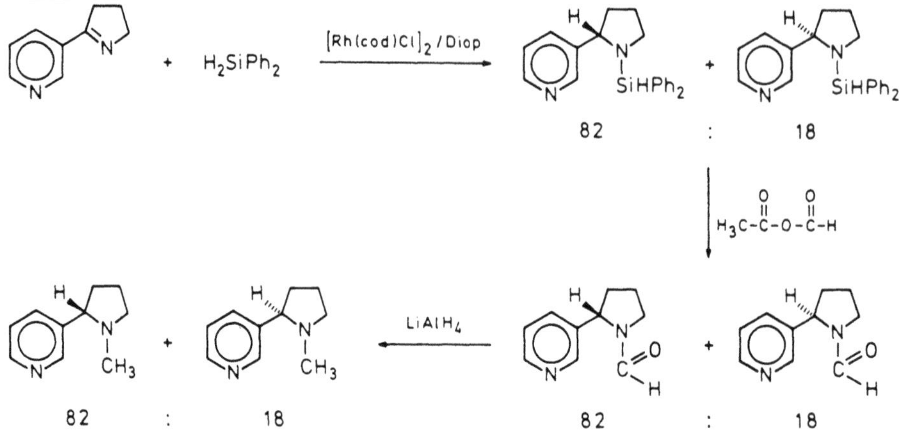

Scheme 6

starting materials, whereas lengthy multistep syntheses are necessary for the optically active phosphines.

Similar to prochiral ketones, prochiral imines can be reduced by catalytic hydrosilylation and subsequent hydrolysis or another derivatization. In Scheme 6 a short enantioselective nicotine synthesis is shown based on the hydrosilylation of a pyridine-substituted cyclic imine, which occurs with 64% ee using the in situ catalyst $[Rh(cod)Cl]_2$/Diop.[10] Derivatization of the hydrosilylation product with acetic formic anhydride yields the *N*-formyl derivative (Scheme 6), which can be reduced with $LiAlH_4$ to give nicotine without changing the enantioselectivity achieved in the hydrosilylation step.

Whereas the reactions discussed up to now belong to reaction types for which enantioselective catalysis is well established, the following examples introduce new reaction types to enantioselective catalysis.

The hydrosilylation of oximes with diphenylsilane can be catalyzed by rhodium complexes (Scheme 7). Three moles of silane is

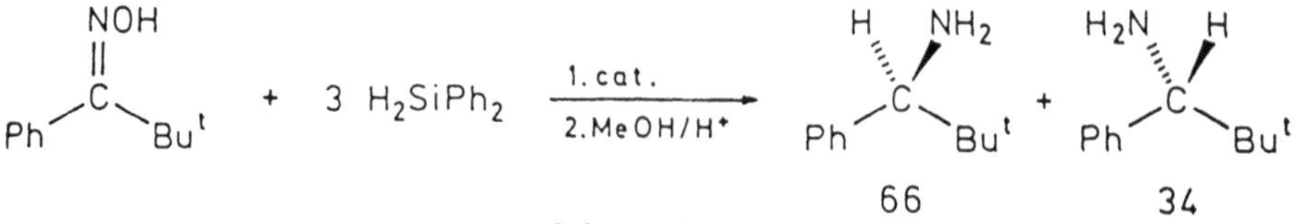

Scheme 7

consumed. Two moles ends up in the disiloxane $Ph_2HSi\text{-}O\text{-}SiHPh_2$ containing the oxime oxygen atom, and the third in the silylamine, which is subsequently hydrolyzed to the corresponding primary amine.[11] With the in situ catalyst $[Rh(cod)Cl]_2$/Diop, 32% ee is achieved in the hydrosilylation of *tert*-butylphenylketoxime with diphenylsilane (Scheme 7).

The oximes of 12 other prochiral ketones give similar optical inductions.[11] The reaction allows the transformation of a prochiral ketone to the corresponding optically active primary amine by enantioselective catalytic hydrosilylation of its readily accessible oxime.

The Michael addition of 1,3-dicarbonyl compounds to α,β-unsaturated ketones can be catalyzed by transition metal compounds such as $Ni(acac)_2$. Due to the mild reaction conditions such catalyses may give better yields than the usual base catalysis. Scheme 8 shows an example, the reaction of methyl indanonecarboxylate and methylvinylketone, in which an asymmetric quaternary carbon atom is formed. An in situ catalyst consisting of the components $Co(acac)_2$ and 1,2-diaminodiphenylethane gave the Michael adduct of Scheme 8 in 66% ee, provided the reaction is carried out at -50°C.[12]

Meso-diols such as *cis*-cyclopentanediol in Scheme 9 can be monophenylated with $Ph_3Bi(OAc)_2$. In this reaction two new asymmetric centers are formed. Catalyses with copper acetate and ligands of the pyridineoxazoline type (Scheme 9) provide enantioselection, in the case of *cis*-cyclopentanediol 50% ee.[13] The reaction is applicable to a variety of *meso*-diols; also, kinetic resolutions of racemic diols are possible.[14]

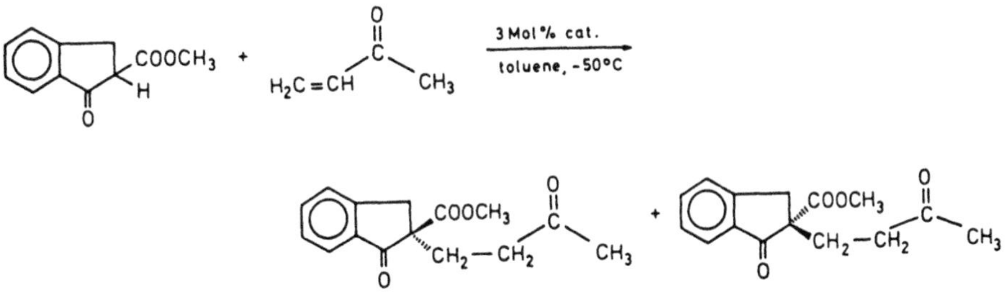

COOCH₃ H O + H₂C=CH C O CH₃ → 3 Mol % cat. / toluene, -50°C

COOCH₃ CH₂—CH₂ C O CH₃ O + COOCH₃ CH₂—CH₂ C O CH₃ O

Scheme 8

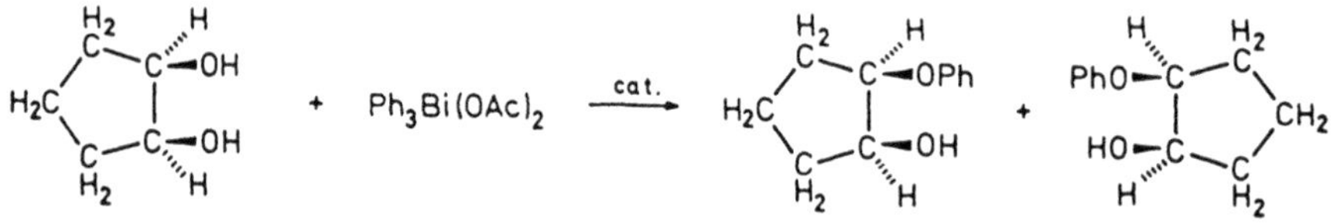

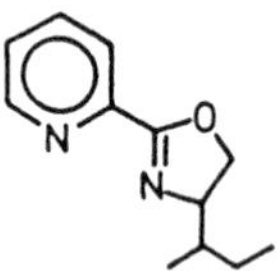

Scheme 9

REFERENCES

1. H. B. Kagan and T. P. Dang, *J. Am. Chem. Soc., 94,* 6429 (1976).
2. H. Brunner and W. Pieronczyk, *Angew. Chem. Int. Ed. Engl., 18,* 620 (1979).
3. H. Brunner, W. Pieronczyk, B. Schönhammer, K. Streng, I. Bernal, and J. Korp, *Chem. Ber., 114,* 1137 (1981).
4. H. Takaya, K. Mashima, K. Koyano, M. Yagi, H. Kumobayashi, T. Taketomi, S. Akutagawa, and R. Noyori, *J. Org. Chem., 51,* 629 (1986).
5. T. Minami, Y. Okada, R. Nomura, S. Hirota, Y. Nagahara, and K. Fukuyama, *Chem. Lett.,* 613 (1986).
6. H. Brunner and W. Leitner, *Angew, Chem. Int. Ed. Engl., 27,* 1180 (1988).
7. H. Brunner and M. Kunz, *Chem. Ber., 119,* 2868 (1986).
8. H. Brunner, R. Becker, and G. Riepl, *Organometallics, 3,* 1354 (1984).
9. H. Brunner and A. Kürzinger, *J. Organomet. Chem., 346,* 413 (1988).
10. H. Brunner, A. Kürzinger, S. Mahboobi, and W. Wiegrebe, *Arch. Pharm. (Weinheim), 321,* 73 (1988).
11. H. Brunner, R. Becker, and S. Gauder, *Organometallics, 5,* 739 (1986).
12. H. Brunner and B. Hammer, *Angew. Chem. Int. Ed. Engl., 23,* 312 (1984).
13. H. Brunner, U. Obermann, and P. Wimmer, *J. Organomet. Chem., 316,* C1 (1986).
14. H. Brunner, U. Obermann, and P. Wimmer, *Organometallics, 8,* 821 (1989).

2

Catalytic Enantioselective Hydrogenation of Ketones and Imines Using Platinum Metal Complexes

BRIAN R. JAMES, AJEY M. JOSHI, PAL KVINTOVICS, ROBERT M. MORRIS,† and IAN S. THORBURN*

Department of Chemistry, University of British Columbia, Vancouver, British Columbia, Canada

ABSTRACT

A brief overview is presented on systems that effect catalytic enantioselective hydrogenation of ketones and imines using H_2, or transfer from 2-propanol, and platinum metal complexes. Recent developments from my own group are then presented, and these describe Ir and Rh systems with inexpensive chiral ligands based on naturally occurring carboxylic acids (mandelic acid, methionine); up to 80% ee has been achieved for certain alkylarylketone substrates using *N*-acetyl-*S*-methionine (*R,S*)-sulfoxide as ligand. Nonfunctionalized chiral sulfoxides at Rh and Ir centers provide ineffective catalyst systems. Recently reported Ru-chiral phosphine catalysts realize enantioselectivities up to 100% for certain ketones (M. Kitamura et al., *J. Am. Chem. Soc., 110,* 629 (1988)); our work on closely related systems shows that trimeric hydrides are present, and these effect catalytic H_2 hydrogenation of acetophenone, imines, and acetonitrile.

---

*Research Group for Petrochemistry of the Hungarian Academy of Sciences, University of Chemical Engineering, Veszprèm, Hungary
†Department of Chemistry, University of Toronto, Toronto, Ontario, Canada

I. INTRODUCTION

The basic principles of catalytic asymmetric hydrogenation of prochiral organic substrates (exemplified generally by Eq. 1) are now well established[1-4] and, in the case of reduction of certain carbon-carbon double bonds (X = C<), complete enantioselectivity has been attained with a range of platinum metal complexes, typically rhodium[3,5-7] or ruthenium[8-11] species containing chiral ditertiary chelating phosphine ligands; some of these systems are mechanistically understood at a highly sophisticated level.[3,5-7] The success and degree of understanding in hydrogenation of prochiral ketones

$$R^1R^2C{=}X + H_2 \rightarrow R^1R^2CH\text{-}XH \quad (1)$$

(Eq. 1, X = O) to give optically active alcohols have been generally less notable,[2,12-16] although recent communications report enantiomeric excesses (ee values) of up to 100% for a range of functionalized ketones with some Ru-chiral phosphine systems.[17,18] The story with prochiral imines (Eq. 1, X = N-) is generally quite dismal,[2,13,19] but we are aware of a patent and a report in which amines have been generated in >90% ee from selected imines using Rh-chiral phosphine catalysts.[20] There is certainly a direct correlation between the degree of success in catalytic asymmetric hydrogenation of the three types of prochiral substrates and the volume of literature on systems describing catalytic hydrogenation of analogous achiral olefins,[21,22] ketones,[13,22,23] and imines.[13,22-24] One should also note that there is also an indirect method to chiral alcohols from prochiral ketones, and to chiral amines from imines, using a catalytic asymmetric hydrosilylation-hydrolysis procedure; Rh-chiral phosphine catalysts dominate and optical yields of up to ~85% ee have been attained.[1,2,13,25-27] Extremely effective stoichiometric reductions of prochiral ketones and imines have also been realized using chiral borohydride derivatives.[28,29]

The above brief overview includes lead references for an entry into the established but still expanding literature on catalytic asymmetric hydrogenation. It is evident that the vast majority of catalysts have involved platinum metal complexes containing chiral phosphine ligands. The syntheses of such ligands often require lengthy, multistep procedures[2,30] and, if available commercially, the phosphines are expensive, sometimes even more per gram than the corequisite platinum metal precursor reagent. The high cost of the catalysts certainly prejudices against their commercial use, but does not preclude it, as exemplified by the development of the Monsanto L-dopa synthesis.[31] The use of naturally occurring, inexpensive, chiral ligands (without chemical modification, e.g., by incorporation of a phosphine moiety) is attractive but (outside of enzymatic systems!) has met with limited success.[32-34] This present paper will mainly describe our efforts to utilize readily available, cheap, and naturally occurring carboxylates and amino acids, or simple derivatives thereof, as chiral ligands in iridium and rhodium systems for enantioselective hydrogenation of ketones and imines. Also, because of the recent spectacular success using ruthenium/chiral phosphine catalysts,[8-10,17,18] some findings in this area will be mentioned.

II. IRIDIUM CATALYSTS

A. Sulfoxide Systems

About a decade ago, we reported work on the use of *trans*-dichloro-*mer*-tris(*S*-dimethylsulfoxide)hydridoiridium(III), $IrHCl_2(DMSO)_3$, **1**, for the hydrogen-transfer hydrogenation (using 2-propanol as the source of hydrogen) of α,β-unsaturated aldehydes to the unsaturated alcohols[35]:

$$RCH = CHCHO \rightarrow RCH = CHCH\ OH; \qquad R = C_6H_5,\ CH_3 \tag{2}$$

Related catalysts containing tertiary phosphine ligands invariably reduce the olefinic bond in such substrates, and the unusual selectivity for the sulfoxide system may be attributable to a difference in the acidity of the coordinate "hydride" compared to that in the phosphine systems. The mode of addition of a metal hydride across a carbonyl moiety, a key step in hydrogenation of aldehydes or ketones, remains poorly defined; formation of an alkoxide (Eq. 3) or an

$$RR'C{=}O \cdots HM \longrightarrow \left[\begin{array}{c} >C{=}O \\ H{-}{-}M \end{array} \right] \longrightarrow >CHOM \qquad (3)$$

$$RR'C{=}O \cdots MH \longrightarrow \left[\begin{array}{c} >C{=}O \\ M{-}{-}H \end{array} \right] \longrightarrow >C(M){-}OH \qquad (4)$$

α-hydroxyalkyl species (Eq. 4) is feasible, and will depend critically on the polarity of the M-H bond. A demonstrated acidic hydrogen within $IrHCl_2(DMSO)_3$, through deuterium exchange data,[36] suggested addition as shown in Eq. 4.[35] However, confusing the mechanistic picture further is that complex **1** also catalyzes the 2-propanol transfer hydrogenation of α,β-unsaturated ketones to the saturated ketones, i.e., hydrometallization is now favored at the olefinic center, although the saturated ketones could be reduced more slowly to the alcohols.[37-39] Further, **1** exists as several isomers,[40] and it is not clear in the earlier work from Henbest's group[37,38] as to which isomer was being used as catalyst.

Nevertheless, even if poorly understood, this Ir-sulfoxide system (**1**) had been demonstrated to effect catalytic solvent transfer hydrogenation of carbonyl functions, and several analogous chiral sulfoxide-iridium systems were subsequently tested as catalysts. The saturated ketones 2-octanone and 2,3-butanedione were slowly reduced to alcohols but with no measurable ee when *R,R*-PTSE, *R*-MPTSO, or *R*-TBPTSO was used as ligand (Table 1 lists the abbreviations and structures of the various ligands); the bulky substrates benzil and 1,3-diphenyl-2-methylpropen-1-one were not reduced at all. Reduction of 3-methyl-3-penten-2-one, an α,β-unsaturated ketone, prochiral at the olefin gave the saturated ketone, but with zero ee. The lack of chiral induction in products from successfully hydrogenated substrates may be due to a concurrent reduction of the chiral sulfoxide because recovered iridium species (not well characterized) contained thioether (sulfide) ligands. We had experienced such a problem earlier with some chiral sulfoxide complexes of Rh[41] (see below) but not Ru.[42]

Table 1 Abbreviations and Some Ligand Structures

General

COE = cyclooctene; COD = 1,5-cyclooctadiene; NBD = norbornadiene, HD = 1,5-hexadiene; DMA = *N,N'*-dimethylacetamide

Sulfoxides (which can be S- or O-bonded)

DMSO = dimethylsulfoxide; DPSO = diphenylsulfoxide; NPSO = di-*n*-propylsulfoxide; MPSO = methylphenylsulfoxide; MBMSO = *S*-2-methylbutyl(methyl)sulfoxide; TMSO = tetramethylene sulfoxide;

MPTSO = methyl-*p*-tolylsulfoxide, $Me-C_6H_4-S(O)Me$

TBPTSO = *t*-butyl-*p*-tolylsulfoxide, $Me-C_6H_4-S(O)^tBu$

PTSE = 1,2-bis(*p*-tolylsulfinyl)ethane, $[Me-C_6H_4-S(O)CH_2]_2$

MSE = *meso*-1,2-bis(methylsulfinyl)ethane, $[MeS(O)CH_2]_2$

DIOS = 2,3-*O*-isopropylidene-2,3-dihydroxy-1,4-bis(methylsulfinyl)butane

(R = R,S-S(O)Me)

Carboxylic acids

mandelic, $PhCH(OH)CO_2H$; phenylalanine, $PhCH_2CH(NH_2)CO_2H$; methionine, $MeSCH_2CH_2CH(NH_2)CO_2H$

Phosphines

diphos = 1,2-bis(diphenylphosphino)ethane, $PPh_2(CH_2)_2PPh_2$

chiraphos = 2,3-bis(diphenylphosphino)butane, $PPh_2CH(Me)CH(Me)PPh_2$

prophos = 1,2-bis(diphenylphosphino)propane, $PPh_2CH(Me)CH_2PPh_2$

diop = 2,2-dimethyl-4,5-bis(diphenylphosphinomethyl)-1,3-dioxolane (analogous to DIOS, but with R = PPh_2)

binap = 2,2'-bis(diphenylphosphino)-1,1'-dinaphthyl

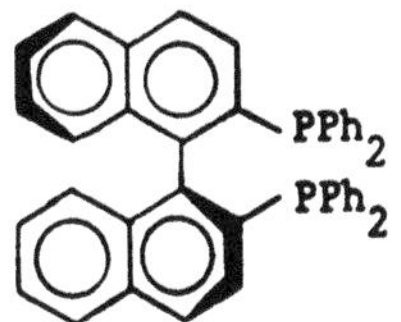

B. Sulfoxide-Carboxylate Systems

The simple route that we had discovered for the synthesis of $IrHCl_2(sulfoxide)_3$ complexes, via oxidative addition of HCl to labile Ir(I) species in the presence of sulfoxide,[40] allowed us to incorporate auxiliary carboxylate ligands, including chiral ones (Eq. 5, X = Cl or carboxylate).[43] A range of carboxylate derivatives was isolated: benzoates (and p-substituted ones) appear from

$$\frac{1}{2}[IrCl(COE)_2]_2 + HX \xrightarrow[CH_2Cl_2]{DMSO} \underset{\mathbf{1} \text{ or } \mathbf{2}}{IrHClX(DMSO)_3} + 2COE \qquad (5)$$

NMR and IR spectroscopic data to be of the configuration shown in **2a**, while systems with the CO_2^- group on an aliphatic carbon (acetate, trifluoroacetate, mandelate, α-methoxy- and α-acetylmandelate, and *N*-acetylphenylalanine) are tentatively considered to have structure **2b**.[43] Unfortunately, the carboxylate species (**2**) under hydrogen

1	**2a**	**2b**	
Ir: H, S, S, S (equatorial); Cl, Cl (axial)	Ir: H, S, O, S (equatorial); O_2CR, Cl (axial)	Ir: RCO_2, S, O, S (equatorial); H, Cl (axial)	S = S-DMSO; O = O-DMSO

transfer conditions (2-propanol, 80°C, under Ar), in contrast to **1**, decompose slowly to metal and no reduction of ketonic substrates is observed. Some preliminary data show that **2b** with R = racemic PhCH(OMe) in ethanol or acetone catalyzes homogeneously the H_2 reduction (80 atm H_2, 50°C) of 1-butanal to 1-butanol; whether effective asymmetric hydrogenation can be accomplished with species such as **2** using H_2 remains to be established.

C. Carboxylate-Phosphite Systems

There is clearly a subtle balance in factors affecting stabilization of Ir(III) hydrides or Ir(I) intermediates in the catalytic cycles (see below) against reduction or disproportionation to metal. A comparison of **1** and **2** shows that an *S*-DMSO (vs. an *O*-DMSO) and a chloride (vs. O_2CR) are beneficial, presumably because of their

Table 2 Enantioselective Hydrogenation of Acetophenone Using (Carboxylato)(phosphite)iridium Transfer Hydrogenation Catalysts[a]

PhCOMe ⟶ PhCH(OH)Me

$P(OMe)_3$:Ir	Conv. (%)	Opt. yield (%)
2.0	10	2.3(*S*)
2.0[b]	75	1.0(*S*)
3.0[c]	15	2.0(*S*)
3.0[c,d]	12	2.0(R)
3.0[c,e]	15	9.0(*S*)
4.0	20	4.0(*S*)
4.0[b]	25	8.0(*S*)
4.0[c,e]	20	12.0(*S*)

[a]Refluxing 2-propanol solutions (10 ml) under Ar, containing 2.5×10^{-3} M Ir added as $[IrCl(COE)_2]_2$, 2.5×10^{-3} M *R*-mandelic acid, and $P(OMe)_3$; reaction time 6 hr.
[b]Added NaOMe : Ir = 5.0.
[c]Added Et_3N : Ir = 5.0.
[d]*S*-Mandelic acid used.
[e]*R*-Acetylmandelic acid used.

"softer" character, a factor known to favor stabilization of hydrides and lower oxidation states.[44] In attempts to stabilize the hydrido(carboxylate)iridium(III) species under hydrogen transfer conditions, the well-known "stabilizers," tertiary phosphines and phosphites, were then used to replace the sulfoxide as ancillary ligands in species akin to **2**.

The procedure of Reaction 5 was used to synthesize or form in situ the required species. For example, use of benzoic acid and triphenylphosphite with the iridium(I) cyclooctene precursor yields the complex $IrHCl(O_2CPh)\{P(OPh)_3\}_3$, **3**, although its geometry (cf. **1** and **2**) is uncertain.[43] This particular isolated complex is completely inactive as a catalyst for hydrogen transfer from 2-propanol to ketones. However, closely related in situ species with a range of carboxylate ligands do effect enantioselective hydrogenation of acetophenone to 1-phenylethanol, but a maximum optical yield of only 12% has been realized so far (Table 2); the catalyst is formed using

$Ir/P(OMe)_3$/*R*-acetylmandelic acid at a ratio of 1:4:1 with some added base. Use of *R*-mandelic acid itself has given a maximum 8% ee of the *S*-1-phenylethanol, and choice of *S*-mandelic acid does lead to the other enantiomer of 1-phenylethanol. Mandelic acid was chosen initially because it was thought that the hydroxy substituent might hydrogen-bond with the substrate carbonyl, a factor that was invoked within a more effective chiral rhodium-hydroxyalkyl(ferrocenyl)phosphine system[45]; however, the replacement of -OH by $-OCCH_3$ in our system improves the ee, and such H bonding is clearly not essential.

The reaction conditions for these (carboxylato)iridium catalysts had been chosen following studies on their use for the stereoselective hydrogenation of *t*-butylcyclohexanone to the *cis*- and *trans*-alcohols[14]; here trialkylphosphites were the most effective ancillary phosphorus-containing ligands for reaction rates [compared to $P(OPh)_3$ and tertiaryphosphines], and the rates were enhanced by added base (Et_3N, NaOMe) and inhibited by added HCl. Conversions of the cyclohexanone were maximized at a phosphite/Ir ratio of ∿4, and the most effective carboxylic acids, including chiral ones, were benzoic, *R*-mandelic, *R*-acetylmandelic, *R*-malic, crotonic and cinnamic acids (less effective were propionic, 1*S*,3*R*,-camphoric, trifluoroacetic, fumaric, and maleic acids). *N*-Benzylidenephenylamine is reduced (Eq. 6, 33% in 6 hr) using conditions similar to those given in Table 2 [with $P(OMe)_3/Ir/PhCO_2H$ = 4:1:1), but conditions have still to be found to give effective conversions of the prochiral substrate $PhC(CH_3){=}NCH_2Ph$.

$$PhCH{=}NPh \longrightarrow PhCH_2NH(Ph) \tag{6}$$

A better understanding of the mechanism of ketone (and imine) hydrogenation (see Sec. II.A) would certainly assist in designing more effective systems for chiral induction. The effect of added acid and base on the rates may indicate loss of HCl in one step of the catalytic cycle (presumably during formation of an isopropoxide species), and Scheme 1, based largely on suggestions in the literature,[13,23,32] outlines plausible pathways for the hydrogen transfer

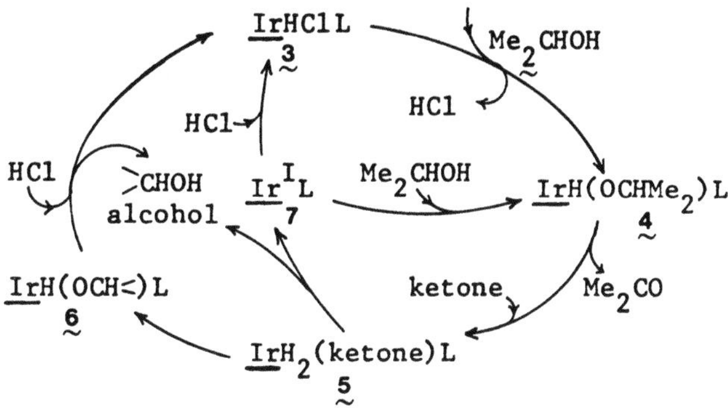

Scheme 1 Plausible catalytic cycles for catalytic hydrogen transfer from 2-propanol to ketones; here Ir stands for Ir{P(OMe)$_3$}*n*, *n* = 2 or 3 and could vary within the cycles, and L = mono- or bidentate carboxylate.

hydrogenation. β-Hydride transfer from the isopropoxide 4, formed from 3, would yield a dihydride 5 containing coordinated ketone (the substrate or acetone, depending on their relative concentrations and binding constants). A single hydride transfer within 5 could yield a hydrido(alkoxide) 6 as shown (analogous to 4) according to Eq. 3, or an α-hydroxyalkyl species (Eq. 4); protonolysis of the Ir-O (or Ir-C) bond then liberates the alcohol product with regeneration of 3. Transfer of both hydrogens within 5 (via 6) would form the alcohol and an Ir(I) species 7 which could oxidatively add HCl to reform 3 or 2-propanol to give 4 directly. Details within the plausible steps of Scheme 1 are essentially nonexistent: e.g., as well as the ambiguity within the 5 → 6 conversion (Eq. 3 or 4), the coordination geometries of the various species are unknown, and end-on ketone coordination (rather than side-on as shown in Eqs. 3 and 4) could be involved. Indeed, a "harder" metal center such as Ir(III) is expected to favor end-on bonding, while the softer Ir(I) should favor the π-bonded, side-on interaction.[46] Fluxionality between mono- and bidentate carboxylate has been shown to be important in dehydrogenation of alcohols catalyzed by Ru(II) carboxylate phosphine complexes[47] and could be critical in these Ir systems. The imine hydrogenation (Eq. 6) could occur via pathways corresponding to those shown in Scheme 1 for the ketones, 6 now being a hydrido(dialkylamido) species $HIrN(Ph)CH_2Ph$. One report (documenting an H_2 hydrogenation) suggests that initial η^2 bonding of the C=N (cf. Eq. 3) is favored in alcohol solvents because of intramolecular hydrogen bonding between coordinated alcohol and the imine nitrogen.[24]

Although the optical inductions (Table 2) are unsatisfactorily low, particularly compared to the spectacular success using chiral phosphine systems (see Sec. I), the iridium systems provide a rare example of producing any chiral induction using "off-the-shelf," naturally occurring ligands (mandelic acids).

It should be noted that these carboxylate/phosphite systems catalyze hydrogen transfer hydrogenation of α,β-unsaturated ketones and aldehydes (e.g., PhCH=CHCOPh, CH_3CH=CHCHO) selectively at the olefinic bond.

III. RHODIUM CATALYSTS

A. Sulfoxide Systems

We have synthesized a large number of rhodium complexes containing S-bonded sulfoxide ligands.[41,48,49] These include $RhCl_3L_3$ (where L = DMSO, MPSO, TMSO, and *R*-MPTSO), $RhCl_3(DMSO)_2L$ (where L = amide, amine oxide, and phosphine oxide), $RhCl_4L_2^-$ (where L = DMSO and NPSO), $[RhClL_2]_2$ (where L = DMSO, DPSO, DIOS), $[RhCl(COE)L]_2$ (where L = DPSO, TMSO), $RhCl_2(DMSO)_2^-$, and $Rh(MSE)_2^+$. Attempts to generate hydrido species via heterolytic cleavage of H_2 by the Rh(III) complexes (Eq. 7), or by oxidative addition of HX (X = H or Cl) to the Rh(I) complexes noted above, or to Rh(I) sulfoxide species (sulfoxide = MSE, DIOS, PTSE) formed in situ (Eq. 8; cf. Eq. 5), have been unsuccessful.[41,50] The Rh(III)/H_2 reactions invariably yield Rh(I) species,

$$\mathrm{Rh(III)Cl} + H_2 \rightarrow \mathrm{Rh(III)H} + H^+ + Cl^- \qquad (7)$$

$$\mathrm{Rh(I)} + HX \rightarrow \mathrm{Rh(III)(H)X} \qquad (8)$$

via presumed Rh(III)H intermediates, i.e., Reaction 7 plus the reverse of Reaction 8; the HX oxidative addition reaction is less favored than for Ir (Eq. 5), a well-established fact for analogous phosphine systems, and attributed to the weaker metal-hydride bond strength (Rh < Ir).[51]

Catalytic hydrogenations of olefinic and ketonic substrates using rhodium sulfoxide systems with H_2 or 2-propanol are limited by decomposition to metal and/or competing catalytic hydrogenation of the sulfoxide to thioether.[39,41] The presence of an ancillary tertiary phosphine ligand, for example, within $Rh(diene)(PPh_3)L^+$ species (diene = COD, NBD; L = O-bonded DMSO, TMSO, NPSO, MBMSO, MPSO, MPTSO, TBPTSO, DIOS),[41,52] proved to be of no benefit for catalytic activity. Under H_2, disproportionation is common (e.g., Eq. 9), and any catalytic activity observed is that of the bis(phosphine) species, and so produces nonchiral products.[41,42]

$$2\ Rh(NBD)(PPh_3)(DIOS)^+ \xrightarrow[S]{5H_2} Rh(H)_2(PPh_3)_2(S)_2^+ + Rh(DIOS)_2^+ + 2\ \text{norbornane} \quad (S = \text{acetone, DMA}) \tag{9}$$

Efforts to use Rh complexes with monodentate or bidentate sulfoxides, chiral at the S or in a carbon backbone but containing no other functional groups, have met with no success in catalytic asymmetric hydrogenation.

B. Carboxylate/Amino Acid Systems

The chemistry of rhodium(II) bridged-carboxylato complexes such as $Rh_2(O_2CR)_4$ and ligated adducts $Rh_2(O_2CR)_2(N\text{-}N)_2X_2$ (**8**) and $Rh_2(O_2CR)_2(L\text{-}L)_2$ is extensive; N-N = 2,2'-dipyridyl-o-phenanthroline, X = halide, L-L = acetylacetonate, dimethylglyoximate, etc.[53] The catalytic activity of the tetraacetate for hydrogenation of olefins has been known for some time,[54] and extension to ketone substrates and

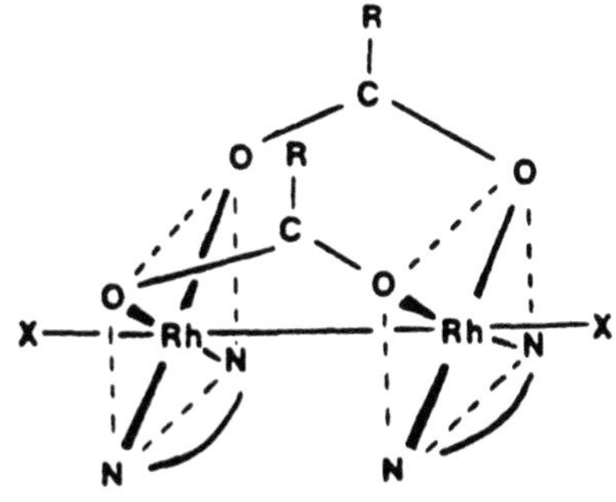

8

chiral systems has been initiated more recently.[34,55] The tetrakis(*R*-mandelato) complex in the presence of PPh_3, needed to maintain solution homogeneity, effects H_2 hydrogenation (at 1 atm) of the prochiral olefinic acid E-α-methylcinnamic acid to $PhCH_2CH(CH_3)CO_2H$ in 15% ee (*S*)[34]; the precursor catalyst is probably $Rh_2[\mu\text{-}O_2CCH(OH)Ph]_4(PPh_3)_2$. The $Rh_2[\mu\text{-}O_2CCH(OH)Ph]_2(N\text{-}N)_2Cl_2$ complexes in basic ethanol also catalyze the H_2 hydrogenation (30°C, 1 atm) of ketones, but data on the degree of optical induction are not yet available.

Markó's group at Veszprem was probably the first to report on the use of amino acids with Rh for catalytic hydrogenation, in species of the type $Rh(COD)[NH_2CH(R)CO_2]$ and acylated bridged carboxylate derivatives $Rh_2(COD)_2[\mu\text{-}O_2CCH(R)NHCOMe]_2$[33]; removal of coordinated diene in Rh species by treatment with H_2 (cf. Eq. 9) has become a standard procedure for generating sites for substrate and/or ligand binding.[32] The amino acid systems are poor hydrogenation catalysts, however, and even in the presence of tertiary phosphines to stabilize against reduction to the metal, only a 7% ee has been realized (using an acyl species for hydrogenation of olefinic acids).[33] We had earlier considered utilizing the "two-centered" chirality ligand *S*-methionine-*S*-sulfoxide at Rh for asymmetric hydrogenation, following a report of a simple stereospecific oxidation of *S*-methionine by chloroauric acid,[56] but we were unable to reproduce the procedure.[41] A testing of *S*-methionine-*R*,*S*-sulfoxide, racemic at the sulfoxide center but readily made by H_2O_2 oxidation of commercially available *S*-methionine,[57] was then carried out, but in situ formed Rh(I) species containing this ligand are not effective hydrogenation catalysts. A minor modification (á la Veszprem) to give the *N*-acetylated *S*-methionine-*R*,*S*-sulfoxide (AMSO, **9**) has led, however, to a system that effects enantioselective transfer hydrogenation of alkylarylketones with up to 80% ee.[16]

```
                H
                ▼
Me-S-CH2CH2-C-CO2H
   ||           ▲
   O            NHC(O)Me
```

9

Table 3 Transfer Hydrogenations Using Rhodium-AMSO Catalysts[a]

Substrate	Conv. (%)	Opt. yield (%)
PhCOMe	22	38
	33[b]	41
	45[c]	63
	8[d]	1
	48[e]	22
	40[f]	37
p-MeC_6H_4COMe	38[c]	80
PhCOEt	27[c]	76
2-Heptanone	6	—
PhCH=NPh	58	—

[a]Refluxing 2-propanol solutions (10 ml) under Ar, containing 2.5×10^{-3} M Rh(I) precursor, added as $[RhCl(HD)]_2$, and base (usually KOH):AMSO:Rh = 5:1:1, unless stated otherwise; 1.0 M substrate, reaction time 8-10 hr. The AMSO from *S*-methionine gives *R*-alcohol products, and vice versa.
[b]KOH : Rh = 4.0.
[c]AMSO : Rh = 2.0.
[d]KOH added as aqueous solution (2 ml).
[e]NaOMe added as base.
[f]$KOBu^t$ added as base.

Table 3 summarizes the findings for the in situ Rh-AMSO systems. Added hydroxide or alkoxide is essential and the system operates optimally with Rh/AMSO/KOH = 1:2:4-5 in refluxing 2-propanol. The ketone must be added prior to the base, otherwise metal precipitates; a required order of addition of reagents is not unusual in either Rh-catalyzed hydrogen transfer[58] or H_2 hydrogenation reactions,[59] the substrate stabilizing Rh(I) against disproportionation and/or hydrogenation to the metal. Dialkylketones are not effectively reduced under corresponding conditions, and this has been observed also with Rh-phosphine systems.[13] It is of interest that the nature of the Rh(I) precursor affects markedly the observed ee, effectiveness decreasing in the order $[RhCl(HD)]_2$(38%, see Table 3) > $[RhCl(COD)]_2$ (33%) >> $[RhCl(COE))]_2$(11%), $[RhCl(NBD)_2]_2$(7%), implying that the diene remains coordinated within the catalytic intermediates. Indeed, AMSO is unlikely to displace the diene, as judged by formation of $[RhCl(COE)(R_2SO)]_2$ complexes (R = $[CH_2]_2$ or Ph), with S-bonded sulfoxide, when the precursors are treated with a large excess of the sulfoxide.[48] Further, as methionine sulfoxide itself is an ineffective ligand (see above), the acylation probably prevents binding via the nitrogen and promotes chelation through the sulfoxide and carboxylate. That neither nonfunctionalized sulfoxides (Sec. III.A) nor simple amino acids generate catalytic systems implies that AMSO becomes effective via chelation. The role of added base is poorly understood generally but, of several possibilities including generation of polynuclear species with bridging alkoxides (see below),[13] removal of Cl^- almost certainly occurs, which assists coordination of the isopropoxide. A plausible key intermediate, consistent with the observations, is $Rh(HD)(AMSO)(ketone)^+$, **10**. Within a mechanistic cycle such as that shown for the Ir systems (Scheme 1), **10** would be "equivalent" to species **7**; the postulated steps (e.g., corresponding to **7** → **4** → **5** → **7**) could again be accommodated by changes in denticity of the diene and/or AMSO ligands. Addition of PPh_3 (2 equiv./Rh) to the system gives somewhat higher conversions than those shown in Table 3, but no optical induction occurs, implying that the AMSO has been replaced by the phosphine ligands. It is worth noting that with the addition of Et_3N as base (a commonly used cocatalyst in ketone hydrogenations[13]), no acetophenone reduction occurs; whether complications result from dehydrogenation of the amine, a facile process that occurs in the presence of certain Rh (and Ru) complexes,[49] is unknown. Use of analogous Ir(I)-AMSO systems under comparable conditions gives somewhat higher conversions of acetophenone to the alcohol, but ee values were essentially zero. A nonchiral imine (Eq. 6) is again reduced (Table 3).

The Rh-AMSO systems are thus reasonably effective for alkylaryl ketones. A Rh(I)-chiral phosphine system, using 2,4-bis(diphenylphosphine)pentane, gives somewhat higher ee values (up to 82%) for analogous substrates, but these are high-pressure (70 atm) H_2 hydrogenations.[12] The Ru(II)-binap systems that effect H_2 hydrogenations of functionalized ketones with 100% enantioselectivity also utilize 50-100 atm H_2, but of interest are ineffective for acetophenone itself (<1% chemical yield, 74% ee).[18] Related Rh(I) and Ir(I) species that catalyze H_2 transfer from basic 2-propanol to alkylarylketones are $M(COD)(N-N)^+$, where N-N is a chiral Schiff base readily made, e.g., from pyridine-2-carbaldehyde and chiral amines such as *R*- or *S*-1-phenylethylamine; maximum optical yields were 50% for ketones, including acetophenone (42) and propiophenone (36%).[60] Some $Rh(diene)(P-P)^+$ complexes (P-P = chiraphos and prophos) effect similar base-promoted transfer reduction of acetophenone with up to 13% ee.[15] In this work, ^{31}P NMR and CD data suggested that trimeric species of the type $Rh_3(P-P)_3(O^iPr)_2^+$ [based on analogy with data for a known triply bridged methoxide complex, $Rh_3(diphos)_3(\mu_3\text{-}OMe)_2$, isolated from basic methanol solutions of $Rh(NBD)(diphos)^+$][61] and other polynuclear species were formed, and these were considered to be the active catalysts; synthesized $Rh_3(P-P)_3(OMe)_2^+$ complexes were then used catalytically in the basic 2-propanol solutions, and optical yields were increased to ~25%.[15] Mechanistic schemes, analogous to Scheme 1, can be envisaged for a trinuclear species, perhaps involving more than one metal site. The difficult problem again arises of proving whether catalysis occurs at one or several sites within a multinuclear species that maintains its integrity, or at a mononuclear species that may be nondetectable.[21,22,62] Neither the Ir carboxylate/phosphite nor the Rh-AMSO system is ligand-deficient, and intuitively catalysis via mononuclear species is favored. Kinetic together with spectroscopic data are sorely needed for some of the more well-defined ketone hydrogenation systems.

IV. RUTHENIUM CATALYSTS

The recent success in obtaining up to 100% ee for enantioselective H_2 hydrogenation of certain olefinic acids[8-11] and functionalized ketones[17,18] using complexes of empirical formula $RuX_2(P-P)$ (where P-P = chiraphos, diop, and binap) deserves comment, especially in view of the possible involvement of polynuclear species. We have

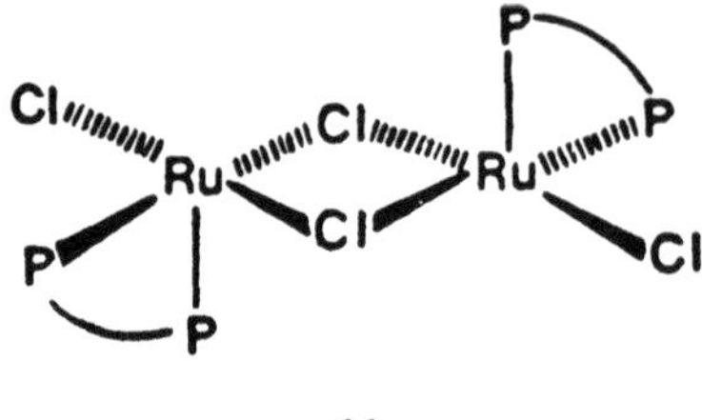

11

shown that the chiraphos and diop complexes are chloride-bridged dimers (11) and that treatment with H_2, at least in the presence of Et_3N base which removes HCl as $Et_3NH^+Cl^-$, gives the trimeric hydride $[RuHCl(P\text{-}P)]_3$.[10] The *S*,*S*-chiraphos structure shown (12) is notably unsymmetric showing one metal-metal bond (2.80 Å), with three bridging chlorides between Ru1 and Ru3, while Ru2 is bonded to only a single chloride that bridges all three Ru atoms; ^1H and ^{31}P NMR data allow for tentative placing of a terminal hydride on both Ru3 and Ru2, and a bridging one across Ru1-Ru2.[10] Whether such trimers are involved in the asymmetric hydrogenations remains to be established.

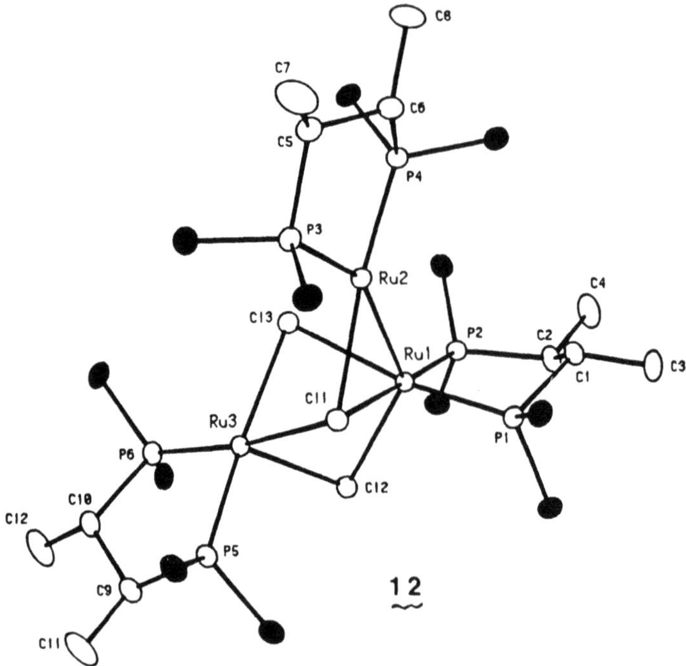

12

The trimer also catalyzes under mild conditions (1 atm H_2, 50°C, 5 x 10^{-3} M Ru, 1.0 M substrate in DMA) the hydrogenation of acetophenone, several imines, including prochiral ones (Eq. 10), and

$$PhCR^1{=}NCH(R^2)(R^3) \longrightarrow PhCH(R^1)\text{-}NHCH(R^2)(R^3) \quad (R^1 = H, Me;\ R^2 = H, Me;\ R^3 = Me, Ph) \tag{10}$$

acetonitrile, with turnovers in the 3-10 hr^{-1} range. Tests using $[RuCl_2(chiraphos)]_2$ for acetophenone reduction (20°, 80 atm H_2, several days) gave low conversions (as do the reported binap systems[18]), typically <2% with ee values of ~25%. The relationship of these systems to the H_2 reduction of imines and acetonitrile catalyzed by $RuCl(PPh_2(CH_2)_4PPh_2)(MeCN)_3^+$ under comparable conditions[63] is also undetermined; data on chiral analogs of this cation, and on the trimer **12**, with ketone and imine substrates are not yet available. Preliminary kinetic data on the hydrogenation of MeCN catalyzed by the nonchiral $[RuHCl(PPh_2(CH_2)_4PPh_2)]_3$ show increasing turnover per Ru with increasing concentration of trimer, evidence that implies catalysis at an "aggregated" site[62]; there are good synthetic models for reduction of acetonitrile at a trinuclear site,[64] but we are unaware of supporting kinetic data.

V. EXPERIMENTAL SECTION

The procedures used for the H_2 and transfer hydrogenation and for the determination of optical yield by rotation and/or the use of chiral shift reagents can be traced through the literature.[14,16,34,35] The details of the synthesis and characterization of sulfoxides, including AMSO, and the complexes, isolated or in situ, are generally available in the appropriate reference quoted; details on the Ir/carboxylato(hydrido)sulfoxide (**2**) and -phosphite (**3**) species will be published elsewhere.

The ketone and imine substrates are generally available commercially or are readily synthesized by standard procedures.

VI. CONCLUDING REMARKS

Effective catalysts are now being reported for enantioselective hydrogenation of ketones and imines, but the systems have developed largely from empirical studies, very much akin to the early work on asymmetric hydrogenation of alkenes. Again, chiral phosphines appear optimal, but our less expensive AMSO system appears promising (at least for ketones). There is considerable scope for such functionalized sulfoxide ligands, where possible synergic effects of chirality at both the sulfur and, for example, the amino acid moiety may be realized.

Careful spectroscopic and kinetic data on reproducible systems (H_2 and transfer hydrogenation) are desperately needed to define reaction mechanisms, in particular the chiral recognition step; the roles of "cocatalysts," especially base and halides, and the solvent, which is often critical, also require elucidation.

ACKNOWLEDGMENTS

We thank the Natural Sciences and Engineering Research Council of Canada for grants; the Guggenheim Foundation for a fellowship (BRJ); and Johnson, Matthey Ltd., for a loan of Ir, Rh, and Ru.

REFERENCES

1. B. Bosnich (ed.), *Asymmetric Catalysis*, Martinus Nijhoff, Dordrecht, 1986.
2. (a) H. B. Kagan, in *Asymmetric Synthesis*, Vol. 5 (J. D. Morrison, ed.), Academic Press, New York, 1985, Chap. 1; (b) in *Comprehensive Organometallic Chemistry*, Vol. 8 (G. Wilkinson, ed.), Pergamon Press, Oxford, 1982, Chapter. 53.
3. J. Halpern, in *Asymmetric Synthesis*, Vol. 5 (J. D. Morrison, ed.), Academic Press, New York, 1985, Chap. 2.
4. K. E. Koenig, in *Asymmetric Synthesis*, Vol. 5 (J. D. Morrison, ed.), Academic Press, New York, 1985, Chap. 3.
5. J. M. Brown, *Angew. Chem. Int. Ed. Engl., 26,* 190 (1987).
6. J. Halpern, *Pure Appl. Chem., 55,* 99 (1983); in *Catalysis of Organic Reactions* (R. L. Augustine, ed.), Marcel Dekker, New York, 1985, p. 3.

7. C. R. Landis and J. Halpern, *J. Am. Chem. Soc., 109,* 1746 (1987).

8. T. Ikariya, Y. Ishii, H. Kawano, T. Arai, M. Saburi, S. Yoshikawa, and S. Akutagawa, *J. Chem. Soc. Chem. Commun.,* 922 (1985).

9. R. Noyori, M. Ohta, Yi Hsiao, M. Kitamura, T. Ohta, and H. Takaya, *J. Am. Chem. Soc., 108,* 7117 (1986).

10. B. R. James, A. Pacheco, S. J. Rettig, I. S. Thorburn, R. G. Ball, and J. A. Ibers, *J. Mol. Catal., 41,* 147 (1987).

11. H. Takaya, T. Ohta, N. Sayo, H. Kumobayashi, S. Akutagawa, S. Inoue, I. Kasahara, and R. Noyori, *J. Am. Chem. Soc., 109,* 1596 (1987).

12. J. Bakos, I. Toth, B. Heil, and L. Marko, *J. Organometal. Chem., 279,* 23 (1985).

13. B. Heil, L. Marko, and S. Toros, in *Homogeneous Catalysis with Metal Phosphine Complexes* (L. H. Pignolet, ed.), Plenum Press, New York, 1983, p. 317.

14. B. Heil, P. Kvintovics, L. Tarszabo, and B. R. James, *J. Mol. Catal., 33,* 71 (1985).

15. R. Spogliarich, J. Kaspar, and M. Graziani, *J. Catal., 94,* 292 (1985).

16. P. Kvintovics, B. R. James, and B. Heil, *J. Chem. Soc. Chem. Commun.,* 1810 (1986).

17. R. Noyori, T. Ohkuma, M. Kitamura, H. Takaya, N. Sayo, H. Kumobayashi, and S. Akutagawa, *J. Am. Chem. Soc., 109,* 5856 (1987).

18. M. Kitamura, T. Ohkuma, S. Inoue, N. Sayo, H. Kumobayashi, S. Akutagawa, T. Ohta, H. Takaya, and R. Noyori, *J. Am. Chem. Soc., 110,* 629 (1988).

19. S. Vastag, J. Bakos, S. Toros, N. E. Takach, R. B. King, and B. Heil, *J. Mol. Catal., 22,* 283 (1984).

20. W. R. Cullen, M. D. Fryzuk, B. R. James, G-J. Kang, J. P. Kutney, R. Spogliarich, and I. S. Thorburn, U.S. Patent 079,625, filed July 30, 1987; G-J. Kang, W. R. Cullen, M. D. Fryzuk, B. R. James, and J. P. Kutney, *J. Chem. Soc. Chem. Commun.,* 1466 (1988).

21. B. R. James, in *Comprehensive Organometallic Chemistry,* Vol. 8 (G. Wilkinson, ed.), Pergamon Press, Oxford, 1982, Chap. 51.

22. B. R. James, *Homogeneous Hydrogenation,* John Wiley and Sons, New York, 1973.

23. G. Mestroni, A. Camus, and G. Zassinovich, in *Aspects of Homogeneous Catalysis,* Vol. 4 (R. Ugo, ed.), Reidel, Dordrecht, 1981, p. 71.

24. C. J. Longley, T. G. Goodwin, and G. Wilkinson, *Polyhedron, 5,* 1625 (1986).

25. I. Ojima, K. Yamamoto, and M. Kumada, in *Aspects of Homogeneous Catalysis,* Vol. 3 (R. Ugo, ed.), Reidel, Dordrecht, 1977, p. 185.

26. A. Karim, A. Mortreux, F. Petit, G. Buono, G. Peiffer, and C. Siv, *J. Organometal. Chem., 317,* 93 (1986).

27. H. Brunner, G. Riepl, and H. Weitzer, *Angew. Chem. Int. Ed. Engl., 22,* 331 (1983).

28. M. Srebnik and P. V. Ramachandran, *Aldrichimica Acta, 20,* 3 (1987).

29. K. Yamada, M. Takada, and T. Iwakuma, *J. Chem. Soc., Perkin Trans. I,* 265 (1983).

30. R. S. Dickson, *Homogeneous Catalysis with Compounds of Rhodium and Iridium,* Reidel, Dordrecht, 1985, pp. 86-99.

31. W. S. Knowles, M. J. Sabacky, and B. D. Vineyard, *Ann. N.Y. Acad. Sci., 295,* 274 (1977); W. S. Knowles and M. J. Sabacky, U.S. Patent 3,849,480 (1974).

32. B. R. James, *Adv. Organometal. Chem., 17,* 319 (1979).

33. Z. Nagy-Magos, P. Kvintovics, and L. Marko, *Transition Met. Chem., 5,* 186 (1980); L. Marko, Z. Nagy-Magos, and P. Kvintovics, *Proc. IXth Int. Conf. Organomet. Chem.,* Dijon, 1979, Abstract P41T.

34. F. Pruchnik, B. R. James, and P. Kvintovics, *Can. J. Chem., 64,* 936 (1986).

35. B. R. James and R. H. Morris, *J. Chem. Soc. Chem. Commun.,* 929 (1978).

36. Y. M. Y. Haddad, H. B. Henbest, and J. Trocha-Grimshaw, *J. Chem. Soc. Perkin I,* 592 (1974).

37. H. B. Henbest and J. Trocha-Grimshaw, *J. Chem. Soc., Perkin I,* 601 (1974).

38. Y. M. Y. Haddad, H. B. Henbest, J. Husbands, T. R. B. Mitchell, and J. Trocha-Grimshaw, *J. Chem. Soc., Perkin I,* 596 (1974).

39. M. Gullotti, R. Ugo, and S. Colonna, *J. Chem. Soc. (C),* 2652 (1971).

40. B. R. James, R. H. Morris, and P. Kvintovics, *Can. J. Chem., 64,* 898 (1986).

41. R. H. Morris, Ph.D. Dissertation, University of British Columbia, Vancouver, 1978.

42. (a) B. R. James, R. S. McMillan, R. H. Morris, and D. K. W. Wang, *Adv. Chem. Series, 167,* 123 (1978); (b) B. R. James and R. S. McMillan, *Can. J. Chem., 55,* 3927 (1977).

43. B. R. James and P. Kvintovics, *Proc. XXII Int. Conf. Coordin. Chem.,* Budapest, 1982, Abstract p. 732, to be published.

44. F. A. Cotton and G. Wilkinson, *Advanced Inorganic Chemistry,* 4th ed., John Wiley and Sons, New York, 1980, p. 1113.

45. T. Hayashi, A. Katsumura, M. Konishi, and M. Kumada, *Tetrahedron Lett.*, 425 (1979).

46. S. Toros, L. Kollar, B. Heil, and L. Marko, *J. Organometal. Chem.*, *255*, 377 (1983).

47. A. Dobson and S. D. Robinson, *Inorg. Chem.*, *16*, 137 (1977).

48. (a) B. R. James, R. H. Morris, F. W. B. Einstein, and A Willis, *J. Chem. Soc. Chem. Commun.*, 31 (1980); (b) B. R. James and R. H. Morris, *Can. J. Chem.*, *58*, 399 (1980).

49. S. N. Gamage, R. H. Morris, S. J. Rettig, D. C. Thackray, I. S. Thorburn, and B. R. James, *J. Chem. Soc. Chem. Commun.*, 894 (1987).

50. S. Gamage, Ph.D. Dissertation, University of British Columbia, Vancouver, 1985.

51. Ref. 22, p. 258.

52. B. R. James, R. H. Morris, and K. J. Reimer, *Can. J. Chem.*, *55*, 2353 (1977).

53. (a) E. B. Boyar and S. D. Robinson, *Coord. Chem. Rev.*, *50*, 109 (1983); (b) T. R. Felthouse, *Progr. Inorg. Chem.*, *29*, 73 (1982).

54. B. C. Y. Hui, W. K. Teo, and G. L. Rempel, *Inorg. Chem.*, *12*, 757 (1973).

55. H. Pasternak, E. Lancman, and F. Pruchnik, *J. Mol. Catal.*, *29*, 13 (1985).

56. G. Natile, E. Bordignon, and L. Cattalini, *Inorg. Chem.*, *15*, 246 (1976).

57. B. W. Christensen and A. Kjaer, *Chem. Commun.*, 225 (1965), and references therein.

58. D. Beaupere, L. Nadjo, R. Uzan, and P. Bauer, *J. Mol. Catal.*, *20*, 185 (1983); *14*, 129 (1982).

59. Ref. 22, p. 275.

60. (a) G. Zassinovich and F. Grisoni, *J. Organometal. Chem.*, *247*, C24 (1983); (b) G. Zassinovich, C. Del Bianco, and G. Mestroni, *J. Organometal. Chem.*, *222*, 323 (1981); (c) G. Zassinovich and G. Mestroni, *J. Mol. Catal.*, *42*, 81 (1987).

61. J. Halpern, D. P. Riley, A. S. C. Chan, and J. J. Pluth, *J. Am. Chem. Soc.*, *99*, 8055 (1978).

62. R. M. Laine, *J. Mol. Catal.*, *14*, 137 (1982).

63. I. S. Thorburn, S. J. Rettig, and B. R. James, *J. Organometal. Chem.*, *296*, 103 (1985).

64. M. A. Andrews and H. D. Kaesz, *J. Am. Chem. Soc.*, *99*, 6763 (1977).

3

Catalytic Asymmetric Hydrogenation of α-Acetamidocinnamic Acids

JUAN G. ANDRADE AND GUENTER PRESCHER

Degūssa Corporation, Ridgefield Park, New Jersey

ADOLF SCHAEFER

Degūssa AG, Hanau, Federal Republic of Germany

ULRICH NAGEL

Institute of Inorganic Chemistry, University of Munich, Munich, Federal Republic of Germany

ABSTRACT

The homogeneous catalytic asymmetric hydrogenation of substituted cinnamic acids by [Rh(1-benzyl-(3R,4R)-bis(diphenylphosphino)pyrrolidine)(COD)]BF_4 as catalyst prepared from L-tartaric acid has been studied. Contrary to other catalysts, this rhodium complex affords very high chemical and optical yields of *N*-acetylated amino acids, under mild conditions and using substrate-to-catalyst ratios as high as 16,000. The method has been used for the preparation of L-phenylalanine and L-dopa.

There are several processes suitable for the production of amino acids: extraction and isolation from natural protein hydrolysates, chemical synthesis, fermentation, enzymatic catalysis, and asymmetric hydrogenation of amino acid precursors.

The work described herein is the result of an effort to combine Degūssa's experience in precious metal catalysts with its know-how in chemical synthesis of amino acids, and was primarily developed with the goal of providing cheap access to enantiomerically pure amino acids like, for example, L-phenylalanine, an important raw material for the production of Nutrasweet® (trademark of Nutrasweet Company). As will be shown later, the method with some limitations has also been applied to some other types of compounds.[1]

A major drawback of any available standard chemical method used for amino acid synthesis is the fact that a racemic mixture of the desired amino acids is obtained. Considering the fact that in most cases only one of the enantiomers is biologically active and therefore the important one, this can be a major obstacle to be resolved. A classical approach for overcoming this problem is shown in Fig. 1.

By reaction with a stereospecific enzyme, the desired L-amino acid can be isolated from a mixture of previously prepared racemic amino acid derivatives. However, for qualifying as an attractive route, the remaining untouched 50% of D-amino acid derivative in this method has to be first racemized in order to be recycled and used again (Fig. 1). This is not always straightforward since in some cases, side reactions such as polymerization can take place.

After the discovery of the Wilkinson hydrogenation catalyst in the late 1960s, attention was focused on trying to modify this and other related catalysts by introducing chiral phosphines instead of the common triphenylphosphine ligands. With chiral ligands diastereomeric transition states can be formed on coordination of suitable

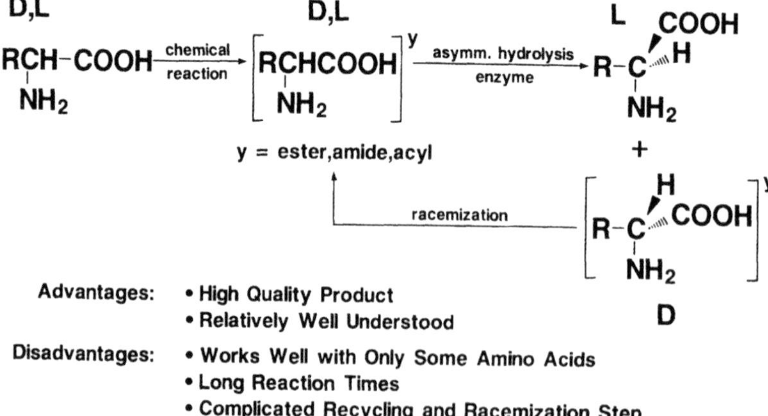

Fig. 1 Commonly used chemical L-amino acids production process.

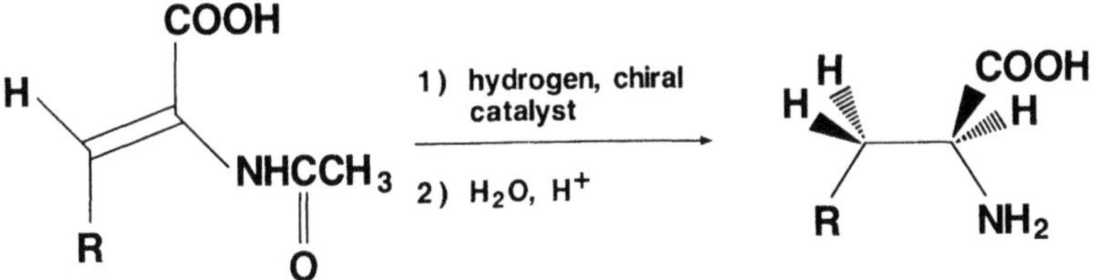

Fig. 2 Catalytic asymmetric hydrogenation.

substrates to the metal chelate. Then the initial substrate coordination complex leads to enantiomeric hydrogenations products. When related to amino acid synthesis, the general idea is depicted in Fig. 2.

A potential amino acid precursor as, for instance a substituted acetamidoacrylic acid, is hydrogenated in the presence of a chiral catalyst, giving an optically active acetylated amino acid which in turn can be cleaved to the desired optically pure amino acid.[2] Some of the most efficient and commonly used phosphines for pursuing this concept are the bidentate phosphines shown in Fig. 3. Ligands like Diop discovered earlier by Kagan,[3] have a chiral carbon backbone, whereas others, like Dipamp developed at

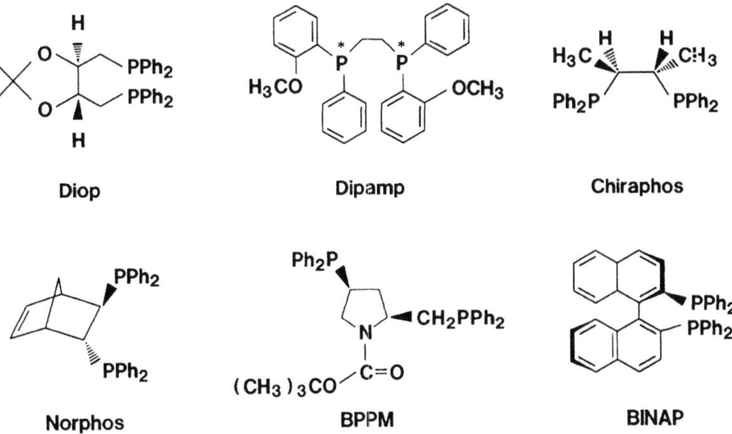

Fig. 3 Examples of some bidentate phosphines used in catalytic asymmetric hydrogenation.

Table 1 Asymmetric Hydrogenation on α-*N*-Acylaminoacrylic Acids; Literature Examples of Some Rh Catalysts

Substrates	Chiral diphosphines, optical purity e.e.(%)				
	DIPAMP	DIOP	CHIRAPHOS	BPPM	BINAP
$CH_2=C(CO_2H)NHCOCH_3$	94	73	91	98.5	67
$PhCH=C(CO_2H)NHCOPh$	96	64	99	-	96
$PhCH=C(CO_2H)NHCOCH_3$	95	81	89	91	84
$(AcO)(OCH_3)C_6H_3CH=C(CO_2H)NHCOCH_3$	94	84	83	86	84

Source: Ref. 2.

Monsanto,[4] the phosphorus atoms themselves carry the chiral information. Some of these ligands have been used quite extensively in the asymmetric synthesis of amino acid precursors (Table 1).
success (Table 1).

Although in some cases good results are obtained, apart from Rh-Dipamp, none of the catalysts prepared with these ligands have industrial significance. The reasons for this are many: high catalyst-to-substrate ratios, catalysts are difficult to recycle, and quite often the synthesis of the chiral phosphine is itself expensive and cumbersome. Under these conditions asymmetric hydrogenation cannot compete with enzymatic or classical methods.

Using our own experience[5] as well as some literature data, we started to develop an industrially efficient catalyst for asymmetric hydrogenation. The following requirements were found to be important: the catalyst should possess a bidentate phosphine ligand which upon coordination to the metal could provide a five membered chelate ring with C_2 symmetry. In addition, the ligand should be accessible from

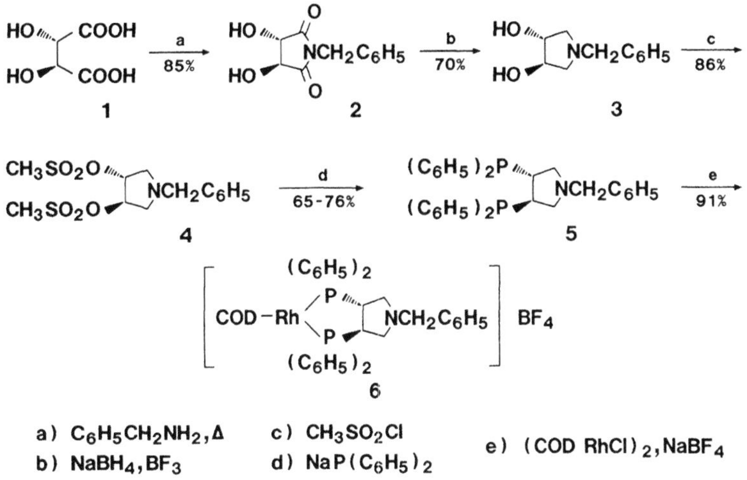

a) $C_6H_5CH_2NH_2$,Δ c) CH_3SO_2Cl e) $(COD\ RhCl)_2$,$NaBF_4$
b) $NaBH_4$,BF_3 d) $NaP(C_6H_5)_2$

Fig. 4 Synthesis of the Degussa rhodium catalyst Deguphos.

cheap commercially available starting materials, and the catalyst should be recoverable. The result of this effort was the synthesis of Deguphos® (Degussa AG), shown in Fig. 4.

The preparation of the diphosphine 6 starts with (*R*,*R*)-tartaric acid and benzylamine; these are refluxed in xylene until two moles of water are obtained. The resulting imide 2 is then reduced with a mixture of sodium borohydride/borontrifluoride in diglyme to give the dihydroxypyrrolidine 3. Reaction of 3 with methanesulfonyl chloride in dichloromethane in presence of triethylamine affords 4. The mesylate 4 is then transformed to 6 by reaction with sodium diphenylphosphide.[6] Other derivatives of 6 can be prepared by hydrogenolysis of the benzyl group in 4 with Pd/C in acetic acid. The corresponding pyrrolidine acetate can be used then as precursor for a variety of related N-substituted diphosphine ligands.[1b] Contrary to the experience with other catalysts as, for example, Chiraphos (Fig. 2), the enantioselectivity (ee) is independent of the hydrogen pressure (up to 70 bar). Substrate/catalyst ratio (s/c) can be as high as 50000/1. The reaction rates ($0.008\ sec^{-1}$)

$R_4(R_3)C{=}C(CO_2R_1)(NHCOR_2)$ (7 a-d) $\xrightarrow{H_2\ /\ \text{Deguphos}}$ $R_4(H)(R_3)C{-}C(H)(CO_2R_1)(NHCOR_2)$ (8 a-d)

	R_1	R_2	R_3	R_4
7,8a	H	CH_3	Ph	H
7,8b	H	CH_3	H	H
7,8c	H	CH_3	3-Indolyl	H
7,8d	H	CH_3	$(3\text{-}CH_3O, 4\text{-}C_2H_5O)\text{-}C_6H_3$	H

Fig. 5 Substrates of asymmetric hydrogenation.

at 1 atm are comparable to those of Norphos (Fig. 2). A raise in pressure up to 50 bars increases the rate 50 times (Fig. 5; Table 2). Reactions carried out in a pilot plant, in which up to 40 kg of α-acetamidocinnamic acid (s/c = 10,000) was used, showed similar results compared to laboratory experiments. The chemical and optical yields were excellent for this precursor and very good for the other substituted derivatives. One should point out that in the case of L-phenylalanine for 1 metric ton of starting material only 400 g of the catalyst are necessary!

Table 2 Catalytic Asymmetric Hydrogenation of Amino Acid Precursors with Deguphos

Substrate	(S)/(C)	Ph_2(bar)	T (°C)	React. time (hr)	Yields	e.e.(%)$^{(1)}$
7a	10,000$^{(2)}$	10-15	50	3-4	100	99.5
7a	13,000	40-50	50	2	100	98.5
7a	16,000	40-50	60	7	100	97.0
7b	4,400	50	25	0.5	>99.5	89.7
7b	10,900	20	21	2	>99.5	88.4
7c	2,150	50	50	6	95	68.9
7d	5,000	25-50	50	3	>99.5	95.5

Solvent:methanol
(1) Based on optical rotation ee% = (R - S/R + S) x 100 ≈ (a/a_0) x 100.
(2) 16.6 g catalyst, 40-kg substrate, 25% in methanol.

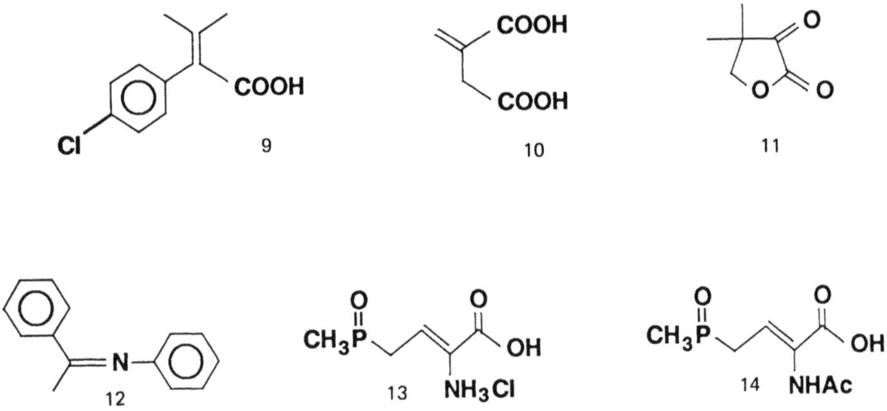

9 10 11

12 13 14

Fig. 6 Asymmetric hydrogenation of other types of double bonds.

Most important is the fact that the catalyst was prepared in quantities up to 10 kg and can therefore be manufactured on an industrial scale. Furthermore, using proprietary technology, the catalyst and ligands can be easily recovered from reaction mixture. After the hydrogenation the enantiomerically pure amino acids can be obtained in a trivial manner by cleavage of the *N*-acetyl group with hydrogenchloride/water.

Preliminary experiments aimed at the utilization of these catalysts for the hydrogenation of other types of olefins indicate that this is possible with some limitations. The olefins which we have studied are shown in Fig. 6. We tried to investigate the effects of olefin substitution and the hydrogenation C=X (X = N, O) double bonds. The results of these experiments are shown in Table 3.

It can be stated that the results were not as promising as those of the asymmetric hydrogenation of amino acid precursors.

Finally, it is interesting to mention that the hydrogenation of amino acid precursors can also be carried out in water.[8] For that purpose of course, a slight modification of the catalyst is necessary. Water solubility can be achieved by quaternization of the amino group of the pyrrolidine ring (Fig. 7). The enantioselectivity of the hydrogenation drops only slightly to 90% ee, independent of whether water or methanol is used (Fig. 7).

Table 3 Asymmetric Hydrogenation of Other Types of Double Bonds

Catalyst	Substrate	(S)/(C)	Solvent	Pressure (bar)	Temp. (°C)	React. time (hr)	Yield (%)	e.e (%)
Deguphos	9	200	methanol	50	80	4	>98.0	28.3
Deguphos/NEt_3	10	4,300	methanol	8	50	2	100.0	34.4
Deguphos	11	200	toluene	50	50	26	37.5	26.4
Deguphos	12	505	methanol	50	50	21	>99.0	5.7
(COD)Rh-(NBZDPP) BF_4[a,b]	13	400	water	50	30	24	92.3	85.0
(COD) Rh-(NBZDPP) BF_4[a,b]	14	400	methanol	15	30	24	96.0	90.0

[a]NBZDPP = (3*R*,4*R*)-*N*-benzoylbis(diphenylphosphino)pyrrolidine.
[b]Isolated yields.[10]

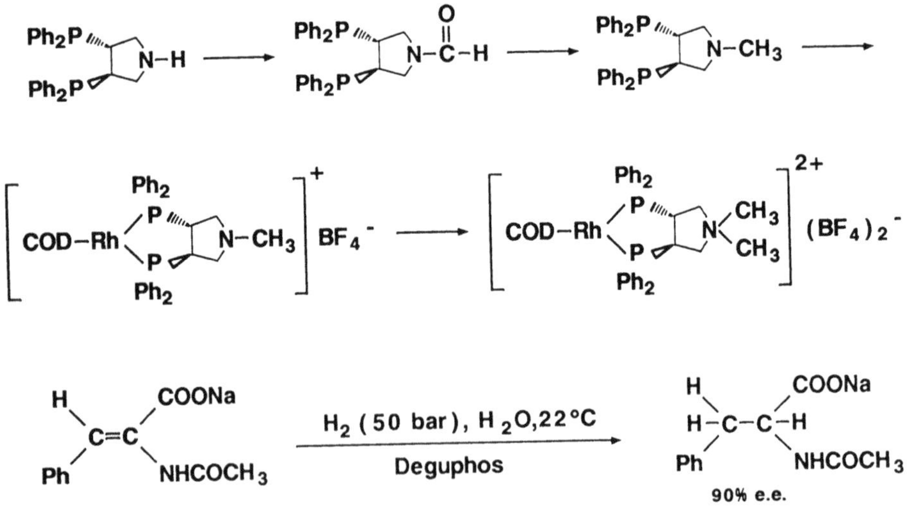

Fig. 7 Asymmetric hydrogenation in aqueous solution (from Ref. 8).

of the hydrogenations drops only slightly to 90%, independent of whether water or methanol is used (Fig. 7).

Finally, it is also important to mention that Nagel and Kinzel at Munich recently showed that it is possible to bind 3,4-(*R*,*R*)-bis(diphenylphosphino)pyrrolidine to silica, but activity and selectivity decrease considerably when reusing the catalyst three times.[9] We hope that further experiments with Deguphos will lead to new applications for this catalyst.

REFERENCES

1. (a) Part of this work was presented at the Third IUPAC Symposium on Organometallic Chemistry Directed Toward Organic Synthesis, Kyoto, Japan, 1985; and (b) at X International Conference on Phosphorus Chemistry, Bonn, FRG, 1986; Parts also have been published; U. Nagel, E. Kinzel, J. Andrade, and G. Prescher, *Chem. Ber., 119*, 3326 (1986).
2. For recent reviews see K. E. Koenig, in *Asymmetric Synthesis*, Vol. 5 (J. D. Morrison, ed.), Academic Press, New York, 1985, p. 71.
3. H. B. Kagan and T. P. Dang, *J. Am. Chem. Soc., 94*, 6429 (1972).
4. B. D. Vineyard, W. S. Knowles, M. J. Sabacky, G. L. Bachmann, and D. J. Weinkauff, *J. Am. Chem. Soc., 99*, 5946 (1977).
5. W. Bergstein, A. Kleemann, and J. Martens, *Synthesis*, 76 (1981).
6. DE-3446303, Degussa AG (1984); DE-3403194, Degussa AG (1984).
7. H. Brunner, *Angew. Chem., 95*, 921 (1983).
8. U. Nagel and E. Kinzel, *Chem. Ber., 119*, 1731 (1986).
9. U. Nagel and E. Kinzel, *J. Chem. Soc. Chem. Commun.*, 1098 (1986).
10. DE-3609818, Hoechstag (1986).

4

Cyclopentane Construction by Rh-Catalyzed Intramolecular C-H Insertion: Scope and Selectivity

DOUGLASS F. TABER, MICHAEL J. HENNESSY,[*] R. SCOTT HOERRNER,[†] KRISHNA RAMAN,[‡] ROBERT E. RUCKLE, JR.,[§] and JONATHAN S. SCHUCHARDT[¶]

Department of Chemistry and Biochemistry, University of Delaware, Newark, Delaware

ABSTRACT

On exposure to a catalytic amount of rhodium acetate (5 mol % or less), an α-diazoketone will smoothly undergo 1,5 C-H insertion to give the corresponding cyclopentane. This reaction presumably proceeds via the corresponding metallocarbene. In substrates having more than one site where C-H insertion could occur, steric and electronic considerations govern selectivity. Strategies for making this process enantioselective are discussed.

I. INTRODUCTION

Carbon-carbon bond formation by carbene insertion has been known for many years.[1] However, the first development of intramolecular C-H insertion as a synthetic method was by Wenkert.[2] This work, based

Current affiliations:
[*]Jackson Laboratories, E. I. du Pont de Nemours & Co., Inc., Deepwater, New Jersey
[†]Department of Chemistry, Northwestern University, Evanston, Illinois
[‡]Himont, Inc., Wilmington, Delaware
[§]Department of Chemistry, The Pennsylvania State University, State College, Pennsylvania
[¶]ARCO Chemical Co., Newtown Square, Pennsylvania

mostly on Cu-based carbenoid generation, incorporated in its late stages the observation by Teyssie[3] of $Rh_2(OAc)_4$-mediated insertion of ethyl diazoacetate into alkane C-H bonds.

At the inception of the work described here, there were two particular limitations to this approach: the diazocarbene precursors were not readily available, and there had been little investigation of what selectivity these inherently hot "carbenes" might show. As outlined below, it is apparent that α-diazo-β-keto esters, readily prepared by diazo transfer, are effective precursors for $Rh_2(OAc)_4$-mediated C-H insertion. There is a substantial preference for cyclopentane formation in this process, and the method can show significant diastereo-, enantio-, and chemoselectivity. It is also apparent that simple α-diazoketones, even those having β-hydrogen, can be induced to cyclize efficiently.

II. CYCLIZATION OF α-DIAZO-β-KETO ESTERS

A. Substrate Preparation

Initial investigations[4] focused on the cyclization of α-diazo-β-keto esters. The precursor β-keto esters are readily prepared *inter alia* by alkylation of acetoacetate dianion,[5] by acylation of lithiomethyl acetate with an acid chloride,[6] or by carbomethoxylation of a methyl ketone with sodium hydride and dimethyl carbonate. Triethylamine-mediated diazo transfer from a sulfonyl azide[7] then gives the α-diazo-β-keto ester. The acyclic ketoesters were attractive precursors both because they were readily prepared and because the cyclized ketoesters would be versatile precursors to highly substituted cyclopentanes.

B. Cyclization of Simple Acylic α-Diazo-β-Keto Esters

Commercially available rhodium acetate is a bluish green powder that is scarcely soluble in CH_2Cl_2. On dropwise addition of a solution of diazoketone in *dry* (filtered through anhydrous K_2CO_3) CH_2Cl_2 to a suspension of a catalytic amount (5 mol % is sufficient) of $Rh_2(OAc)_4$ in CH_2Cl_2, one observes a clear green solution with a white layer of foam on top (N_2 evolution). The reaction is very fast at room temperature, being finished as soon as diazo addition is complete. It should be noted, however, that on chilling, even to 0°, the reaction stops altogether.

An initial screen (Ref. 4; Table 1) indicated that C-H insertion proceeded selectively, in a 1,5 sense, to give cyclopentane derivatives. Insertions into methylene C-H and methine C-H both proceeded

Table 1 Cyclization of Diazoketones

Diazoketone	Product	Yield %
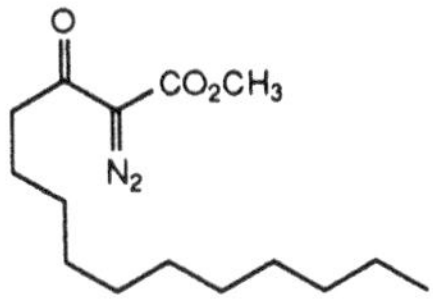 1	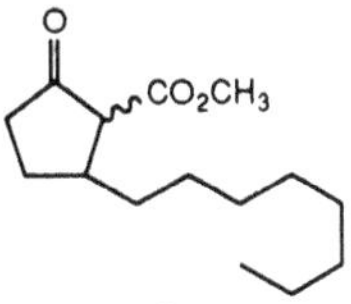 2	68
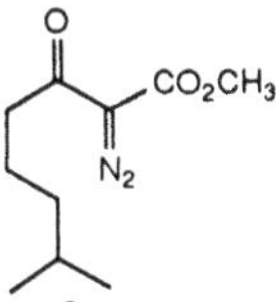 3	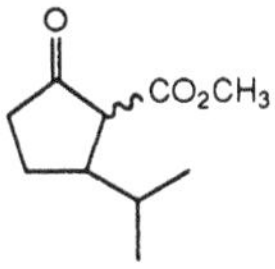 4	55
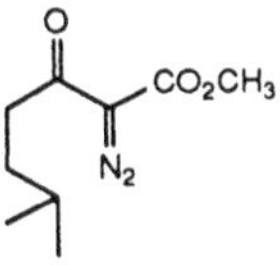 5	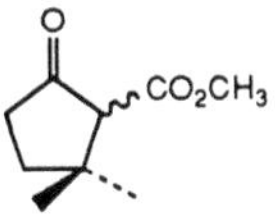 6	64
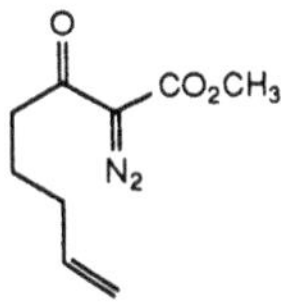 7	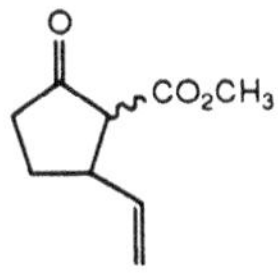 8	48
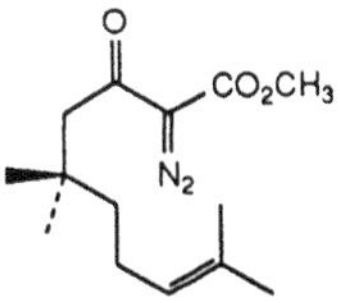 9	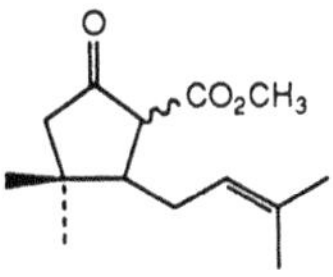 10	77

readily. The cyclization of 5 indicated that there was some hope for selectivity in this process. Despite a 3:1 statistical preference for 1,5 insertion into a methyl group, only the product of methylene insertion, 10, was observed.

C. Steric and Electronic Effects

More recently,[8] steric and electronic effects on the selectivity of this cyclization have been explored (Ref. 8; Table 2). After correcting for the relative number of methine vs. methylene C-H sites available for insertion, it is apparent (entry 1) that methine C-H is more reactive than methylene C-H by a factor of 4.6:1. From Table 1, it is also apparent that methylene C-H is much more reactive than methyl C-H. This is the order one would expect for any electrophilic process.

More striking is the observation (Table 2, entry 2) that benzylic C-H is *less* reactive than aliphatic C-H. In a similar competition reported by Gilbert et al.,[9] a simple thermally generated vinyl carbenoid showed a substantial *preference* for benzylic insertion. A possible rationale for the difference in the Rh-mediated process is that reactivity is a function of the availability of electron density in the C-H bond, and that inductively the phenyl group is electron withdrawing. Such electronic selectivity, which has since been confirmed in other cases,[10] is one of the more attractive features of this Rh-mediated process.

It is apparent (entries 4-7) that steric factors also play an important role in governing selectivity. The ligands on the rhodium also influence selectivity. For entry 6, rhodium acetate gave a 3:1 ratio of cis[25] to trans.[26] With rhodium octanoate the ratio was 5:1, whereas with tetraphenylporphyrin rhodium chloride it was 15:1.

III. APPLICATIONS TO NATURAL PRODUCT SYNTHESIS

A. Pentalenolactone E Methyl Ester

All the examples of Rh-mediated C-H insertion to this point had been for unstrained systems. It was not clear that insertion would proceed as well if cyclization led to an increase in torsional strain. Especially worrisome was the indication from theoretical calculations[11] that the activation energy for C-H insertion would depend

Table 2 Steric and Electronic Effects

	Starting Diazoketone	Product(s)			Yield %
1	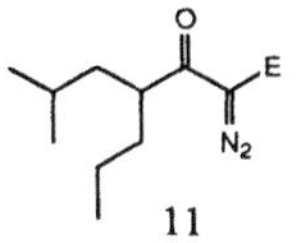 11	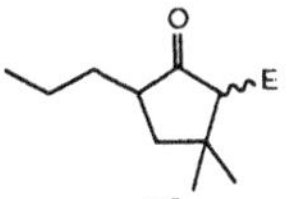 12	2.3 : 1 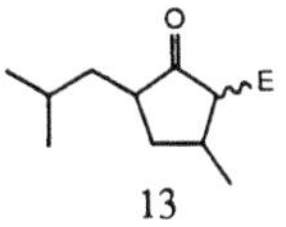	13	84
2	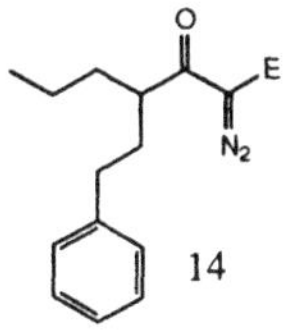 14	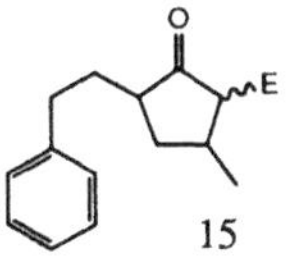 15	2.3 : 1 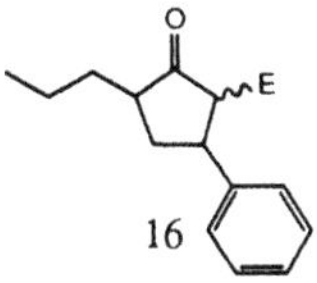	16	75
3	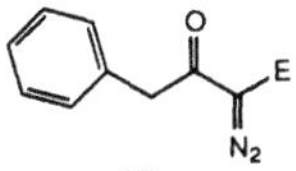 17	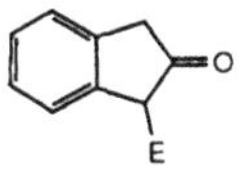 18			76
4	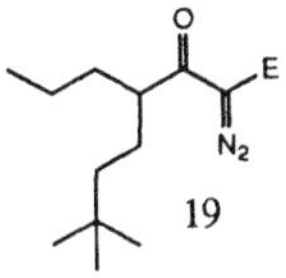 19	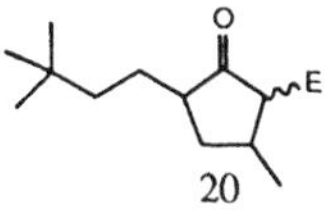 20	3.4 : 1 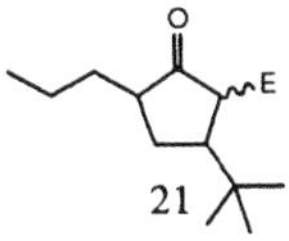	21	73
5	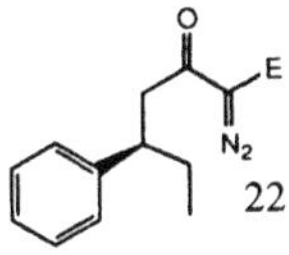 22	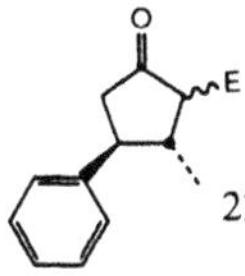 23			77
6	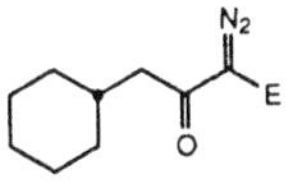 24	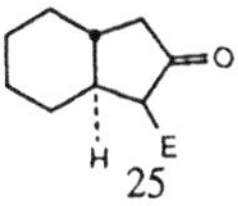 25	see text 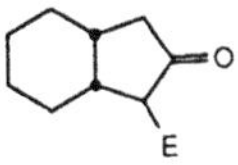	26	63
7	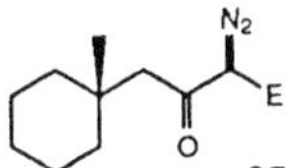 27	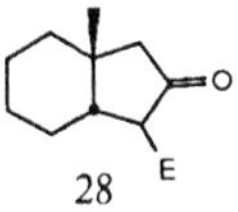 28			93

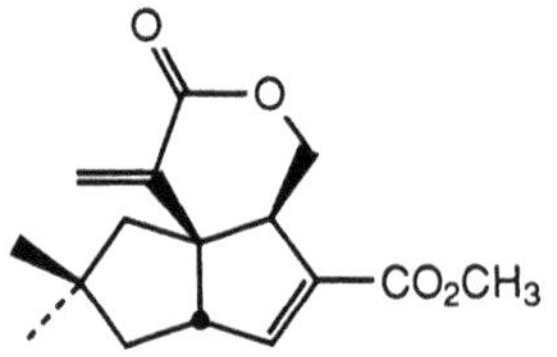

29

Fig. 1

on the angle of approach of the intermediate Rh complex to the target C-H bond. An attractive retrosynthetic dissection of the sesquiterpene antibiotic pentalenolactone E, **29** (Ref. 12; Fig. 1), revealing an appealing symmetry,[13] prompted the exploration of this question.

Cyclization of the sterically constrained diazoketone **33** (Scheme 1) was the key to the success of this synthetic approach. In the event, **33** cyclized smoothly to give **34**. This suggests that even for systems incorporating torsional strain, Rh-mediated intramolecular C-H insertion may be effective. The increase in molecular complexity[14] on going from **33** to **34** is striking.

B. (+)-α-Cuparenone

It seemed likely that the Rh-mediated insertion was in fact proceeding with retention of absolute configuration at the C-H site (for more on the mechanism of the insertion process, see Sec. IV below). If such were the case, then insertion in an enantiomerically pure ternary stereogenic center would provide a method for the enantiospecific construction of cyclic quaternary centers.[15,16]

The requisite enantiomerically pure iodide **37** (Scheme 2) was prepared by alkylation of 4-methylphenylacetic acid, following the Evans procedure.[17] Alkylation of the dianion of methyl acetoacetate[4] with **37** gave **38**, which on diazo transfer and cyclization gave **39**. That ketone **39** so prepared was indeed of high optical purity was confirmed by carrying it on to (+)-α-cuparenone **40**.

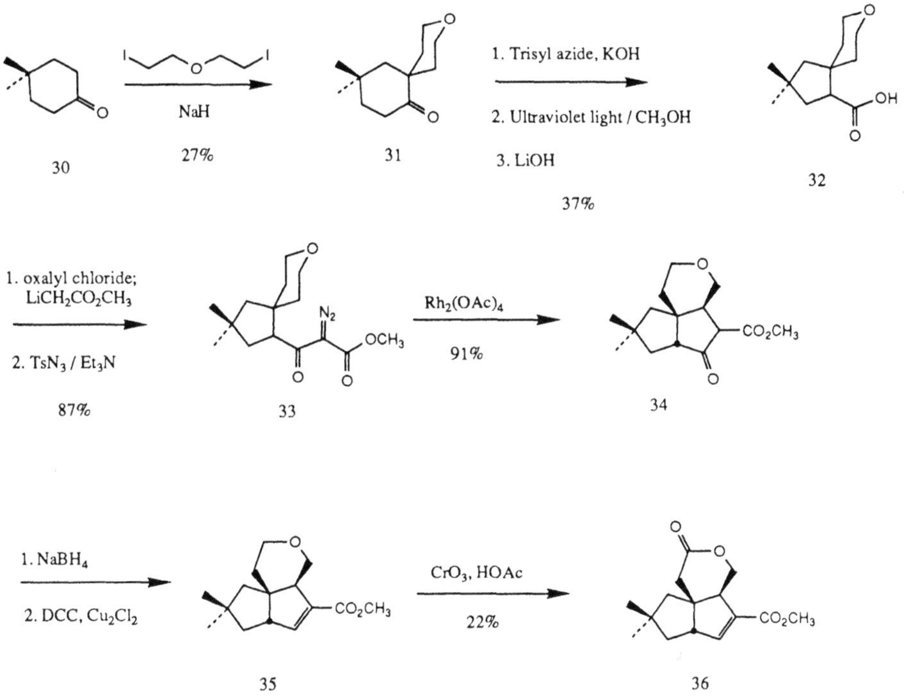

30 → (I–CH2CH2–O–CH2CH2–I; NaH; 27%) → 31 → (1. Trisyl azide, KOH; 2. Ultraviolet light / CH_3OH; 3. LiOH; 37%) → 32

→ (1. oxalyl chloride; $LiCH_2CO_2CH_3$; 2. TsN_3 / Et_3N; 87%) → 33 → ($Rh_2(OAc)_4$; 91%) → 34

→ (1. $NaBH_4$; 2. DCC, Cu_2Cl_2) → 35 → (CrO_3, HOAc; 22%) → 36

Scheme 1 Synthesis of pentalenolactone E methyl ester.

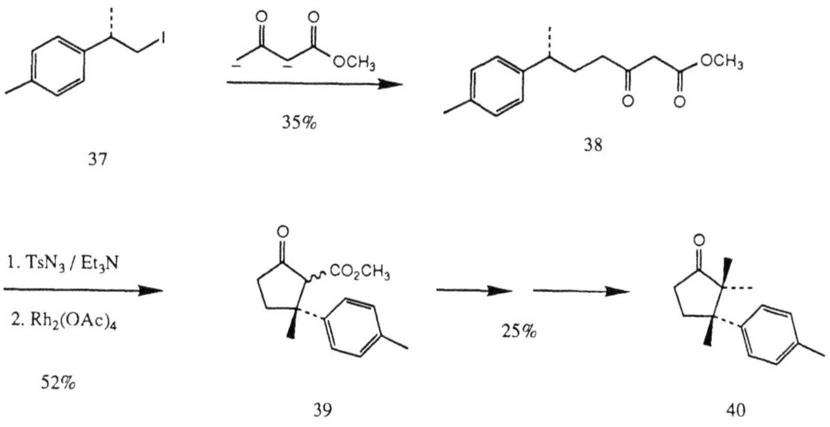

37 → (dianion of methyl acetoacetate; 35%) → 38

→ (1. TsN_3 / Et_3N; 2. $Rh_2(OAc)_4$; 52%) → 39 → → (25%) → 40

Scheme 2 Synthesis of (+)-α-cuparenone.

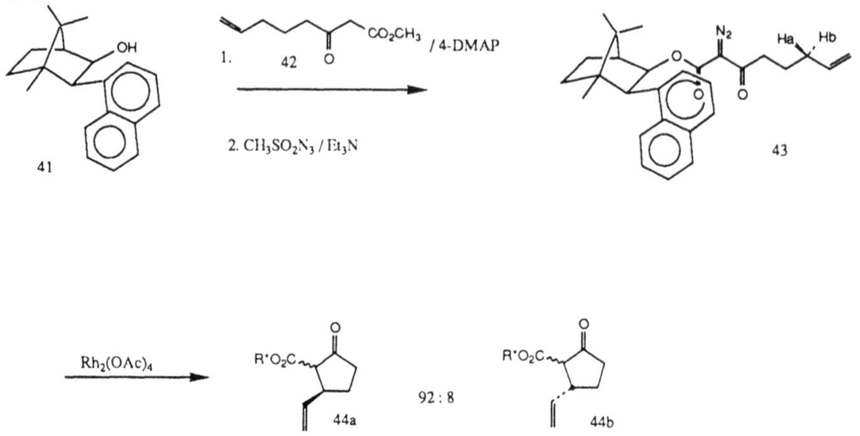

41

1. 42 / 4-DMAP

2. $CH_3SO_2N_3$ / Et_3N

43

$Rh_2(OAc)_4$

44a 92 : 8 44b

Scheme 3 Enantioselective C-H insertion.

C. (+)-Estrone Methyl Ether

As the transition state leading to C-H insertion appeared to be highly ordered, and as insertion proceeded with retention of configuration, it seemed that design of a chiral auxiliary that would direct the cyclization selectively toward one of the two diastereotopic methylene hydrogens on the hydrocarbon chain might be feasible. The naphthylborneol 41 (Ref. 18; Scheme 3) was designed to this end. In fact, cyclization of the diazo ester 43 derived from 41 proceeds with excellent diastereoselectivity, to giving 44a and 44b in a ratio of 92:8.

The synthetic utility of this approach was demonstrated by carrying the 44a/44b mixture onto estrone methyl ether, 47 (Ref. 19; Scheme 4), of high enantiomeric purity. Thus, methylation followed by heating with anhydrous sodium methoxide gave, by retro-Dieckmann rearrangement, keto ester 45, at the same time returning naphthylborneol 41 for recycling. Alkylation of the dianion of 45 with the *racemic* iodide, followed by decarbomethoxylation and pyrolysis, then gave (+)-estrone methyl ether 47.

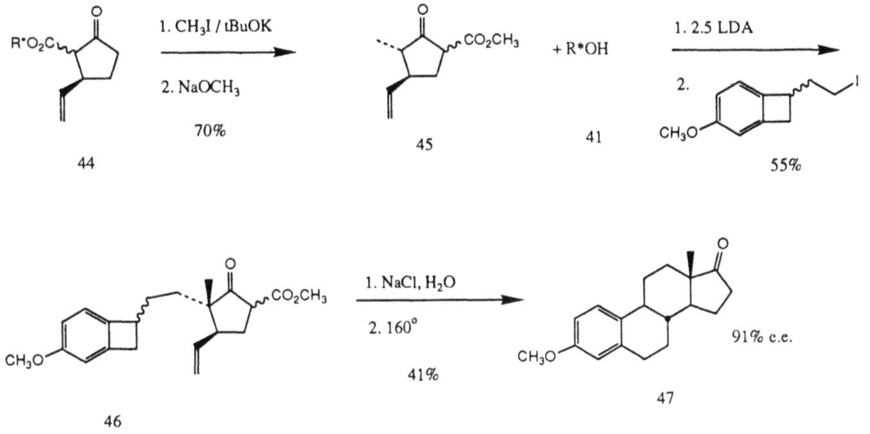

Scheme 4 Synthesis of (+)-estrone methyl ether.

IV. MECHANISM OF Rh-MEDIATED C-H INSERTION

A variety of transition metal complexes have been screened[20] for their ability to catalyze intramolecular C-H insertion. While some complexes show marginal reactivity, none [including mononuclear Rh(I) and Rh(III) species] comes close to the efficiency of the Rh(II)-Rh(II) dimers. It is therefore reasonable to think that *both* Rh centers may be involved in the insertion process.

A possible mechanism for Rh-mediated C-H insertion is outlined in Scheme 5. The diazoketone 48 could be considered to be polarized as shown. Donation of electron density to the open coordination site on one of the rhodium atoms, followed by loss of two of the bridging oxygen ligands, would give 50, in which electron density has been concentrated on the distal rhodium. Donation of this electron density back toward carbon, with concomitant loss of N_2, would then give carbene complex 51.

Oxidative addition of the electrophilic Rh(I) center of 51 into the remote C-H bond would give Rh(III) carbene complex 52. Carbene rearrangement could then give the hydride complex 53, which should undergo reductive elimination to give $Rh_2(OAc)_4$ 49 back again, along with the product 54.

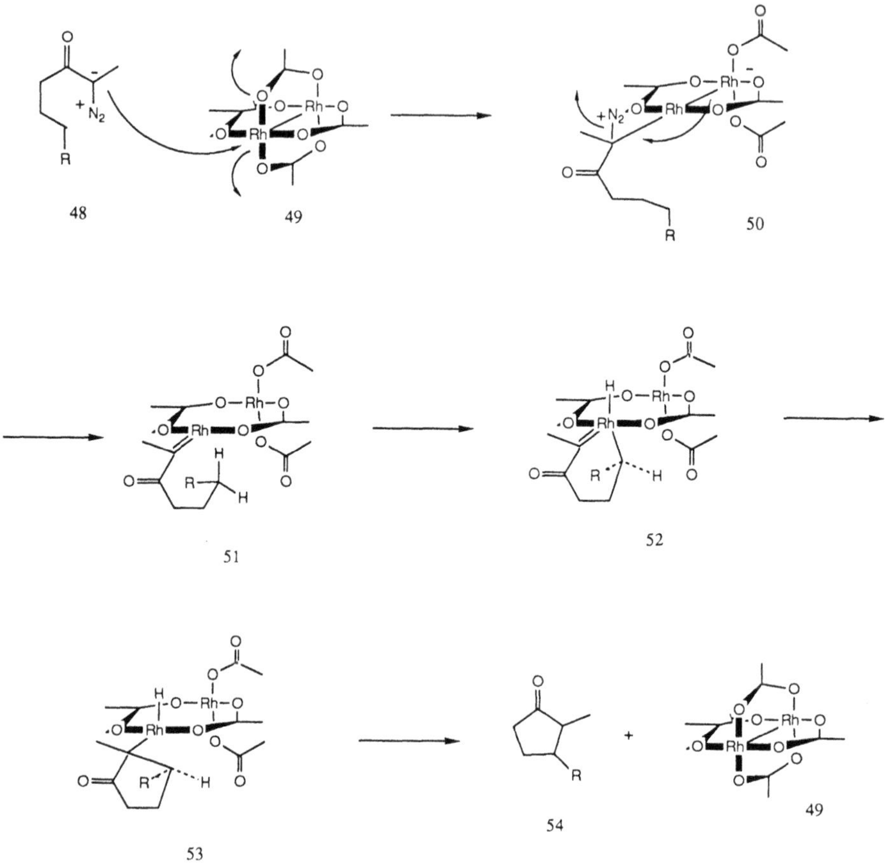

Scheme 5 Mechanism of Rh-mediated C-H insertion.

Whatever mechanism is operative, it is apparent that the last step would likely be reductive elimination of a highly reactive, coordinatively unsaturated Rh species. An obvious advantage that an intermediate such as **53** would have is that it could potentially wrap back around the reactive Rh center before that center could enter into detrimental side reactions with other species in solution.

V. SIMPLE α-DIAZO KETONES AND ESTERS

With one exception,[21] the diazo precursors used in these and other[22] studies have been constructed without β-hydrogens. This was partly because these precursors were readily available, but also partly because it was recognized that if a β-hydrogen were available, β-hydride elimination from the intermediate metallocarbene could compete with the desired intramolecular insertion.

To assess the feasibility of cyclizing a simple α-diazoketone, 56 (Scheme 6) has been prepared, using an improved procedure for diazo transfer recently developed in this laboratory.[7b] Carbene generation from 56 could lead to three products: 57, 58, and/or 59. In fact, on exposure to a catalytic amount of $Rh_2(OAc)_4$ at room temperature, 56 smoothly cyclizes[10] to a mixture of cyclopentane product 57 (62%) and β-hydride elimination product 58 (23%; exclusively cis). The alternative cyclopentane product 59 has been independently prepared; no trace of it is found in the reaction mixture. These results might suggest that 60, rather than 61 (Scheme 6), is the preferred rotamer in the transition state leading to cyclization.

Diazoketone 62 and diazo ester 64 also cyclize efficiently. $Rhodium_2(trifluoroacetate)_4$ has given the best yields for the cyclization of 62 to 63. $Rhodium_2(benzoate)_4$ is the most effective catalyst so far found for the cyclization of 64 to 65.

For such cyclizations to be useful synthetically, practical procedures for the assembly of *unsymmetric* α-diazoketones were needed. Two complementary approaches have been developed.[10] In the first (Scheme 7), it was demonstrated that the enolate of an alkyl/aryl β-diketone 66, on exposure to methanesulfonyl azide, smoothly fragments to give the *alkyl* diazoketone 67.

An alternative approach was to use the known coupling of aliphatic diazoalkanes with acid chlorides (71 + 72 → 73). To take advantage of this method, an acceptable procedure for the construction of the requisite diazoalkane 71 was needed. Of the several methods in the literature,[23] the one that has proven most effective so far is nitrosation of the urea, followed by fragmentation (69 → 70 → 71).

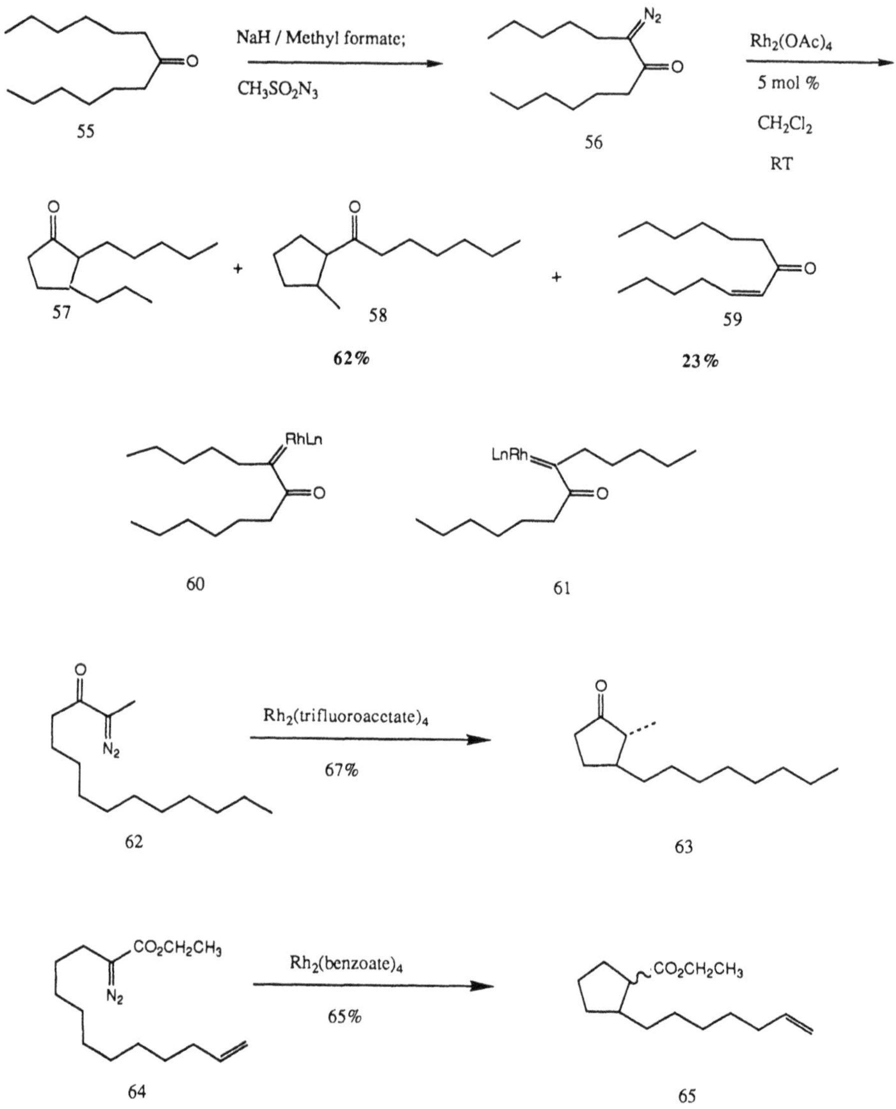

NaH / Methyl formate; $CH_3SO_2N_3$

55 → 56

$Rh_2(OAc)_4$ 5 mol % CH_2Cl_2 RT

57 + 58 + 59

62% 23%

60 61

Rh_2(trifluoroacetate)$_4$ 67%

62 → 63

Rh_2(benzoate)$_4$ 65%

64 → 65

Scheme 6 Cyclization of simple diazoketones and esters.

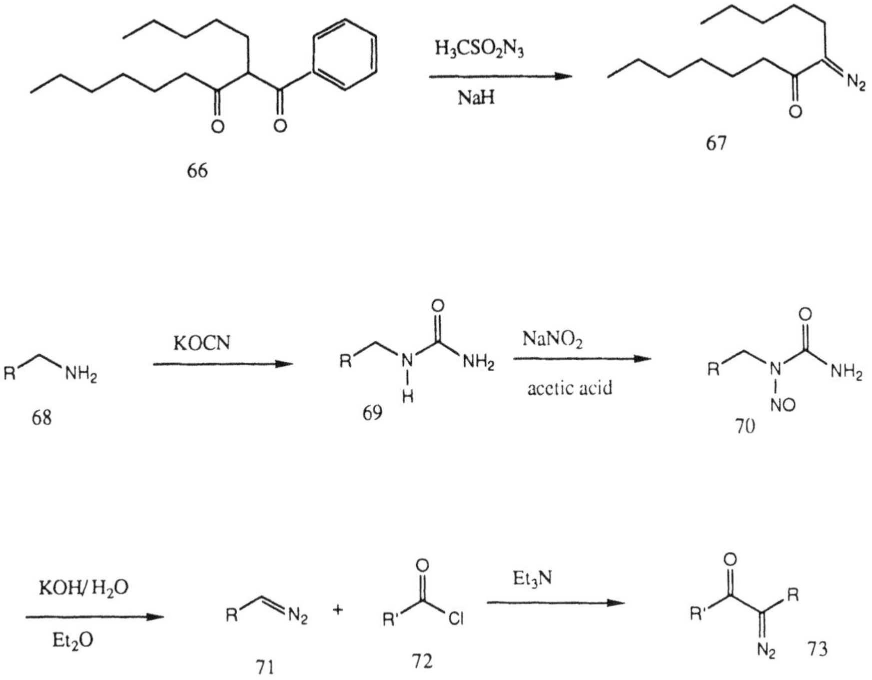

Scheme 7 Preparation of diazoketones.

VI. DIRECTIONS FOR THE FUTURE

A. Enantioselective C-H Insertion

A primary objective in investigating the mechanism of Rh(II)-mediated C-H insertion has been to design a Rh-Rh dimer that would catalyze intramolecular C-H insertion with a synthetically useful degree of enantioselection. Based on the results above, that tetra*carboxylates* are most efficient for catalyzing the desired 1,5 C-H insertion, one can speculate that complexes related to 74 (Scheme 8) might show synthetically useful enantioselectivity.

The R,R diacid needed for the preparation of 74 should be readily available by Evans alkylation.[17] Ligand exchange with $Rh_2(OAc)_4$ would then give 74. If the Rh-Rh dimer indeed opens up but maintains two bridging carboxylates, trans[24] one to another, as outlined in Scheme 5, then 74, on reaction with a diazoketone, 75 (Scheme 8), would give 76.

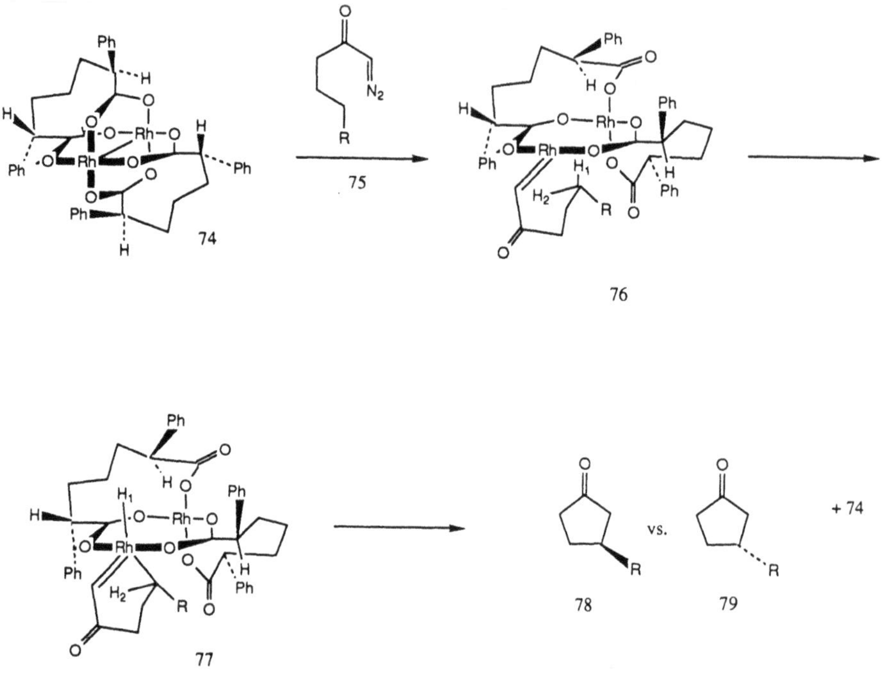

Scheme 8 Proposed enantioselective catalysis of C-H insertion.

It is apparent that in 76 the right-hand phenyl has swung up and the left-hand phenyl has swung down, establishing C_2 symmetry around the rhodium bound to the carbene. As the reaction proceeds, then, insertion into H_1 to give 78 could be substantially preferred over insertion into H_2 to give 79, as the latter would necessitate swinging the pendant alkyl group *through* the left-hand phenyl.

B. Intramolecular Cyclopropanation: Synthesis of Prostaglandin E_2

Total synthesis of the primary prostaglandins, despite the many advances that have been recorded over the last 20 years,[25] is still a lengthy undertaking. The observation that even with the availability of β-hydride, the rhodium carbenoid derived from a simple α-diazoketone will efficiently do 1,5 C-H insertion, suggested that such simple rhodium carbenoids might also be effective for intramolecular cyclopropanation. This has now been shown[26] to be the case.

The starting point for this investigation was the readily available decadienal 80 (Scheme 9), an autooxidation product of fatty acids that is a byproduct of commercial margarine production. Enantioselective assembly of the C-11 stereogenic center[27] followed by saponification and silylation provides acid 81. Coupling of the derived acid

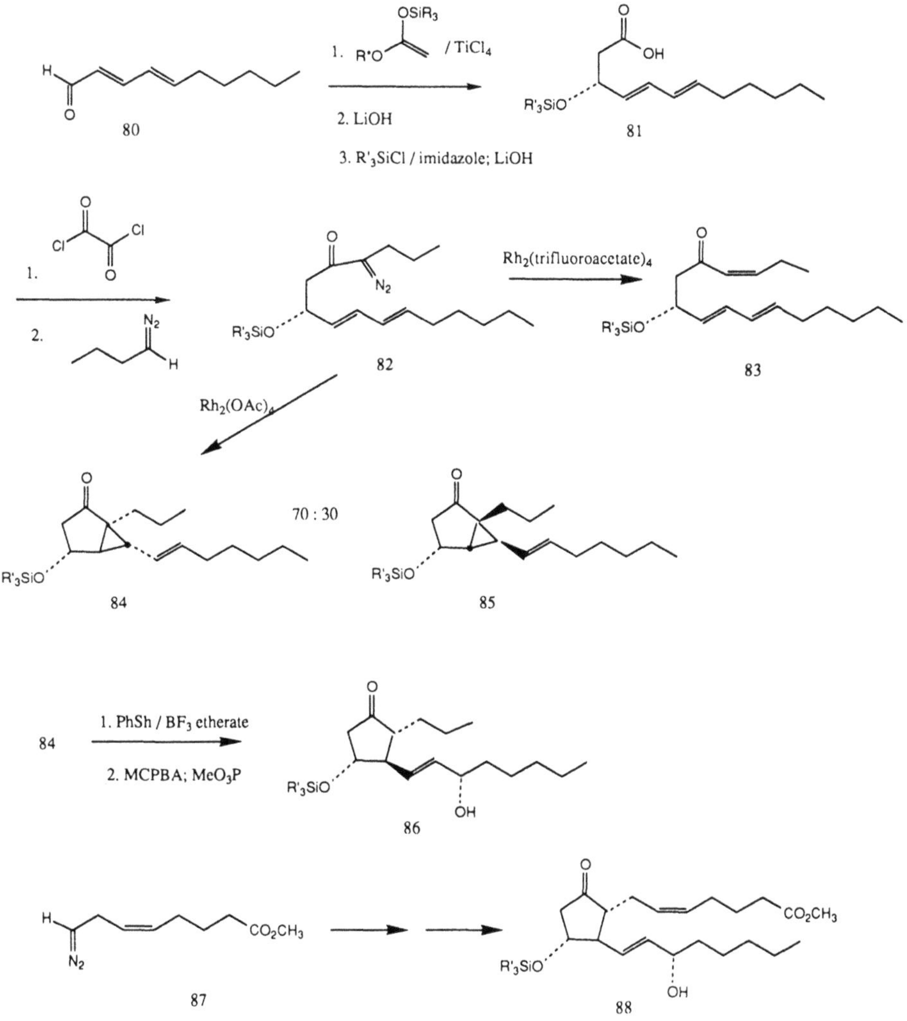

Scheme 9 Proposed synthesis of prostaglandin E_2.

chloride with 1-diazobutane then gives the requisite diazo ketone **82**. Exposure of **82** to a catalytic amount of rhodium trifluoroacetate in CH_2Cl_2 gives almost exclusively the product, **83**, of β-hydride elimination. Rhodium acetate is more favorable, returning **84** and **85** (70:30), with only a trace of enone **83**. Opening of **84** with thiophenol, followed by oxidative rearrangement,[28] then cleanly gives **86**. Completion of the synthesis of PGE_2 **88**, by way of diazoester **87**, is under active investigation.

VII. CONCLUSION

At the outset of this investigation, the scope and selectivity of Rh-mediated intramolecular C-H insertion had been little explored. It has now been demonstrated that the precursor diazoketones and esters are readily prepared, that the cyclizations can show significant regio-, diastereo-, and chemoselectivity, and that simple α-diazoketones, even those having β-hydrogens, can be induced to cyclize efficiently.

There are many potential applications of these cyclizations in natural product synthesis. Only a few of these have yet been explored. Rh-mediated intramolecular C-H insertion should also have wide applicability in the synthesis of pharmaceuticals and other fine organic chemicals.

In addition, it should be noted that this reaction opens up a whole new way of thinking about organic reactivity. The preponderance of carbon-carbon bond-forming reactions are based on either nucleophilic addition or displacement, or on electrophilic addition to an alkene. In contrast, the Rh-mediated insertion proceeds, in what appears to be an electrophilic process, by direct addition to an unactivated C-H bond. This Rh-mediated process should be of continuing interest as a tool for probing the availability of electron density in a sigma framework.

REFERENCES

1. S. D. Burke and P. A. Grieco, *Org. React. (N.Y.)*, *26*, 361 (1979).
2. E. Wenkert, L. L. Davis, B. L. Mylari, M. F. Solomon, R. R. daSilva, S. Shulman, R. J. Warnet, P. Ceccherelli, M. Curini, and R. Pellicciari, *J. Org. Chem.*, *47*, 3242 (1982) and references cited therein.
3. (a) A. J. Anciaux, A. J. Hubert, A. F. Noels, N. Petinot, and P. Teyssie, *J. Org. Chem.*, *45*, 695 (1980). (b) A. Demonceau, A. F. Noels, A. J. Hubert, and P. Teyssie, *J. Chem. Soc. Chem. Commun. (1981)*, 688.
4. D. F. Taber and E. H. Petty, *J. Org. Chem.*, *47*, 4808 (1982).
5. M. W. Rathke and J. Deitch, *Tetrahedron Lett.*, *12*, 2953 (1971).
6. L. Weiler and S. N. Huckin, *J. Am. Chem. Soc.*, *96*, 1082 (1974).
7. (a) M. Regitz, J. Hocker, and A. Liedhegener, *Organic Synthesis*, John Wiley and Sons, New York, 1973, Collect. Vol. V, p. 197. (b) D. F. Taber, R. E. Ruckle, Jr., and M. J. Hennessy, *J. Org. Chem.*, *51*, 4077 (1986).
8. D. F. Taber and R. E. Ruckle, Jr., *J. Am. Chem. Soc.*, *108*, 7686 (1986).
9. J. C. Gilbert, D. H. Giamalva, and M. E. Baze, *J. Org. Chem.*, *50*, 2557 (1985).
10. M. J. Hennessy, Ph.D. Dissertation, Univ. of Delaware, 1988.
11. J. Y. Saillard and R. Hoffman, *J. Am. Chem. Soc.*, *106*, 2006 (1984).
12. D. F. Taber and J. S. Schuchardt, *J. Am. Chem. Soc.*, *107*, 5289 (1985).
13. D. F. Taber and J. S. Schuchardt, *Tetrahedron*, *43*, 5677 (1987).
14. S. H. Bertz, *J. Am. Chem. Soc.*, *103*, 3599 (1981).
15. D. F. Taber, E. H. Petty, and K. Raman, *J. Am. Chem. Soc.*, *107*, 196 (1985).
16. *Copper*-catalyzed intramolecular C-H insertion had previously been shown to proceed with retention of absolute configuration, albeit in mediocre yield: H. Ledon, G. Linstrumelle, and J. Julia, *Tetrahedron Lett.*, *14*, 25 (1973).
17. D. A. Evans, M. D. Ennis, and D. J. Mathre, *J. Am. Chem. Soc.*, *104*, 1737 (1982).
18. D. F. Taber and K. Raman, *J. Am. Chem. Soc.*, *105*, 5935 (1983).
19. D. F. Taber, K. Raman, and M. D. Gaul, *J. Org. Chem.*, *52*, 28 (1987).

20. J. C. Amedio, Jr., unpublished observations, Univ. of Delaware.

21. L. D. Cama and B. G. Christensen, *Tetrahedron Lett.*, *19*, 4233 (1978).

22. For a recent comprehensive review of the chemistry of α-diazo ketones and esters, see G. Maas, *Topics in Current Chemistry*, Vol. 137, Springer-Verlag, Berlin, 1987, p. 75.

23. For a review of diazoalkane preparation, see S. Patai (ed.), *The Chemistry of Diazonium and Diazo Groups*, John Wiley and Sons, New York, 1978.

24. M. J. Doyle, M. F. Lappert, G. M. McLaughlin, and J. McMeeking, *J. Chem. Soc. Dalton Trans. (1974)*, 1494.

25. For alternative synthetic approaches to the prostaglandins, see G. Stork, P. M. Sher, and H.-L. Chen, *J. Am. Chem. Soc.*, *108*, 6834 (1986) and references cited therein.

26. R. Scott Hoerrner, unpublished observations, Univ. of Delaware.

27. G. Helmchen, U. Leikauf, and I. Taufer-Knopfel, *Angew. Chem. Int. Ed. (Eng.)*, *24*, 874 (1985).

28. For examples of *base*-catalyzed cyclopropane opening by thiophenol, see: (a) D. F. Taber, *J. Am. Chem. Soc.*, *99*, 3513 (1977); (b) K. Kondo, T. Umemoto, Y. Takahatake, and D. Tunemoto, *Tetrahedron Lett.*, *15*, 113 (1977); (c) K. Kondo, T. Umemoto, Y. Kumiko, and D. Tunemoto, *Tetrahedron Lett.*, *16*, 3927 (1978).

Part II

New Catalysis Technologies

5

Membrane-Enclosed Enzymatic Catalysis (MEEC): A Practical Method for the Use of Enzymes as Catalysts in Organic Synthesis

MARK D. BEDNARSKI

Department of Chemistry, University of California at Berkeley Berkeley, California

ABSTRACT

This paper describes a particularly convenient technique for manipulating enzymes to be used in organic synthesis, in which soluble protein is enclosed in dialysis membranes. This type of containment facilitates separation of the enzyme from the reaction medium and allows its reuse. In many instances, enzyme stabilities equaled those observed with gel- or solid-immobilized enzymes. Enzymes remained active in the presence of water-miscible and water-immiscible organic solvents. Regeneration of both nicotinamide and nucleotide triphosphate cofactors was also possible using this technique. Modification of the membrane and how these changes relate to transport rates of different materials were also investigated. The use and limitations of using membrane-enclosed enzymes for practical organic synthesis is discussed.

I. INTRODUCTION

Enzymes catalyze many synthetically useful organic reactions.[1-3] Despite their diversity and catalytic power they are still not commonly used in organic chemistry. One of the major limitations of enzymatic reactions is the availability of simple methods to manipulate proteins.[4-6] The focus of this chapter is to demonstrate a new method to use enzymes in organic chemistry called membrane enclosed enzymatic catalysis (MEEC). This technique has been useful in a variety of reactions commonly used by organic chemists.[7]

II. SYNTHETIC APPLICATIONS OF MEMBRANE REACTORS

Many modern synthetic methods in organic chemistry focus on reactions that form carbon-carbon bonds.[8-10] Two classes of enzymes (aldolases and synthetases) can be used to perform similar reactions enzymatically; both classes of enzymes operate in membrane-enclosed reactors.[7] Equation 1 demonstrates the use of N-acetylneuraminate pyruvate-lyase (NeuAc-aldolase; E.C. 4.1.3.3) as an aldol catalyst to synthesize complex carbohydrates.[11] The enzyme forms a carbon-carbon bond between pyruvic acid (2) and N-acetylmannosamine (1) to give neuraminic acid (3). The enzyme can be recovered and reused for several

HO HO NHAc -O HO OH **1** + H_3C O CO_2H **2** NeuAc -aldolase → AcNH O OH CO_2H HO OH OH OH **3** (1)

reactions when it is enclosed within a membrane and isolation of the product from the protein is also straightforward.

Fructose-1,6-diphosphate aldolase (FDP-aldolase; E.C.4.1.2.13) condenses dihydroxyacetone phosphate (4) with a variety of aldehydes (shown below is the reaction of 2-phenyl propanal (5)) to give polyhydroxylated aldol adducts (Eq. 2).[12] This reaction demonstrates the

HO O OP **4** + H O Me H Ph **5** 1) FDP-Aldolase/DMSO-H_2O 2) Acid Phosphatase (E.C. 3.1.3.2) → Me H H HO O OH Ph OH H **6** (2)

use of the MEEC system with miscible organic solvents such as dimethyl sulfoxide (DMSO). Acid phosphatase, also enclosed within a membrane, is used to dephosphorylate these products.[13] Since acid phosphatase has a low specific activity, large amounts of protein must be used in these reactions. The MEEC system can be used to isolate the protein from this reaction mixture which simplifies isolation of the product.

The enzyme 3-Deoxy-D-manno-octulosonate-8-phosphate synthetase (KDOS; E.C.4.1.2.16) forms a carbon-carbon bond between phosphoenol pyruvate **8** (PEP) and arabinose-5-phosphate 7.[14] This enzyme is not

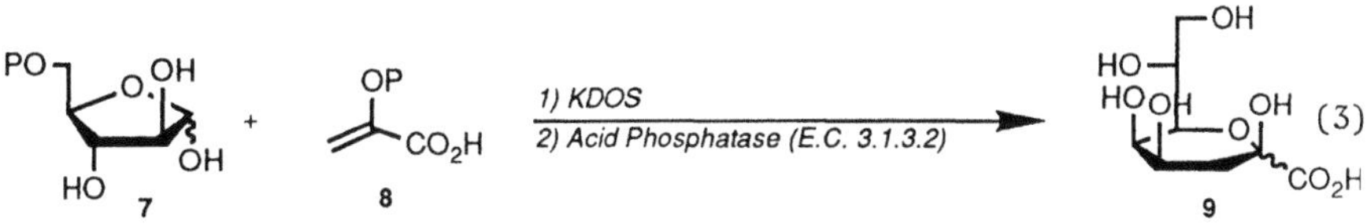

(3)

commercially available but it can be easily isolated as a crude preparation from E. coli K-12 cells.[15] The arabinose-5-phosphate (**7**) used in this reaction is synthesized using hexokinase. Hexokinase, pyruvate kinase, and KDOS are also enclosed in a membrane.[7] The membrane reactor system, therefore, allows the use (and reuse) of crude enzyme solutions without the products being contaminated with residual proteins. Crude enzyme preparations can, therefore, be easily assayed for synthetic utility for a particular organic transfromation using MEEC.

III. OXIDATIONS, REDUCTIONS AND PHOSPHORYLATIONS

Eqs. 4-6 demonstrate the use of MEEC in oxidation, reduction and phosphorylation reactions. These reactions require the use of cofactors such as NADH, NAD and ATP. Equation 4 shows the synthesis of optically pure 2-hydroxy butyric acid **12** from keto acid **10** using D-lactate dehydrogenase (D-LDH; E.C.1.1.1.28).[16] Formate dehydrogenase (FDH; E.C.1.2.1.2) is used to recycle the NADH cofactor via

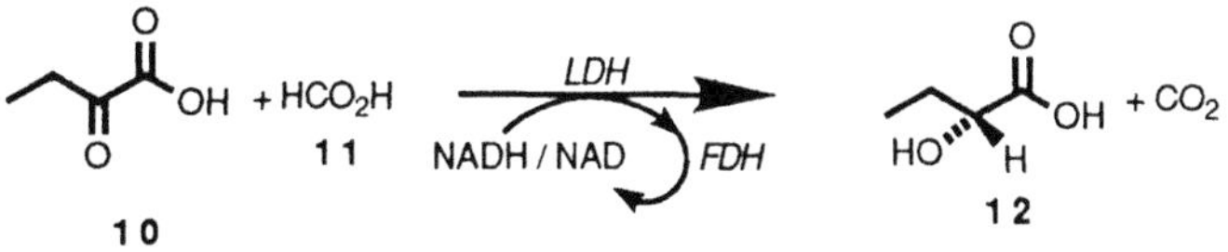

(4)

13 + Pyruvate 2 —HLADH, Water/Hexane; NAD / NADH; LDH→ 14 + Lactate (5)

the oxidation of formic acid 11 to carbon dioxide. Eq. 5 represents a typical example for the use of horse liver alcohol dehydrogenase (HLADH; E.C.1.1.1.1) in the synthesis of optically pure lactones such as compound 14 from symmetrical diol 13.[17] The enzyme L-lactate dehydrogenase (LDH; E.C.1.1.1.27) was used to recycle the NAD cofactor by reducing pyruvate to lactate.

Eq. 6 demonstrates the use of ATP and hexokinase (HK; E.C.2.7.1.1) to phosphorylate glucose.[18] The enzyme is specific for the C_6 hydroxyl group of the sugar resulting in the conversion of glucose (15) into

Glucose 15 + PEP 5 —HK; ATP / ADP; PK→ Glucose-6-phosphate 16 + Pyruvate 2 (6)

glucose-6-phosphate (16). The ATP is recycled using pyruvate kinase (PK; E.C.2.7.1.40) and PEP 5. The recycling of cofactors and their transport rates in membrane reactors are discussed further in Sections IV and V.

IV. HYDROLYTIC ENZYMES

The use of the MEEC system with hydrolytic enzymes is demonstrated in Eqs. 7 and 8. Membrane enclosed enzymes can be used to hydrolyze both amide and ester bonds with high enantiofacial selectivity. Acylase I, for example, was used to resolve α-methylmethionine 17 to give compounds 18 and 19 on a 10 mmol scale.[7] The optical purity of amino acid 19 was greater than 98%.[19] The poor reactivity of this substrate required the use of large amounts of enzyme (five grams of

17 —Acylase I→ 18 + 19 (7)

$$\mathbf{20} \xrightarrow{\text{Lipase}} \mathbf{21} + \mathbf{22} \qquad (8)$$

enzyme for approximately the same amount of substrate. The product could, however, easily be recovered from the reaction mixture and the enzyme reused in other reactions using the MEEC system.[7]

The use of MEEC is less effective with lipases (Eq. 8).[7] The reaction of epoxyester 20 with porcine pancreatic lipase (E.C. 3.1.1.3) to give compounds 21 and 22 proceeded at only one-fifth of the rate expected on the basis of the usual emulsion-based procedures.[20] Lipases require a water-organic interface for activity and the membrane may limit the interfacial area and transport of the substrate and products.

MEEC also was studied using sensitive enzymes such as the 2,6 sialyl transferase (α 2,6ST; E.C.2.4.99.1).[21] CMP-NeuAc 23 was reacted with N-acetyllactosamine 24 to give the trisaccharide 25

$$\underset{\mathbf{23}}{\text{CMP-NeuAc}} + \underset{\mathbf{24}}{\text{N-acetyllactosamine}} \xrightarrow{\alpha\,2,6ST} \underset{\mathbf{25}}{\alpha\,\text{DNeuAc(2,6)}\,\beta\text{DGal(1,4)DGlcNAc}} \qquad (9)$$

(Eq. 9). The enzyme could easily be recycled. This makes the use of sensitive and expensive proteins in organic synthesis more economical.[7]

V. STABILITY AND RECOVERY OF MEMBRANE-ENCLOSED ENZYMES

Membrane reactors can be used to simplify the separation of the protein from the reaction media while allowing reactants and products to contact the catalyst.[22] In all of the cases studied using MEEC, separation of the protein from the reaction mixture was straightforward and recovery of enzymatic activity was high.[7] In reactions summarized in Eqs. 1-4, for example, we demonstrated that the recovered enzymes could be stored at 4°C and reused in other reactions. In cases 1 and 4 we compared the stability of membrane-enclosed enzymes to immobilized enzymes on PAN gel.[23] The stability of the membrane-enclosed enzymes equalled or exceeded the immobilized proteins. The increased stability of enzymes sequestered in membranes is probably due to the high concentration of protein within the membrane which mimics the environment of a cell. The addition of bovine serum albumin (BSA) to the enzyme solution also improves enzyme stability, especially in cases where dilute solutions of enzymes are used.[7]

The use of MEEC to prevent interfacial deactivation of enzymes with immiscible organic solvents was also investigated. Confinement of soluble horse liver alcohol and lactate dehydrogenase in dialysis tubing prevented its contact with hexane during the oxidation of diol **13** to lactone **14** (Eq. 4). Another advantage of MEEC is its use with crude enzyme preparations (Eq. 2). In this case the crude protein mixtures isolated from a bacterial source could be used in synthetic applications without purification and the protein recovered without significant loss of enzymatic activity. The products could also be easily isolated from the reaction mixtures avoiding difficult schemes to remove the protein from the reaction mixture.

Most synthetic applications demand the use of enzymes with unnatural substrates. These substrates usually bind poorly to the enzyme and have poor turnover rates (i.e., kcat/Km is significantly less than the value for the natural substrate). To overcome these problems large amounts of protein can be isolated within a membrane to give a reactor having a high volume activity.[24] We have demonstrated the use of the MEEC system in the enzymatic resolution of α-methyl methionine (Eq. 7), a poor substitute for acylase.[7] A similar reactor has also been used by Ray and coworkers to resolve meso esters by enzymatic reactions using proteases.[25]

VI. COFACTOR REGENERATION

Cofactors are necessary in many enzymatic reactions useful in organic synthesis.[1-3] Since most cofactors are expensive (e.g., NADH costs approximately $\$5 \times 10^4$/mole from Sigma chemical company) it is necessary to regenerate cofactors.[26] The regeneration of cofactors has been extensively reviewed and most problems have been solved using enzymatic regeneration systems.[27] NADH, for example, can easily be recycled using lactate dehydrogenase and NAD is recycled using formate dehydrogenase. ATP can also be regenerated using pyruvate kinase and PEP. We have tested the regeneration of nicotinamide and nucleoside cofactors using membrane enclosed reactors. Both NAD (Eq. 4), NADH (Eq. 5), and ATP (Eq. 6) could easily be recycled in the MEEC reactor.[7] The MEEC system also the advantage that the enzymes necessary for the cofactor recycling do not have to be coimmobilized with the protein used to catalyze the desired reaction. All enzymes are simply added to the membrane reactor with the desired cofactor and reagents.

VII. RATES OF TRANSPORT

The transport of most organic molecules is sufficient for synthetic applications using MEEC. In cases where unnatural substrates are used the reaction rates are usually determined by the rate of the reaction and not membrane diffusion. We have measured the rate of transport for a variety of small organic molecules through commercially available dialysis membranes. Our results show that amino acids such as alanine and phenylalanine transport at a rate of approximately $3 \times 10^{-2}\ \text{min}^{-1}\ \text{cm}^{-1}$. Carbohydrates and carbohydrate derivatives such as glucose-1-phosphate transport at a rate of $4 \times 10^{-2}\ \text{min}^{-1}\ \text{cm}^{-1}$ and cofactors such as NADH transport at a rate of $2 \times 10^{-3}\ \text{min}^{-1}\ \text{cm}^{-1}$. The slower transport rates for large cofactors can, in principle, be used to sequester the cofactor (in the presence of the enzyme) from the reaction media.

VIII. SUMMARY: ADVANTAGES AND DISADVANTAGES OF MEEC IN ENZYMATIC CATALYSIS

The major advantages of MEEC relative to immobilization is its simplicity and convenience. The MEEC system also makes it possible to obtain high-volume activities of enzyme in the reaction vessel and eliminates the loss of enzymatic activity on immobilization. In addition, MEEC is applicable to crude enzyme preparations. Its disadvantages are that reaction rates may be slow under circumstances in which mass transport across the membrane is rate limiting and that enzyme deactivation due to protease contaminants is not prevented (although adding bovine serum albumin within the membrane seems to stabilize some enzymes).

REFERENCES

1. J. B. Jones, *Tetrahedron, 42,* 3351 (1986).
2. G. M. Whitesides and C.-H. Wong, *Angew. Chem. Int. Ed. Engl., 24,* 617 (1985).
3. R. Porter and S. Clark (eds.), *Enzymes in Organic Synthesis,* Ciba Foundation Symposium 111, Pitman, London (1985).
4. I. Chibata, T. Tosa, and T. Sato, *J. Mol. Cata., 37,* 1 (1986).
5. a) A. M. Klibanov, *Science, 219,* 722 (1983); b) A. M. Klibanov, *T.I.B.S., 14,* 141 (1989).
6. I. Chibata, *Immobilized Enzymes—Research and Development,* Halstead Press, New York, 1978.
7. M. D. Bednarski, H. K. Chenault, E. S. Simon, and G. M. Whitesides, *J. Am. Chem. Soc., 109,* 1283 (1987).
8. W. Bartmann and K. B. Sharpless (eds.), *Stereochemistry of Organic and Bioorganic Transformations,* VCH, New York, 1987.
9. J. D. Morrison (ed.), *Asymmetric Synthesis*, Academic, New York, 1983, Vols. 2 and 3.
10. L. Eliel and S. Otsuka (eds.), *Asymmetric Reactions and Processes in Chemistry,* ACS Symposium Series 185, American Chemical Society, Washington D.C., 1982.
11. C. Auge, S. David, C. Gautheron, and A. Veyrieres, *Tetrahedron Lett., 26,* 2439 (1985).
12. a) C.-H. Wong, F. P. Mazenod, and G. M. Whitesides, *J. Org. Chem., 48*, 3199 (1983); b) M. D. Bednarski, E. S. Simon, N. Bischofberger, W. D. Fessner, M.-J. Kim, W. Lees, T. Saito, H. Waldmann, and G. M. Whitesides, *J. Am. Chem. Soc., 111,* 627 (1989).
13. G. Schmidt, *The Enzymes, 5,* 37 (1961).
14. M. D. Bednarski, D. C. Crans, R. DiCosimo, E. S. Simon, P. D. Stein, and G. M. Whitesides, *Tetrahedron Letters, 29,* 427 (1988).
15. P. H. Ray, *J. Bacteriol., 193,* 635 (1980).
16. B. L. Hirschbein and G. M. Whitesides, *J. Am. Chem. Soc., 104,* 4458 (1982).

17. I. J. Jakovac, H. B. Goodbrand, K. P. Lok, and J. B. Jones, *J. Am. Chem. Soc.*, *104*, 4659 (1982).

18. B. L. Hirschbein, F. P. Mazenod, and G. M. Whitesides, *J. Org. Chem.*, *47*, 3765 (1982).

19. The absolute configuration is based on the analogy with the previously determined stereoselectivity of acylase I: J. P. Greenstein and M. Winitz (eds.), *Chemistry of the Amino Acids*, Vol. 2, John Wiley and Sons, 1961, New York, pp. 17353-1767.

20. W. E. Ladner and G. M. Whitesides, *J. Am. Chem. Soc.*, *106*, 7250 (1984).

21. S. Sabesan and J. C. Paulson, *J. Am. Chem. Soc.*, *108*, 2068 (1986).

22. a) W. Pusch, Walch, *Angew. Chem., Int. Ed. Engl.*, *21*, 660 (1982); b) R. E. Kesting, *Synthetic Polymer Membranes*, McGraw-Hill, New York, 1971.

23. A. Pollack, H. Blumenfeld, M. Waz, R. L. Baughn, and G. M. Whitesides, *J. Am. Chem. Soc.*, *102*, 6324 (1980).

24. The amount of enzyme that can be covalently immobilized on most solid supports or gels is limited; it is difficult to obtain high-volume catalytic activities unless the enzyme preparation itself has a high specific activity. These types of immobilization procedures are thus less convenient than MEEC for enzymes with intrinsically low specific activities and for crude enzyme preparations.

25. R. Roy, A. W. Rey, *Tetrahedron Letters*, *28*, 4935 (1987).

26. D. C. Crans, R. J. Kazlauskas, B. L. Hirschbein, C.-H. Wong, O. Abril, and G. M. Whitesides, *Methods Enzymol.*, in press.

27. H. K. Chenault and G. M. Whitesides, *Appl. Biochem. Biotechnol.*, *14*, 147 (1987).

6

Kinetics of Hydrogenation of *p*-Nitrobenzenesulfonamide in a Three-Phase Slurry Reactor

MAKARAND G. JOSHI, WILLIAM E. PASCOE, and DENNIS E. BUTTERFIELD
Chemical Engineering Research Laboratory
Eastman Kodak Company, Rochester, New York

ABSTRACT

In catalytic hydrogenations the interaction between the transport processes and the reaction kinetics can play an important role in determining reactor performance. It is essential to know the intrinsic kinetics of the reaction under consideration for proper reactor design, scale-up, and troubleshooting. A reactor system equipped with a PC-based data acquisition system capable of measuring the intrinsic kinetics of catalytic hydrogenations was assembled. Hydrogenation of *p*-nitrobenzenesulfonamide to *p*-aminobenzenesulfonamide over Raney cobalt in ethanol was chosen as the model system to be studied. The experimental conditions were chosen to ensure that the reaction was not mass transfer-limited. The hydrogenation was modeled as a three-step series reaction. The kinetics of each of the three steps was represented by the power law model. The orders of the steps with respect to organic substrates and hydrogen were determined from the experimental data. The activation energy for the reaction was also determined. A kinetic expression for the reaction was developed.

I. INTRODUCTION

Aromatic amines are major intermediates in the manufacture of a wide variety of products such as polymers, antioxidants, herbicides, insecticides, dyes, pharmaceuticals, and photographic chemicals. These amines are usually made by catalytic hydrogenation of corresponding nitro compounds. Batch slurry reactors are the reactors of choice for these reductions. Since there are three different phases present in the reactor and the nitro functions are reduced quite readily, the mass transfer resistances can play an important role in determining the overall rate of the reaction and the selectivity.[1-5] Furthermore these reductions are also highly exothermic and a careful consideration has to be given to the heat removal from the reactor at least from the safety viewpoint. Thus it is not surprising that the scale-up of these reactions is often difficult. This motivated us to investigate the interaction of transport processes and the reaction kinetics in these reactions. The main objective of the study was to develop a methodology to evaluate the mass and heat transfer effects in production reactors used for these reductions and suggest ways to improve the efficiency of these reactors and/or develop new reactor strategies to carry out these reductions.

A laboratory reactor capable of measuring intrinsic kinetics without interference from the mass transfer limitations was assembled as a first step toward our objective. It was decided to study the reduction of *p*-nitrobenzenesulfonamide (NBS) over Raney cobalt as a model reaction to test the reactor. The overall reaction is shown in Fig. 1. The catalyst was selected due to its activity. The dehalogenation can be a major side reaction in the reduction of halonitro compounds and the role of mass transfer in controlling this undesired side reaction will be looked at in the future.

A literature search aimed at finding reports of past studies of the kinetics of reduction of NBS showed that this particular reduction has not been studied before. However, many similar reductions have been investigated. The actual mechanism of a particular reaction may depend on the catalyst used and other reaction conditions. The purpose

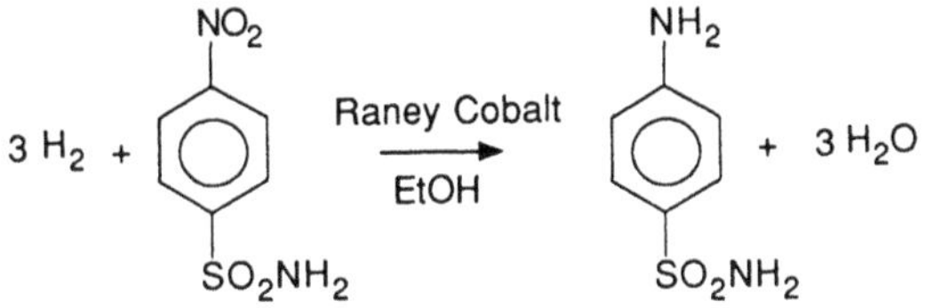

$$3\,H_2 + NO_2C_6H_4SO_2NH_2 \xrightarrow[\text{EtOH}]{\text{Raney Cobalt}} NH_2C_6H_4SO_2NH_2 + 3\,H_2O$$

Fig. 1 The overall reaction.

of the present study was not to determine the actual mechanism of the reaction but to develop an expression which will describe the rate of reaction taking place on the catalyst surface.

II. EXPERIMENTAL

A. Materials

p-Nitrobenzenesulfonamide and its reduction product *p*-aminobenzenesulfonamide (ABS) were reagent grade chemicals obtained from Eastman Kodak Company. The amine (ABS) was procured for developing the analytical method and its calibration. These chemicals were used as received without any further purification. Hydrogen gas supplied by Air Products and Chemicals was reaction grade (99.97% pure). The catalyst, Raney cobalt no. 2427, was manufactured by Davison Specialty Chemical Company, which is a division of W. R. Grace and Company. The catalyst was in a fine-powder form with particle sizes between 40 and 80 μm. The catalyst is pyrophoric and was stored under water. It was used in the wet form and the amounts reported are wet weights. Absolute anhydrous ethanol was used as the reaction solvent. It was procured from Eastman Kodak Company and was spectroscopic grade.

B. Apparatus

A schematic diagram of the apparatus used is shown in Fig. 2. Experiments were carried out in a 500-ml stainless steel autoclave obtained from Autoclave Engineers Inc. It was equipped with a gas-inducing (hollow-shaft) turbine agitator driven by compressed air. The turbine had eight blades and an outside diameter of 25 mm. A removable stainless steel double baffle with a width of 12.7 mm (1/2 in.) was placed

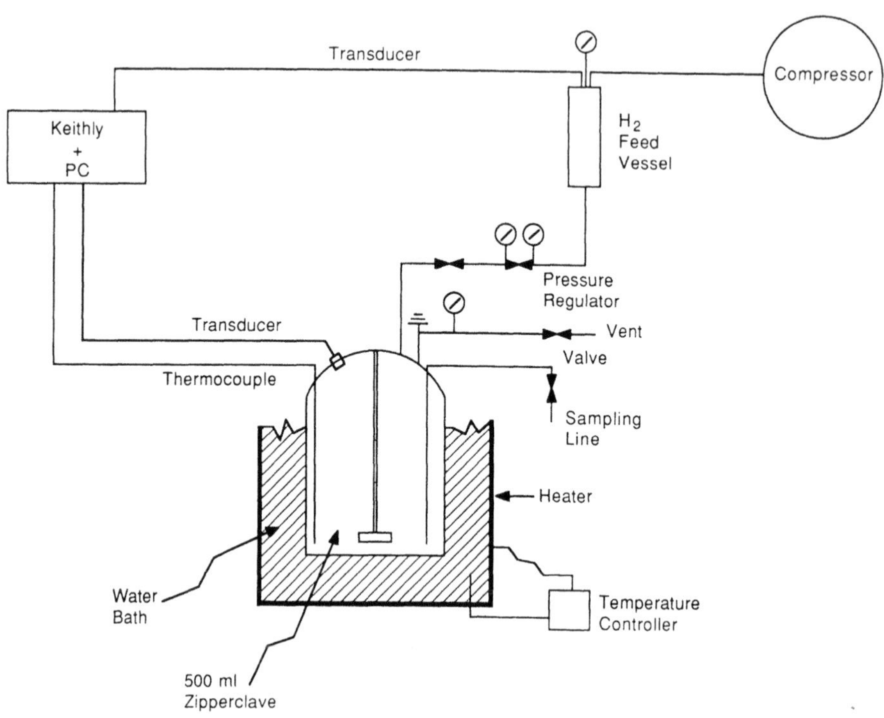

Fig. 2 Schematic drawing of apparatus.

in the reactor to improve mixing. The temperature of the reactor contents was measured by a type J thermocouple dipping directly in the liquid. Hydrogen was fed in the reactor through an opening in the top of the reactor. A pressure gage and a transducer were attached to another opening in the top of the reactor. The reactor was partially submerged in a constant-temperature bath to control the temperature of its contents. The rates of heat generation in the reactor were moderate in all cases and it was possible to control the reaction temperature within 1°C of the setpoint quite easily. The temperature of water in the bath was measured using a thermocouple. The pressure in the reactor was held constant during an experimental run. Hydrogen consumed by the reduction was replenished by supplying it to the reactor from a feed vessel through a pressure regulator. The void volume of the feed vessel and connecting lines was 145 ml. The pressure in the feed vessel was measured by a pressure gage and a transducer. The outputs of the transducers and thermocouples were monitored by a data acquisition system (Keithley DAC Inc.) controlled by a personal computer (Compaq Computer Corporation). The samples of the reaction mixture were analyzed on a Hewlett-Packard model 1090 liquid chromatograph using a Supelcosil LC-8 column obtained from Supelco Inc.

C. Procedure

In a typical experimental run 2 g of wet catalyst (50% water) was suspended in 150 ml of ethanol. This mixture was poured in the empty reactor. After the closing of the reactor, the water bath was raised to surround the reactor partially. The temperature of catalyst suspension was brought up to the desired reaction temperature and hydrogen was introduced into the reactor bringing it to the desired pressure. The stirrer was started and its speed was measured by a strobe light. The stirring speed was adjusted to the desired value by varying the pressure of air being supplied to the stirrer motor. In all reactions the stirrer speed was 2000 rpm. The catalyst suspension was stirred under hydrogen pressure for 1 hr; then the stirrer was stopped and the reactor was vented. This prereduction of catalyst eliminated variability caused by the catalyst induction period. The headspace of the reactor was purged with nitrogen and a solution of 5 g of NBS in 150 ml of ethanol was added to the catalyst suspension in the reactor. An initial sample of the reaction mixture was taken before the reactor was closed and its contents were allowed to come up to the desired temperature set by the water bath. Once the mixture was in thermal equilibrium with the bath, hydrogen was once again introduced in the reactor and its pressure adjusted to the desired value. The reaction was then started by switching on the stirrer. As hydrogen was consumed in the reactor it was replenished by hydrogen from the feed vessel, which was charged to a pressure much higher than that in the reactor (1500 psi). The progress of the reaction was followed by monitoring the drop of hydrogen pressure in the feed vessel with time using the data acquisition system. Hydrogen pressure in the feed vessel was converted to moles of hydrogen using the van der Waals equation. Thus it was possible to convert the rate of pressure drop in the feed vessel to the rate of hydrogen consumption in the reactor. The pressure in the feed vessel was monitored once per second to yield the desired accuracy in the rate of hydrogen consumption. The temperature of reaction mixture and hydrogen pressure in the reactor were also recorded at the same frequency. Once it was evident that the reaction was over (no further drop in pressure in the feed vessel), the stirrer was stopped and the reactor vented. The reactor was opened and a final sample of the reaction mixture was obtained. The initial and final samples of the reaction mixture were analyzed using a liquid chromatograph (LC). The LC method is described later in this section.

In a few experiments samples of the reaction mixture were obtained during the reaction. A sampling tube was installed in the reactor for these experiments. The procedure for these experiments was the same as for other experiments except that the hydrogen pressure in the feed vessel was not monitored. To take a sample of the reaction mixture the stirrer was stopped and a small (2-ml) sample was obtained by opening the needle valve on the sampling tube. The first couple of milliliters of the liquid coming out of the sampling tube were rejected and then a sample was collected. Once the sample was obtained, the reaction was allowed to proceed by switching on the stirrer. In a typical experiment roughly 10 samples were collected in this way. These samples were analyzed on a LC using a method described below.

A reversed phase ion chromatographic procedure was used to analyze the samples of reaction mixture. A 250 x 4.6 mm ID Supelcosil LC-8 column packed with 5-μm silica particles coated with octyldimethylsilyl was used. The buffer was 0.15 M solution of sodium heptanesulfonate to which 15 ml of 1% acetic acid was added per liter of solution. The second solvent was acetonitrile. A gradient of eluent composition was used. Initially the eluent consisted of 100% buffer. It changed to 80% buffer and 20% acetonitrile in 10 min. The composition of mobile phase was held constant for the next 4 min before it was brought back to 100% buffer in 2 min. The flow rate of mobile phase was 1.5 ml/min. The injection volume was 5 μl. With this method it was possible to get two distinct peaks, one for the starting material (NBS) and a second for the reduction product (ABS). No intermediates were detected. Nevertheless their existence was evident from the mass balance. A linear calibration curve was obtained. The samples of reaction mixture were diluted by a factor of 100 so that their concentration was in the central region of the calibration curve. A new calibration curve was established every time prior to analyzing the unknown samples. The accuracy of the analytic results was checked by preparing a control chart of the liquid chromatograph performance. The data for the control chart were obtained by analyzing solutions of known concentration before and after the samples of reaction mixture. The upper and lower control limits, which are roughly three standard deviations away from the mean, were within 12.5% of the mean.

III. THEORY

The widely accepted reaction mechanism for the reduction of aromatic nitro compounds was deduced by Haber[6,7] in 1898. It is shown for NBS in Fig. 3. If the competing parallel pathways are neglected, the reduction can be approximated by a series reaction:

$$NO_2C_6H_4SO_2NH_2 \longrightarrow NOC_6H_4SO_2NH_2 \longrightarrow NHOHC_6H_4SO_2NH_2 \longrightarrow NH_2C_6H_4SO_2NH_2 \qquad (1)$$

which can be represented by following equations

$$A + B \rightarrow E \qquad (2)$$

$$A + E \rightarrow F \qquad (3)$$

$$A + F \rightarrow G \qquad (4)$$

where A is hydrogen, B is the substrate (NBS here), E and F are intermediates, and G is the reduction product (ABS in the present case).

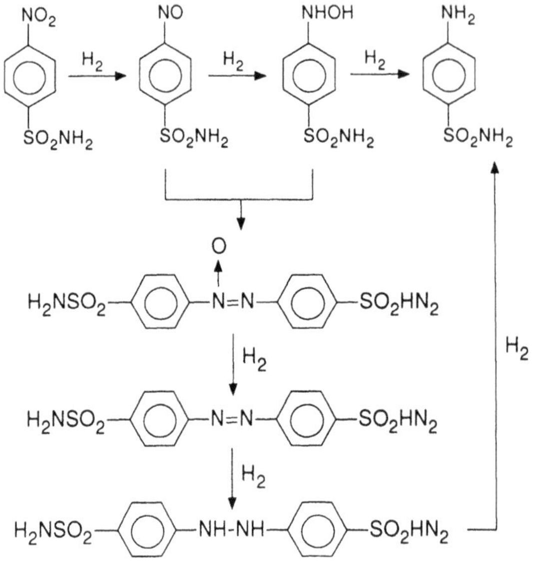

Fig. 3 Reduction mechanism as per Haber.[6,7]

Since hydrogen is usually supplied as a gas, the substrate is dissolved in a solvent, and the reaction takes place on the surface of a catalyst, it is clear that the reactants have to cross one or two phase boundaries in order to react. If the catalyst is porous and the active sites are located in the pores, then there is an additional intraparticle resistance which has to be overcome by the reactants. This intraparticle resistance is negligible if either the active sites are located on the external surface (e.g., if catalyst particles are impervious) or the diffusion in the pores is fast. If it is assumed that the intrinsic kinetics of the reactions shown in Eqs. 2-4 can be represented by the power law model and the intraparticle resistance is negligible, then the rate of change of concentrations of various species is given by:

$$-\frac{dB_\ell}{dt} = wk_1 A_s^{\beta_1} B_s^{\alpha_1} \tag{5}$$

$$-\frac{dE_\ell}{dt} = -wk_1 A_s^{\beta_1} B_s^{\alpha_1} + wk_2 A_s^{\beta_2} E_s^{\alpha_2} \tag{6}$$

$$-\frac{dF_\ell}{dt} = -wk_2A_s^{\beta_2}E_s^{\alpha_2} + wk_3A_s^{\beta_3}F_s^{\alpha_3} \tag{7}$$

where the symbols B, E, and F denote the concentrations in moles/cm^3 of compounds B, E, and F, respectively, and their subscripts ℓ and s denote liquid and catalyst surface, respectively; w is the amount of catalyst in g; α_i and β_i are the orders of the ith reaction with respect to organic compound and hydrogen, respectively; k_i is the rate constant for the ith reaction in $(\text{cm}^3/\text{mole})^{\alpha_i+\beta_i-1}/\text{g sec}$. The rate of change of product concentration G_1 can be obtained by mass balance.

The surface concentrations of various chemical species are virtually impossible to measure directly but they can be eliminated by equating the rates of mass transfer and reactions:

$$R_B = (k_sa_p)_B(B_\ell - B_s) = wk_1A_s^{\beta_1}B_s^{\alpha_1} \tag{8}$$

$$R_E = (k_sa_p)_E(E_\ell - E_s) = wk_2A_s^{\beta_2}E_s^{\alpha_2} - wk_1A_s^{\beta_1}B_s^{\alpha_1} \tag{9}$$

$$R_F = (k_sa_p)_F(F_\ell - F_s) = wk_3A_s^{\beta_3}F_s^{\alpha_3} - wk_2A_s^{\beta_2}E_s^{\alpha_3} \tag{10}$$

$$R_A = M_A(A^* - A_s) = wk_1A_s^{\beta_1}B_s^{\alpha_1} + wk_2A_s^{\beta_2}E_s^{\alpha_2} + wk_3A_s^{\beta_3}F_s^{\alpha_3} \tag{11}$$

where

R = rate of mass transfer, moles/cm^3 sec. (Subscripts denote chemical species.)

k_s = liquid-solid mass transfer coefficient, cm/sec.

a_p = external surface area of catalyst particles per unit volume of the reaction mixture, cm^2/cm^3. (Subscript on (k_sa_p) denotes species.)

A* = concentration of hydrogen in liquid in equilibrium with the gas, moles/cm^3.

A_s = concentration of hydrogen on the catalyst surface, moles/cm^3, and

M_A is the overall mass transfer coefficient for hydrogen which combines the resistances for gas-liquid and liquid-solid mass transfer, sec^{-1}.

M_A is given by

$$M_A = \left(\frac{1}{K_L a_B} + \frac{1}{k_s a_p}\right)_H^{-1} \tag{12}$$

where K_L is gas-to-liquid mass transfer coefficient in cm/sec and a_B is the gas-liquid interfacial area per unit volume of reaction mixture in cm^2/cm^3.

It is assumed above that all three subreactions (B to E, E to F, and F to G) occur simultaneously. This assumption may not be valid in all cases. For example, if there is competition for the active sites on the catalyst, one or more of the reactions could be dominant at a particular time.

The process in the reactor is completely defined by Eqs. 5-12. These equations have many parameters and unknown concentrations, so that certain simplification is in order. If the mass transfer resistances are negligible, then

$$B_s = B_\ell,\ E_s = E_\ell,\ F_s = F_\ell,\ \text{and}\ A_s = A^* \tag{13}$$

and the Eqs. 5-12 are reduced to

$$-\frac{dB_\ell}{dt} = wk_1 A^{*\beta_1} B_\ell^{\alpha_1} \tag{14}$$

$$-\frac{dE_\ell}{dt} = wk_2 A^{*\beta_2} E_\ell^{\alpha_2} - wk_1 A^{*\beta_1} B_\ell^{\alpha_1} \tag{15}$$

$$-\frac{dF_\ell}{dt} = wk_3 A^{*\beta_3} F_\ell^{\alpha_3} - wk_2 A^{*\beta_2} E_\ell^{\alpha_2} \tag{16}$$

The rate of hydrogen consumption in the reactor in moles/cm^3-sec assuming all the three subreactions (B to E, E to F, and F to G) occur simultaneously, is given by

$$R_{H_2} = wk_1 A^{*\beta_1} B_\ell^{\alpha_1} + wk_2 A^{*\beta_2} E_\ell^{\alpha_2} + wk_3 A^{*\beta_3} F_\ell^{\alpha_3} \tag{17}$$

Continuing the simplification if it is assumed that the order of reactions with respect to organics is zero (i.e., α_i are zero) and that the order of all the reactions with respect to hydrogen is the same (i.e., all β_i are equal to β), then Eqs. 14-17 are further reduced to

$$-\frac{dB_\ell}{dt} = wk_1A^{*\beta}U(B_\ell - 0) \tag{18}$$

$$-\frac{dE_\ell}{dt} = wk_2A^{*\beta}U(E_\ell - 0) - wk_1A^{*\beta}U(B_\ell - 0) \tag{19}$$

$$-\frac{dF_\ell}{dt} = wk_3A^{*\beta}U(F_\ell - 0) - wk_2A^{*\beta}U(E_\ell - 0) \tag{20}$$

$$R_{H_2} = wk_1A^{*\beta}U(B_\ell - 0) + wk_2A^{*\beta}U(E_\ell - 0) + wk_3A^{*\beta}U(F_\ell - 0) \tag{21}$$

where U(x) is the unit step function defined as

$$\begin{aligned} U(x) &= 0 \qquad \text{if } x \leq 0 \\ U(x) &= 1 \qquad \text{if } x > 0 \end{aligned} \tag{22}$$

The special form of these equations arises because even though the order of reactions with respect to organic substrate is zero, a reaction takes place only if the substrate is present (concentration greater than zero). Equations 18-21 can be solved numerically to simulate the reactor performance for a given set of initial conditions and kinetic parameters. The results of one such simulation are shown in Fig. 4 which shows the variation of concentrations of various organic species with time. The kinetic parameters used to obtain the curves in Fig. 4 are listed in the caption. Usually hydrogen is fed to the reactor from a vessel and the change in the moles of hydrogen in this vessel is measured by noting the pressure drop in the vessel. Thus the rate of hydrogen consumption is not measured but its integral, the cumulative hydrogen consumption, is measured. Figure 5 shows the variation of moles of hydrogen in the feed vessel with time under the same conditions as Fig. 4. The plot of moles in feed vessel vs. time is piecewise linear. The discon-

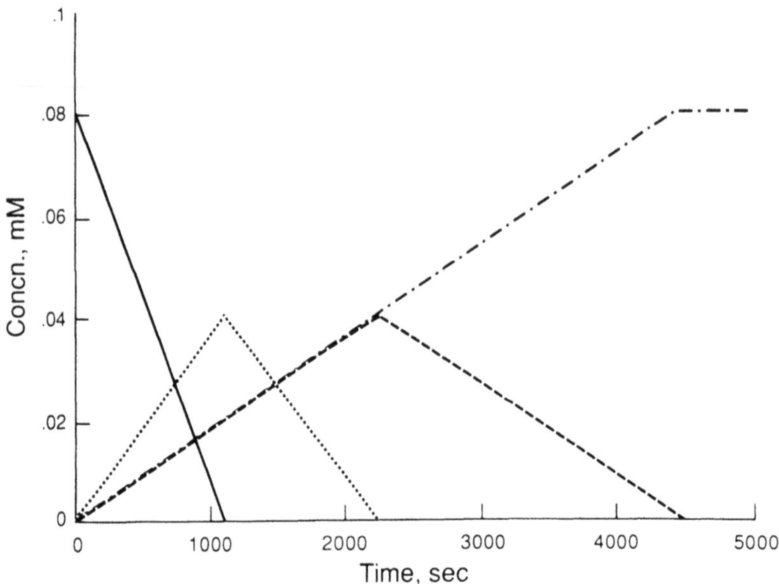

Fig. 4 Simulated variation of concentrations with time in hydrogenation of NBS. Parameters used in simulation: $k_1 = 3.0 \times 10^{-5}$, $k_2 = 1.5 \times 10^{-5}$, $k_3 = 0.75 \times 10^{-5}$ all in $(\text{mole/cm}^3)^{0.5}$/g-sec, $B_i = 0.5$, initial concentration of NBS = 81 mM, hydrogen pressure in reactor = 500 psi, volume of reaction mixture 300 ml.

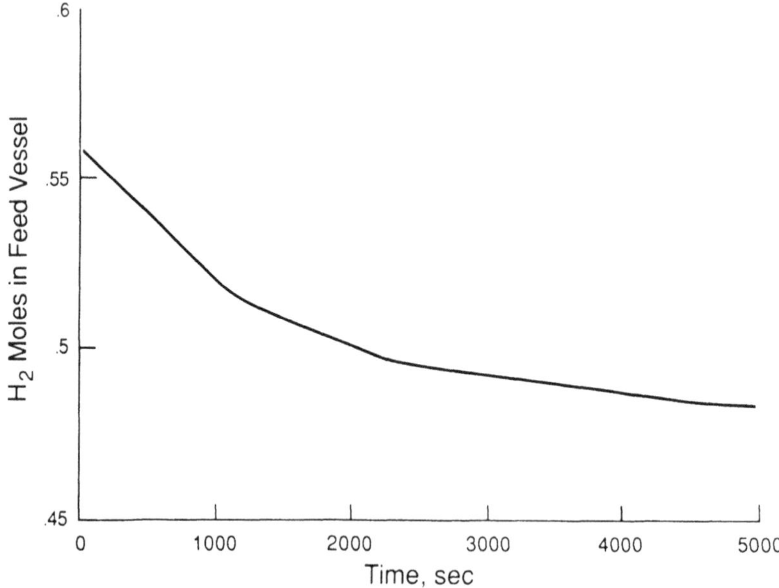

Fig. 5 Simulated decrease in moles of hydrogen in feed vessel using the same parameters as Fig. 4. The volume of feed vessel was 145 ml and initial pressure in feed vessel was 1500 psi.

tinuities in the curve coincide with the disappearance of the substrate and the intermediate.

If it is further assumed that the rate constants for all reactions are equal (i.e., k_i are equal), then the variation of the concentrations of various species with time is shown in Fig. 6. In this case the intermediates are consumed at the same rate as they are produced; hence their concentration is zero at all times. Also, the moles of hydrogen in the feed vessel would drop linearly with time as shown in Fig. 7. It should be noted that the starting material is present in the reaction mixture to the very end of the reaction (Fig. 6). Under these conditions the rate of hydrogen consumption is given by the following equation:

$$R_{H_2} = wkA^{*\beta}[U(B_\ell - 0) + U(E_\ell - 0) + U(F_\ell - 0)] \quad (23)$$

where k is the rate constant for all the reactions (rage constants for reactions are equal). Since the concentrations of the intermediates is zero at all times during the reaction and the starting

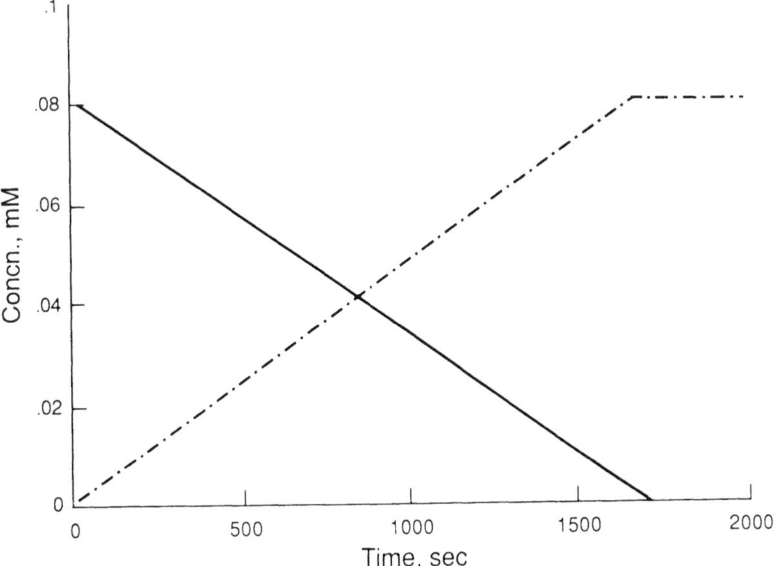

Fig. 6 Simulated variation of concentrations with time in hydrogenation of NBS. The rate constants for all the reactions were assumed to be 2.0 x 10^{-5} $(\text{mole/cm}^3)^{0.5}$/g-sec. The order with respect to hydrogen was 0.5 and the pressure in the reactor was 500 psi.

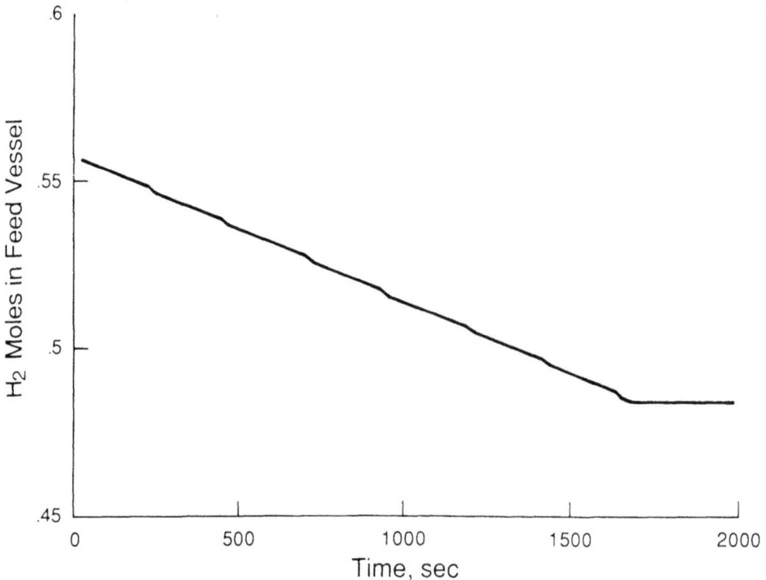

Fig. 7 Simulated drop in moles of hydrogen in feed vessel with the same conditions as Fig. 6.

material is present at all times, the above expression can be simplified to

$$R_{H_2} = wkA^{*\beta} \tag{24}$$

Thus the rate of hydrogen consumption will remain constant during the reduction of NBS if

1. The mass transfer is not limiting.
2. The orders of subreactions are zero with respect to organic compounds.
3. The orders of all subreactions with respect to hydrogen are equal, and
4. The rate constants of all the reactions are the same.

It should be noted that these may not be the only conditions which make the rate of hydrogen consumption constant during the reaction.

IV. RESULTS AND DISCUSSION

The main aim of the present work was to determine the intrinsic kinetics of hydrogenation of *p*-nitrobenzenesulfonamide over Raney cobalt catalyst. As mentioned before, this is a three-phase system; hence it was crucial to establish that the mass transfer limitations were eliminated. The criteria developed by Ramachandran and Chaudhari[8] were used to evaluate the significance of the mass transfer effects. In these criteria the observed rates of reactions are compared with the maximum rates of gas-liquid, liquid-solid, and intraparticle mass transfer. Since the catalyst used had a fine particle size (40-80 μm) and the active sites were located on the external surface of the catalyst, the intraparticle effects were negligible. The rates of gas-liquid and liquid-solid mass transfer were calculated using the correlations of Bern et al.[9] and Sano et al,[10] respectively. The estimated rates of mass transfer were at least 2-3 orders of magnitude higher than the reaction rates calculated from the observed rates of hydrogen consumption under the same conditions. For the above calculations the diffusivity values were estimated by using the Wilke-Chang correlation[11] and the solubility of hydrogen in ethanol at reaction conditions was estimated by using the Henry's law constant reported by Wainwright et al.[12] Thus it was clear that the experiments were being conducted in the kinetic regime. Since the correlation of Bern et al.[9] is based on the data obtained in large-scale equipment, the gas-liquid mass transfer coefficient for hydrogen was measured in the laboratory reactor using the dynamic dissolution method.[13] The measured mass transfer coefficient was three times that obtained from the correlation. This confirmed that the gas-liquid mass transfer of hydrogen is not limiting.

Initially a few experiments were carried out at 50°C with 5 g of NBS dissolved in 300 ml of ethanol. The hydrogen pressure was 500 psi, the catalyst amount was 2 g, and the agitation speed was 2000 rpm. During these experiments it was observed that the moles of hydrogen in the feed vessel decreased linearly with time during the reduction of NBS. A plot of moles of hydrogen in the feed vessel vs. time for

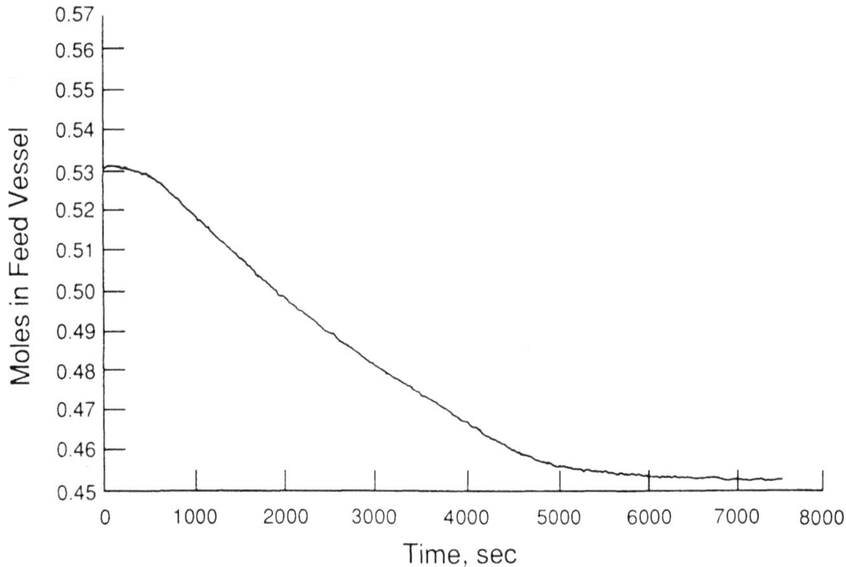

Fig. 8 The drop in moles of hydrogen in feed vessel with time in a typical experiment. The reaction conditions were: temperature = 50°C, hydrogen pressure in reactor = 500 psi, initial amount of NBS = 5 g, amount of catalyst = 2 g, agitation speed = 2000 rpm, solvent = ethanol, volume of reaction mixture = 300 ml.

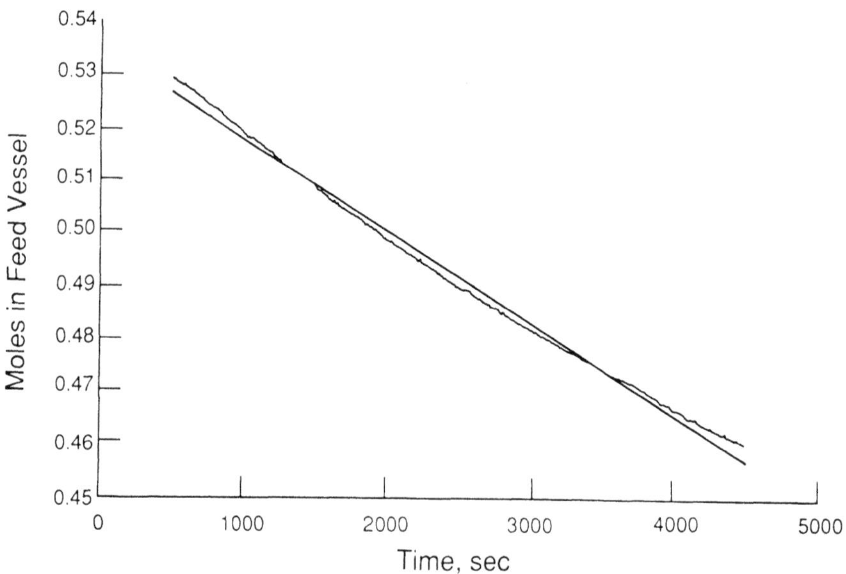

Fig. 9 A linear fit to the data of Fig. 8.

a typical experiment is shown in Fig. 8. The linearity of the data is evident from Fig. 9 where it is plotted along with the least squares line. It was seen in the previous section that the rate of hydrogen consumption will remain constant if the orders of subreactions with respect to organics are zero, the orders with respect to hydrogen are equal, and the rate constants for all the reactions are the same. In order to verify whether the orders with respect to organic species are indeed zero, the reductions were conducted with lower amounts (2 g instead of 5 g) of the starting material (NBS) keeping all other conditions the same. Fig. 10 shows the variation of the moles of hydrogen in the feed vessel with time for one such experiment along with the best straight line fitting the data. With the lower amount of starting material the rate of hydrogen consumption as given by the slope of the line fitting the data was a little lower than that observed with 5 g of starting material. The difference in

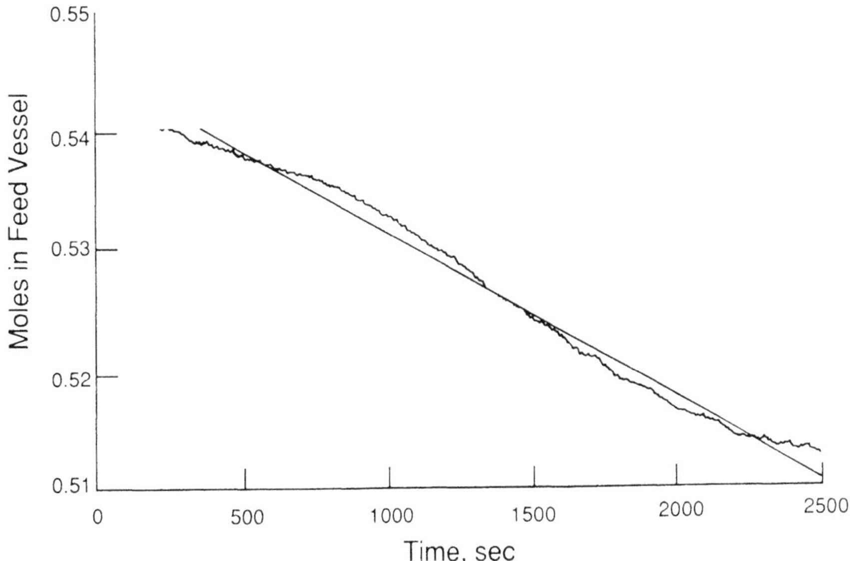

Fig. 10 The drop in moles of hydrogen in feed vessel with the lower (2-g) initial amount of NBS. Other reaction conditions were the same as for Fig. 8.

the rates of hydrogen consumption observed with different amounts of starting material suggested that the orders of reduction reactions with respect to organic substrates were not exactly zero. If the orders were very much different than zero, the variation of the moles of hydrogen in the feed vessel with time would not have been linear. Since it was very much linear (R^2 = 0.995), it was concluded that the orders of the reduction reactions with respect to organic substrates were very close to zero.

In the previous section it was seen that the disappearance of NBS as well as the appearance of ABS will be linear with time and NBS will be present till the end of the reaction if the orders with respect to organics were zero and those with respect to hydrogen were equal and the rate constants for all the reactions were equal. It was also seen that the concentrations of intermediates will be zero at all times under these conditions. A few experiments were conducted in which the samples of the reaction mixture were collected during the course of the reaction to check whether the above predictions were true in the reduction of NBS. The results from one of the experiments are presented in Fig. 11. The experiments showed that NBS disappeared from the reaction mixture much before the reaction was complete but its disappearance was indeed linear with time. The appearance of ABS, however, was not as linear with time. It was not possible to measure the concentrations of the intermediates as the analytic method used did not detect them. Since NBS disappeared before all of ABS was formed, it is obvious that some intermediates were formed and their concentration was not zero for some time during the course of the reaction. Disappearance of NBS before the end of the reaction along with the formation of ABS not being linear with time implied that either the rate constant for the reduction of NBS to the first intermediate (nitroso) is higher than that for other reactions or the order of the first reaction with respect to NBS is not exactly zero. If the first possibility were true, then the rate of hydrogen consumption would not have remained constant during the reaction. Since the decrease in the moles of hydrogen in the feed vessel was linear (R^2

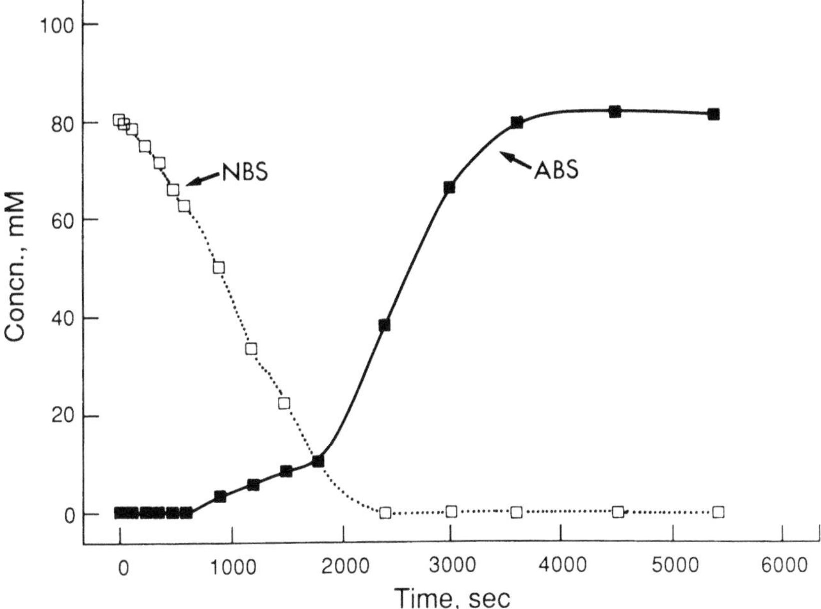

Fig. 11 The plot of NBS and ABS concentration vs. time. The reaction conditions were the same as for Fig. 8.

greater than 0.99 in most experiments), there was reason to believe that the rate constants for the three steps in the reduction were approximately equal to each other. Thus it was concluded that the order of the first step in the reduction may not be exactly zero but is very close to zero.

The primary objective of this study was to determine the kinetics of reduction of NBS to predict the performance of production scale reactors. The study was not aimed at determining the exact mechanism of the reduction. Thus it was felt that, based on the results of the preliminary experiments discussed above, it can be assumed that the orders of the subreactions with respect to organic compounds are zero, the orders with respect to hydrogen are equal, and the rate constants for these reactions are equal. It was realized that these assumptions may not be exactly true. Nevertheless it was felt that the reduction can be modeled with these assumptions to predict the performance of the reactors carrying out the reduction. It was understood that this model of the reduction would be semiempirical in nature but it would still provide some insights into the process.

In the previous section it was shown that the rate of hydrogen consumption during the reduction is given by

$$R_{H_2} = wkA^{*\beta} \tag{25}$$

if the assumptions stated in the last paragraph are true. In any given experiment the pressure of hydrogen in the reactor and thus A* was held constant. The above equation can be rewritten as

$$R_{H_2} = wkHP_{H_2}^{\beta} \tag{26}$$

where P_{H_2} is the partial pressure of hydrogen in the reactor in psi and H is the Henry's law constant for hydrogen in psi. The above equation gives the rate of hydrogen consumption per unit volume in the reactor. The total rate of hydrogen consumption in the reactor, $R_{H_2}^{T}$, in moles/sec, is then given by

$$R_{H_2}^{T} = wVkHP_{H_2}^{\beta}$$

or

$$R_{H_2}^{T} = wKP_{H_2}^{\beta} \tag{27}$$

where V is the volume of the reaction mixture and K is equal to the product (VkH). The temperature dependence of K in the above equation can be expressed by decomposing it as follows:

$$R_{H_2}^{T} = wK_0\, e^{-\Delta E/RT}\, P_{H_2}^{\beta} \tag{28}$$

where ΔE is the activation energy. For predicting the performance of production scale reactors it was desired to determine the constant K_0 in the above equation along with its activation energy ΔE and the order with respect to hydrogen β. The reductions of NBS were conducted at different temperatures keeping the pressure of hydrogen in the reactor the same to determine the activation energy. Figure 12 shows the Arrhenius plot of the total rate of hydrogen consumption.

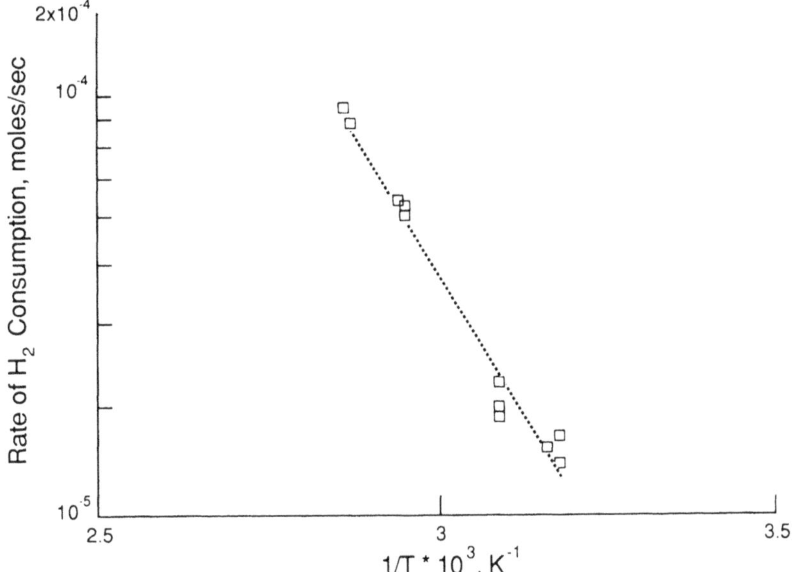

Fig. 12 The Arrhenius plot of the rate of hydrogen consumption. Other conditions which were held constant were the same as for Fig. 8.

The activation energy is 12 ± 0.6 kcal/mole. It should be noted that this value of the activation energy includes the variation of the Henry's law constant H and the volume of reaction mixture V with temperature since the parameter K is equal to the product (VkH). The activation energy for the Henry's law constant for the hydrogen-ethanol system calculated from the data reported in the literature[14] is around -1 kcal/mole (the solubility of hydrogen in ethanol increases slightly with temperature) and the activation energy for the volume of the reaction mixture (which is essentially ethanol) is around 0.3 kcal/mole.[15] Thus the contributions to the measured activation energy from the Henry's law constant and the volume of the reaction mixture are of the same order of magnitude as the uncertainty in the measurement. The activation energy value of 12 kcal/mole compares very well with the activation energy values between 9 and 13 kcal/mole reported by Karwa and Rajadhyaksha for various steps in the hydrogenation of nitrobenzene.[16]

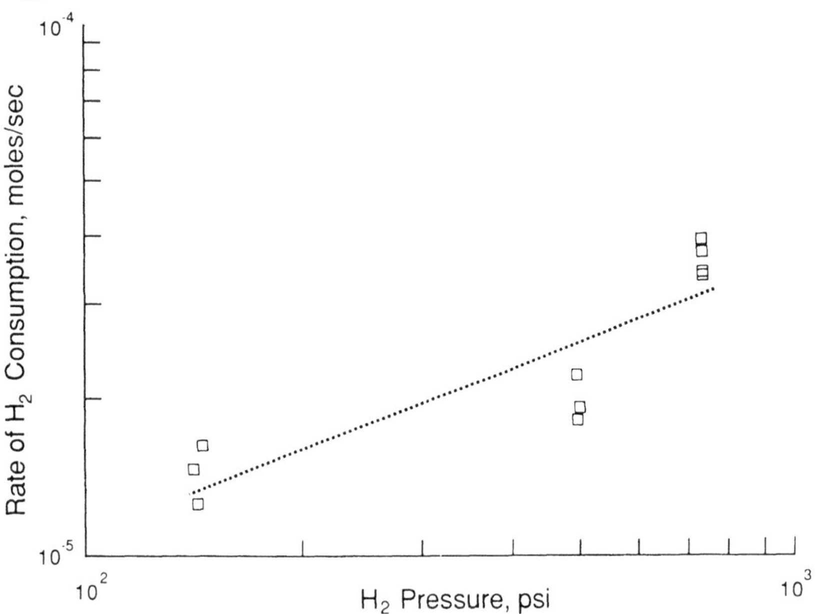

Fig. 13 The log-log plot of rate of hydrogen consumption vs. hydrogen pressure in the reactor. Other conditions were the same as for Fig. 8.

To determine the order of the reactions with respect to hydrogen the reductions were conducted at different pressures of hydrogen in the reactor keeping temperature the same. A log-log plot of the total rate of hydrogen consumption vs. the hydrogen pressure in the reactor is shown in Fig. 13. The order of the reactions with respect to hydrogen is 0.53 ± 0.06. It should be noted here that the rate of hydrogen consumption would have remained constant during the reduction had the mass transfer of hydrogen been limiting. However, in that case the order with respect to hydrogen would have been 1.0. Thus the order being different from 1.0 once again confirms that the mass transfer was not limiting during the reduction of NBS under the conditions used in this work.

Equation 28 can be rewritten as a general linear model as follows:

$$\ln R_{H_2}^{T} = \ln(wK_0) - \frac{\Delta E}{RT} + \beta \ln P_H \tag{29}$$

or

$$\text{Response} = B_0 + B_1*(\text{factor 1}) + B_2*(\text{factor 2}) \tag{30}$$

Table 1 Results of Statistical Analysis

Parameter in Eq. 29	Estimate	Standard error of estimate
$\ln K_0$	6.37	1.002
$-\frac{\Delta E}{R}(°K^{-1})$	-6571.5	297.93
β	0.53	0.058

Thus it was possible to use the regression methods and all the data to determine simultaneously K_0, the activation energy ΔE, and the order with respect to hydrogen β. The results of the statistical analysis, shown in Table 1, confirmed the values obtained earlier. Although the data were obtained by a one-variable-at-a-time approach, which is common in chemical engineering literature, the estimates of various parameters were correct because the independent variables (temperature and pressure) were not correlated (Fig. 14). However,

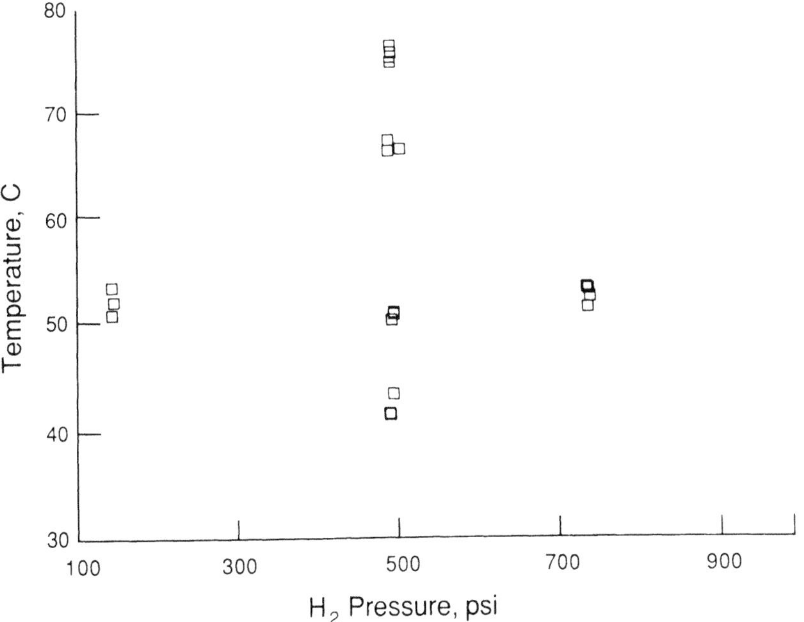

Fig. 14 Plot showing lack of correlation between temperature and pressure.

it should be noted here that it would have been more efficient to use the factorial design to obtain these parameters in this case.

V. CONCLUSIONS

The hydrogenation of *p*-nitrobenzenesulfonamide dissolved in ethanol over Raney cobalt was studied in a laboratory reactor with the objective of understanding the intrinsic kinetics of the reaction. The significance of various mass transfer effects was evaluated which showed that the reaction was not mass transfer-limited under the conditions used in this study. If it is assumed that the overall reaction can be represented as a series reaction and that the kinetics of individual steps can be represented by a power law model, then the orders with respect to organic substrates are close to zero and the orders with respect to hydrogen are 0.5. The rate constants of various steps are nearly equal. The rate of hydrogen consumption during the reduction can be represented by the following equation:

$$R_H = w\ 0.973\ e^{-6571/T}\ P_{H_2}^{0.53} \tag{31}$$

REFERENCES

1. H. C. Yao and P. H. Emmett, *J. Am. Chem. Soc., 81,* 4125 (1959).
2. H. C. Yao and P. H. Emmett, *J. Am. Chem. Soc., 83,* 796 (1961).
3. H. C. Yao and P. H. Emmett, *J. Am. Chem. Soc., 83,* 799 (1961).
4. P. S. Malpani and S. B. Chandalia, *Ind. Chem. J., 8,* 15 (1973).
5. G. W. Roberts, The influence of mass and heat transfer on the performance of heterogeneous catalyst in gas/liquid/solid systems, *Catalysis in Organic Synthesis 1976* (P. N. Rylander and H. Greenfield, eds.), Academic Press, New York, 1976, p. 1.
6. F. Haber, *Z. Electrochem., 4,* 506 (1898).
7. F. Haber and C. Schmidt, *Z. Physik. Chem., 32,* 271 (1900).
8. P. A. Ramachandran and R. V. Chaudhari, *Three-Phase Catalytic Reactors,* Gordon and Breach, New York, 1983, p. 190.
9. L. Bern, M. Hell, and N. H. Schoon, *J. Am. Oil Chem. Soc., 53,* 463 (1976).
10. Y. Sano, N. Yamaguchi, and T. Adachi, *J. Chem. Eng. Jap., 7,* 255 (1974).

11. C. R. Wilke and P. Chang, *AIChE J.*, *1*, 264 (1955).

12. M. S. Wainwright, T. Ahn, D. L. Trimm, and N. W. Cant, *J. Chem. Eng. Data*, *32*, 22 (1987).

13. M. G. Joshi, W. E. Pascoe, and D. E. Butterfield, unpublished results.

14. S. K. Gupta, R. D. Lesslie, and A. D. King, Jr., *J. Phys. Chem.*, *77*, 2011 (1973).

15. *American Institute of Physics Handbook*, McGraw-Hill, New York, 1972, p. 2-157.

16. S. L. Karwa and R. A. Rajadhyaksha, *Ind. Eng. Chem. Res.*, *26*, 1746 (1987).

7

Practical Use and Control of Laboratory Catalytic Reactors

J. M. LAMBERT, JR.

Chemical Data Systems, Erie, Pennsylvania

ABSTRACT

This chapter reviews the mental processes and physical steps involved in performing catalytic investigations in laboratory reactors. The intent is to illustrate the available solutions to the difficulties in obtaining meaningful data. Various types of reactors are reviewed and examples are given of their successful use. Some of the latest developments in the design and control of laboratory reactors are described.

The most commonly practiced methodology for performing research on the catalysis of organic reactions involves the following sequence of steps:

1. Define the problem.
2. Select the catalyst.
3. Obtain the catalyst.
4. Test the catalyst.
5. Ask what next?

Thus, one must first establish and precisely define the goals and objectives. Second, one must then decide what catalyst to use for the first experiment. The chosen catalyst must then be obtained and put into a system for analysis. Upon reaching step 5, the scientist must determine whether to return to step 1, 2, 3, or 4 or conclude that the goals have been achieved. The decision is based on the interpretation of the results of step 4 and the conviction that the assumptions used in steps 1, 2, and 3 are correct.

Positive results of all tests then allow the researcher to proceed with the catalyst to the next larger scale of reactor to repeat the tests. This procedure is iteratively repeated until, ultimately, successful results are achieved on the commercial scale. The process is a tedious one, requiring diligence and careful experimentation.

The importance of step 5 in this methodology is to be stressed. It is here that the culmination of all data and experience is put to the test. Unfortunately, it is also here that the results of tests performed often leave the researcher with little or no more information than was available before the test was performed. Many experiments are performed under conditions or within reactors which will offer little information other than basic "it worked" or "it did not work" data.

Obviously, research performed in well-characterized systems is of distinct advantage. It is also advantageous to perform well-designed and carefully planned experiments in reactors and systems designed to yield the most information. Today's research mandates more information. If experiments do not achieve all expectations, it is highly desirable that they give additional information regarding why they did not.

Since an error in judgment about any of the assumptions or interpretations used in any of the above steps can lead to erroneous conclusions, it is worthwhile to examine each step in detail.

Step 1. Problem definition is a clear, concise statement of the task being undertaken and should include the desired goals. Generally such definitions surround questions such as, "How can one make product X and/or Y?" or "What is another way in which one can increase the value of reactant compounds A and/or B?" It should be recognized that the experimental needs of the chemist and the chemical engineer often differ. Perhaps one of the most key areas in helping the researcher to more precisely define the goals is to make a clear distinction between chemical and physical effects. Although not always easy to perform, it is highly advantageous to design experiments which will permit discernment of the two types of effects. If one is to study chemical effects, it is obviously desirable to strive for obtaining intrinsic kinetic information, free from the limitations of transport or reactor design effects. Conversely, if one is attempting to study various physical effects, experiments should be done with care so that the chemistry is not altered by changes in physical parameters.

Step 2. Catalyst selection involves a mental exercise to decide which catalyst (or catalysts) is (are) to be examined. Often successful catalysts are selected on the basis of "educated guesses" or even lucky hunches. Most often, however, this process relies on more clever intuition guided by literature references, personal experience, and colleagues' advice.

Step 3. Obtaining a chosen catalyst is performed either by purchasing it from a commercial manufacturer, borrowing it from a colleague, or making it from scratch. The latter may involve processes such as crystallization, impregnation, dissolution, incipient wetness and calcination, or ion exchange.

Step 4. Testing of the catalyst is a determination of whether the catalyst meets the intended goals as stated in step 1. This is the point at which one determines whether the catalyst is a good one. Therefore, one must first discern what specific qualities are desirable. A partial list of these qualities is found in Table 1. Depending on the stated purpose of the tests, a catalyst may need to meet only one, or perhaps all, of these criteria before being accepted as the catalyst of choice.

Determination of whether the intended catalyst was, in fact, properly prepared can often involve numerous characterization tech-

Table 1 Criteria for Testing of Catalysts

Chemical influences	Physical influences
Activity	Porosity
Selectivity	Surface area
Lifetime	Particle size and shape
Active site distribution	Crush strength
Promoters/inhibitors	Abrasion resistance
Support interactions	Heat/mass transport

niques. Such examinations might be performed before or after testing of the catalyst for use in the intended reaction.

The second step in testing a catalyst is to determine the conditions of the experiments. One must decide the range to be tested for parameters such as temperature, pressure, partial pressures, contact times, and rotational speed of impellers in stirred vessels. Wide ranges must often be examined, or one may have excellent ideas regarding over what range a catalyst will be active or one may be limited by certain production and/or economic constraints to test and accept catalysts which are operable only within certain ranges of conditions. The researcher must next decide whether single or multiple experiments are to be performed. This decision will be based on the precision and accuracy of the equipment and, of course, constraints of time and financial support. Multiple experiments may need to be performed at selected points within the included parameter space or they may be performed randomly throughout the parameter space in accordance with a statistically designed set of experiments.

The third step in testing a catalyst involves defining the equipment requirements. Within this step one must first determine the scale of operation desirable. In order to properly test all the criteria of Table 1, a full-scale commercial reactor should be used. It is not likely that such luxury will be available. Limited feedstocks and manpower, as well as today's high cost of disposal of undesirable products, usually lead to performing tests in smaller reactors. The fact that a smaller reactor will be used mandates even closer examination of the criteria in Table 1. The desired data obtained from the smaller reactor will eventually need to be scaled up to larger reactors. Since the reactor which is first employed may satisfy only some of the criteria, multiple reactors of increasing size are often necessitated.

Thus one may proceed through the following sequence of reactors: microscale, bench scale, laboratory scale, pilot plant scale, demonstration scale, and semiworks scale before gathering the requisite information allowing production and operation of the commercial scale reactor. The number of reactor scales required is dependent on the validity of data obtained and the confidence of the mathematical models being employed for interpretation of the data.

Systems should be designed so that all chemistry takes place in the reactor. This makes the choice of reactor most important. There are a number of different reactor types available to the researcher. Some are designed to have the fluids move to the catalyst and are referred to as fixed-bed reactors. An example is the common tubular reactor operated in an upflow or downflow manner. Other fixed-bed reactors are known as recycle reactors and may have flow paths which move externally or remain internal to the reactor. Other reactors are designed to have the catalyst move to the fluid and are referred to as moving-bed reactors. Examples of this type of reactor include spinning basket and raining solids reactors. A third type of reactor is designed to move both the fluid and the catalyst. These are collectively referred to as fluid bed reactors and include such types as the transport reactor and the ebullating bed reactor.

Each of the various reactors are designed to provide certain types of data when used with certain types of catalysts and feedstocks. There are often trade-offs to be experienced between operational simplicity and quality of the data.[1] Examples will be subsequently described where the improper choice of reactor led to the improper selection of the catalyst.

Once the reactor has been selected, one must next define the peripheral instrumentation. Sensors of appropriate sensitivity need to be chosen for monitoring and control of the necessary parameters. Analog or digital process control must be selected. Capabilities for maintaining a single set condition or for permitting the ramping of the setpoint must be determined along with whether proportional, integral, derivative, or three-mode (PID) control will be required to achieve the desired performance. Finally, one must decide on the level of supervision required to operate the experimental system. The cost effectiveness of operator control vs. computer control needs to be evaluated. If computer control is deemed appropriate, it must then be decided whether the capabilities for sequential execution of a preprogrammed series of experiments are to be utilized without operator intervention.

Careful attention should next be directed to the instrumentation used to control the fluid flow throughout the system and the various safety devices required for operation with or without operator attendance. Safety features for excursions above or below desired levels of temperature, pressure, and flow need to be specified. A combination of mechanical and electronic devices is preferred. Many electronic safety-monitoring devices permit two safety levels to be set above and below the desired levels. Many also offer a choice of audible and visual alarms, as well as contact closures which can initiate additional safety actions sequences.

Reactant preparation involves selection of valves, pumps, mass flow controllers, mixing and vaporizing equipment. Product handling includes gas/liquid separation, sampling, collection and disposal of the various phases produced.

The final area of equipment definition required is the analytic scheme. One must decide first on the type of analyses required. These might be chromatographic, spectroscopic, or based on a particular physical property of the fluid. Such analyses may be performed off-line, on-line, or in-line, i.e., the sampled fluids may be sent off-line to an analytic instrument not connected to the catalyst test system. Alternatively, the test system may have analytic instruments as an integral part of itself. In this case, either a slip stream or the entire stream is sampled. These are referred to as on-line and in-line sampling, respectively. The latter is obviously preferred as it assuredly is representative of the stream. On-line sampling is required with systems having large flow rates and requires careful design to ensure that each sample is representative. It is easily possible to introduce segregation of the sample due to condensation when passing through small orifices or providing insufficient heating.

With equipment in place and catalyst loaded into the reactor of choice, the experiments can be performed and the analytic data collected. One then needs to interpret the data with proper regard being paid to the accuracy of the instruments and their respective calibration curves. These data can then be entered into a properly devised mathematical model describing the reactor performance.

Reactors might be operated in batch, continuous flow, or pulse flow modes and interpretation may thus be from differential or integral equations. Attention must also be paid to whether the reactor is operated as a plug flow reactor (PFR), a continuously stirred tank reactor (CSTR), or, more likely, as a mathematically described approximation of a combination of PFRs and CSTRs.

In addition to the use of a proper reactor model, it is imperative that a proper reaction model also be used. It is the task of the researcher to separate the physical and chemical effects which are included in the data emanating from any reactor. Careful choice of reactor(s) and experimental conditions can usually offer valuable insight. The chemical information can then be entered into appropriate mathematical equations to permit elucidation of rate constants for each of the steps involved in the often complex reaction pathways described by the assumed reaction model.

One now arrives at step 5: What next? The researcher must decide whether the test was successful and the goals have been satisfied; or whether one must return to step 1 to redefine the problem;

or return to step 2 to select a different type of catalyst; or return to step 3 and load a different catalyst under different conditions with a different reactor, or reanalyze the data.

Should the results of step 4 be positive in all aspects deemed desirable, the researcher proceeds to the next larger scale of reactor along a path leading ultimately to the commercial scale system. Often the cycle begins at step 4 with the new reactor. It needs to be recognized that there are times when the cycle needs to be entered at step 3. Such occurrences are generally mandated by the necessity to change the strength of the catalyst or to change its particle size or shape.

It is all too common, however, that the researcher reaches step 5 only to realize that there is not enough information to guide the decision of where to go next. This is particularly evident when little or no catalyst activity is observed. It is obviously imperative that the above five-step scheme be followed in the order described. The mental exercise of careful design, of experimentation will obviously increase the researcher's chances of being provided with more information at the conclusion of step 4 and permit more intelligent execution of step 5.

The various choices available to the researcher to aid in obtaining appropriate data will next be described in a number of examples. The choice of using a simple tubular reactor or a more complicated stirred reactor is often difficult. It is generally assumed that simple reactions having one or two feeds, can most easily be performed in tubular reactors. Complex feeds are often better evaluated in properly modeled stirred reactors. The nature of stirred reactors is such that the inherent back mixing which occurs will usually lead to less conversion per allowed contact time. But when evaluated with proper mathematical models to allow for the back mixing, the performance of the catalyst can be more accurately compared on the basis of rate data.

Another consideration is the separation of chemical and physical effects. Heat and mass transfer resistances can often obscure the

Table 2 Catalyst Performance in Different Reactors

Catalyst	Percent conversion	
	PFR	CSTR
A	34	73
B	59	55

true chemical performance of a catalyst. Table 2 shows the results of a study[2] comparing two catalysts for cracking activity. The extent of conversion of a gas oil to a gasoline-or-lighter fraction is used to compare two catalysts under similar conditions in tubular, plug flow reactor (PFR) and a continuously stirred tank reactor (CSTR). As can be seen in Table 2, selection of catalyst B, in accordance with the interpretation of the PFR data, would lead to selection of a catalyst with reduced performance. The improved heat and mass transport capabilities of the CSTR reduced the extent of deactivation, thereby allowing for a more accurate appraisal of catalyst performance.

The catalytic oxidation of propylene to acrolein[3] provides another example of varying results when the choice of reactor is changed to better separate chemical and physical effects. A feed consisting of 1% propylene and 6% oxygen in nitrogen was passed through a tubular reactor containing a catalyst at 345°C. A 65% conversion of the propylene was observed with selectivities of 1% acrolein, 1% acrylic acid, 39% acetic acid, and various carbon oxides. It was also found that no more than 1% propylene could be employed. The catalyst was labeled as being "too hot," and was discarded as it did not meet the desired criteria. Later this catalyst was tested in an internal recycle, gradientless (Berty) reactor where it was found that up to 6% propylene could be employed in the feed before thermal runaway was experienced. The reaction products were mostly acrolein, with lesser quantities of acrylic acid and only traces of acetic acid. In commercial scale testing this catalyst was found to produce three times the amount of acrolein than the previous catalyst of choice.

The conceptual design of the Berty reactor was recently modified to provide higher momentum transfer. This was done by using a larger impeller in a wider and less tall reactor. Caldwell,[4] employing studies of naphthalene sublimation and ethanol dehydration over ZSM-5 catalysts, claims the design modifications afford the researcher with improved operability by providing higher mass transfer coefficients.

The use of gradientless reactors has also proven to be valuable in the study of hydrodesulfurization of petroleum products. Early studies of Mahoney et al.[5] employed the use of spinning basket reactors. These reactors led to the design of yet another type of internal recycle reactor which has a stationary, annular, catalyst bed with fluids being recirculated through it by vertically aligned impellers stirred along the reactor center line.

More recently, Carberry et al.[6] designed a new reactor for the study of continuous flow, three-phase catalysis. In their new reactor, the catalyst to be tested is coated on the reactor wall. A well-mixed gas phase is contacted with the liquid phase by a wiper which continuously sweeps the catalytically active wall. It is easily recognized that proper maintenance of the degree of mixing is important in stirred reactors. Less obvious is that proper maintenance of fluid states throughout the entire system can also be critical.

Allowing the indiscriminate condensation of fluids prior to entry to a reactor can lead to nonreproducible catalyst contacting, particularly in systems plumbed with narrow tubing. Additionally, allowing condensation to occur downstream of the reactor can induce two-phase flow or pressure variations into the sampling mechanism leading to nonreproducible analytical results. It has been demonstrated[7] that relative errors in system reproducibility of between 1 and 43% can be introduced from such indiscriminate condensation. Therefore it is advisable to construct catalyst test systems entirely within thermostated ovens to ensure that complex mixtures retain their respective properties at all points within the system.

The unraveling of complex reaction pathways will necessitate the gathering of voluminous amounts of data. This can most effectively be accomplished through the use of systems employing supervisory computers with the ability to conduct sequential experiments in an unattended manner. Tulane University researchers[8] recently added greatly to the understanding of competitive reaction occurring in intrazeolitic media. In their studies employing continuously fed tubular reactors, hexane isomerization was observed to be modified by the cofeeding of aromatics. Control of the experiments was performed by a CDS model 8100 computer-controlled micro-pilot plant. This system was designed for operation in either a continuous flow or pulse flow mode.

The most recent introduction of equipment for the study of catalytic reactions is the TAP (Temporal Analysis of Products) reactor system. Patented by Monsanto[9] and offered through Autoclave Engineers, this transient response system allows investigation of catalytic reactions on a time scale previously unattainable. The TAP system employs very fast (100-µsec) pulse valves to introduce reactants to a heated reaction chamber packed with commercial catalyst. Reaction intermediates and products are analyzed at a rate of 10,000 data points per second with a UTI quadrupole mass spectrometer and an HP high-speed data acquisition system and computer. Results made public to data have shown its application to the catalysis of butane and ethylene oxidation,[10] ammoxidation,[10] and hydrodesulfurization of thiophenic compounds.[11]

REFERENCES

1. W. V. Weekman, *AIChE J., 20,* 833 (1974).
2. W. V. Weekman, *AIChE J., 16,* 397 (1970).
3. J. M. Berty, *Chem. Eng. Prog., 70*(5), 78 (1974).
4. L. Caldwell, CZARS Report M-446, Preterit, South Africa, deck (1982), submitted to *Appl. Catal.*
5. J. A. Mahoney, K. K. Robinson, and E. C. Myers, *Chemtech., 12,* 758 (1978).
6. J. J. Carberry, P. Tipnis, and R. Schmitz, *Chemtech., 5,* 316 (1985).

7. K. J. Kuruc and J. M. Lambert, Jr., "Catalysis, 1987" (J. W. Ward, ed.), Elsevier Science Publishers B. V., Amsterdam (1988).

8. J. Chen, A. M. Martin, et al., *J. Catal., 111,* 425-428 (1988), and *I&EC Res., 27,* 401-409 (1988).

9. J. T. Gleaves and J. R. Ebner (Monsanto Co.), U.S. Patent 46226412 (1986).

10. J. T. Gleaves, J. R. Ebner, and T. C. Kuechler, *Cat. Rev. Sci. Eng., 30*(1), 49-116 (1988), and *Chem. Eng. News, 65*(27), 21 (1987).

11. W. Moser, Proc. 5th Annual University Co-operative Program, College Station, Texas, 1987.

Part III

Mechanisms in Catalysis

8

Heterogeneous Catalysis: A Molecular Perspective

ROBERT L. AUGUSTINE AND PATRICK J. O'HAGAN

Department of Chemistry, Seton Hall University, South Orange, New Jersey

ABSTRACT

A discussion is presented which traces the development of the understanding of olefin hydrogenation on heterogeneous catalysts from the basic Horiuti-Polanyi mechanism, through the proposed analogies with homogeneous catalysts to the consideration of the atomic arrangements of various types of surface sites and the application of organometallic reaction principles to rationalize product formation.

I. HORIUTI-POLANYI MECHANISM

One of the oldest, if not the oldest, presently viable organic reaction mechanism is that for the hydrogenation of double bonds originally put forth by Horiuti and Polanyi[1] in 1934 which is shown in its original form in Fig. 1. This mechanism has been used extensively to provide explanations for the results obtained in a variety of alkene hydrogenations and deuterations including double-bond isomerization, extent of deuterium addition and exchange, and the effect of reaction variables on the product stereochemistry. Over the years this mechanism has been slightly modified to replace the -M with -* to signify an adsorbed species and the inclusion of a

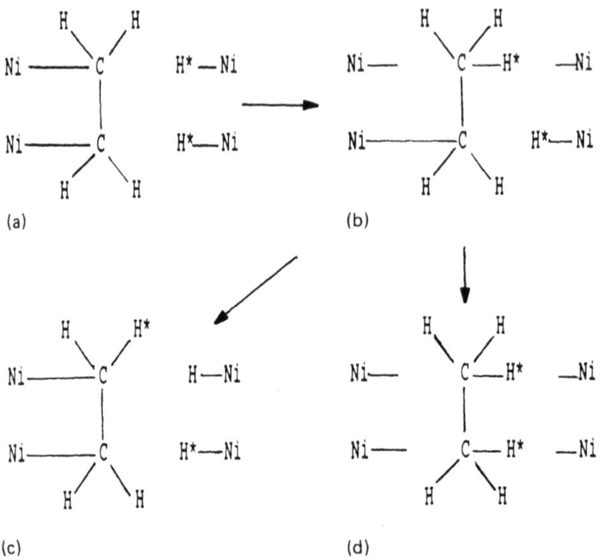

Fig. 1 Original Horiuti-Polanyi olefin hydrogenation mechanism.[1]

π-adsorbed olefin to replace the di-σ-adsorbed species. The modified mechanism is shown in Fig. 2. Comparison with Fig. 1 shows how little this mechanism has changed over the past 40 years or so.

One of the most obvious features of this reaction mechanism is the use of the -* to signify an adsorbed species. This notation is limited in that it does not permit the differentiation between the various catalytically active metals. In fact, the mechanism as written implies that all double-bond hydrogenations take place in the same manner regardless of the nature of the catalyst used. It provides no means of distinguishing the differences noted in the products observed in deuteration[2,3] or stereochemistry[3,4] studies run over the different metal catalysts. All of the variations observed on using different catalysts have been explained by invoking varying degrees of reversibility in the various steps of the mechanism for the different catalysts. Further, even though it is now rather widely recognized that the surface of the dispersed metal catalysts are far from being uniform, the Horiuti-Polanyi mechanism

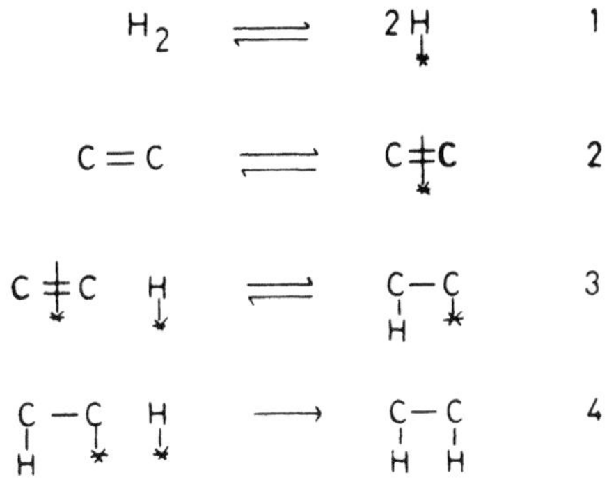

$$H_2 \rightleftharpoons 2\,\underset{*}{H} \qquad 1$$

$$C{=}C \rightleftharpoons \underset{*}{C{\ddagger}C} \qquad 2$$

$$\underset{*}{C{\ddagger}C} \quad \underset{*}{H} \rightleftharpoons \underset{H}{C}{-}\underset{*}{C} \qquad 3$$

$$\underset{H}{C}{-}\underset{*}{C} \quad \underset{*}{H} \longrightarrow \underset{H}{C}{-}\underset{H}{C} \qquad 4$$

Fig. 2 Present version of the Horiuti-Polanyi olefin hydrogenation mechanism.

provides no means for distinguishing between reactions taking place on different types of surface sites. Indeed, it is just this all-purpose generality of the -* which has provided this mechanism with the longevity it enjoys.

II. HOMOGENEOUS-HETEROGENEOUS CATALYST ANALOGIES

A number of attempts have been made to clarify this situation particularly with respect to the different reaction characteristics expected for the various types of surface atoms present on the catalyst. The most significant of these involved the correlation of the reactivity of various homogeneous catalysts with the different types of surface sites that might be expected to be present on a dispersed metal particle.[5,6] This approach assumed that the surface active sites were monoatomic in nature with the different types of sites present on the metal surface distinguished by their degrees of coordinative unsaturation, as shown in Fig. 3. The ^3M sites are corner or kink atoms, the ^2M atoms are on edges or steps, and the ^1M sites are on faces or terraces.

In Fig. 4 is shown the mechanism for the hydrogenation of a double bond over the homogeneous catalyst $(\Phi_3P)_3RhH(CO)$ (**1**).[7] **1** loses a Φ_3P group and adsorbs a double bond to give the five-coordinate intermediate, **3**. Hydrogen transfer to one of the sp^2

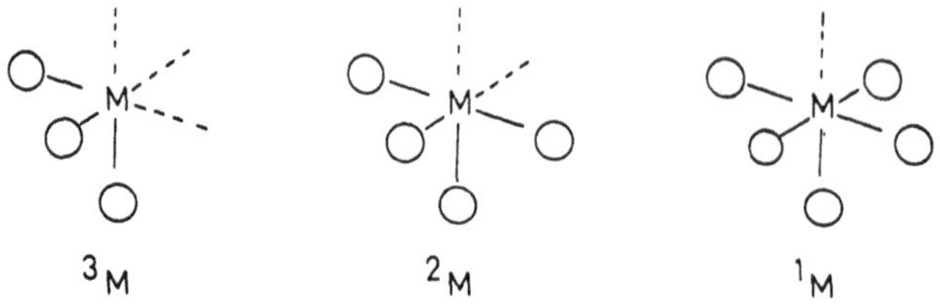

Fig. 3 Catalyst surface site description.[6]

carbons produces the square planar metal alkyl, 4. Addition of dihydrogen to 4 gives the octahedral dihydride, 5, which on reductive elimination liberates the alkane and regenerates the catalytically active diphosphine, 2, which then adsorbs another double bond to initiate another catalytic cycle.

Examination of this mechanism shows that it is the homogeneous analog of the Horiuti-Polanyi mechanism shown in Fig. 2. Siegel[6] equated these two mechanisms by proposing that hydrogenation takes place over ^{3}M or corner atoms which have interacted with hydrogen to produce a ^{3}MH site as shown in Fig. 5. Olefin adsorption and hydrogen transfer give the half-hydrogenated state, 6, which is analogous to the metal alkyl, 4, in Fig. 4. Adsorption of dihydrogen

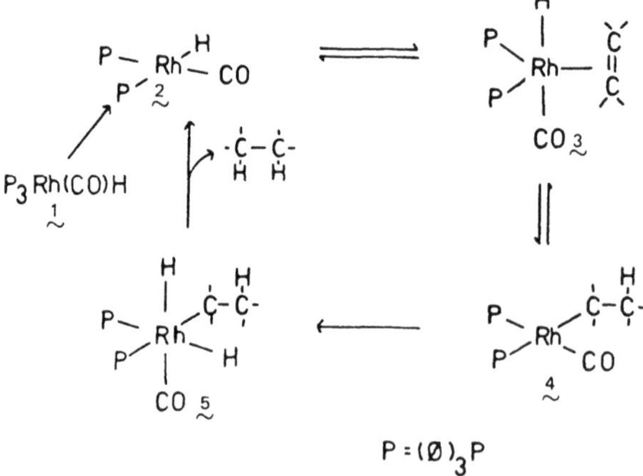

Fig. 4 Mechanism for alkene hydrogenation over $(\Phi_3P)_3Rh(CO)H$.[7]

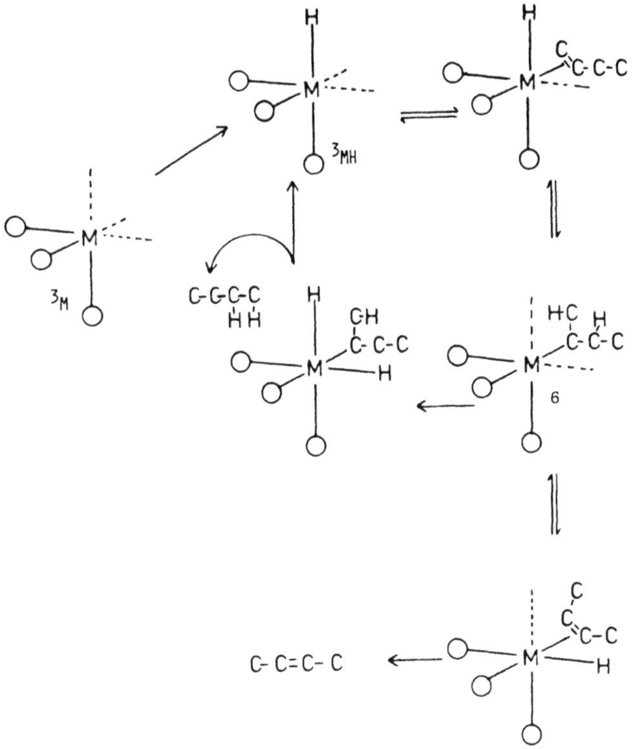

Fig. 5 Mechanism for alkene hydrogenation over ^{3}MH sites.[6]

followed by reductive elimination gives the alkane and regenerates the ^{3}MH site for participation in further catalytic cycles. In the absence of additional hydrogen the species 6 can conceivably undergo a β elimination to produce an adsorbed olefin which can then desorb. Depending on which hydrogen in 6 is involved in the β elimination, the olefin produced can be either the starting material or an isomer. With deuterium instead of hydrogen and having the various steps reverse as needed, one can use this mechanism to account for all of the factors associated with alkene hydrogenation in the same way the Horiuti-Polanyi mechanism does.

However, it was also recognized[6,8] that there are mechanistic pathways other than that shown in Fig. 5 available to the various

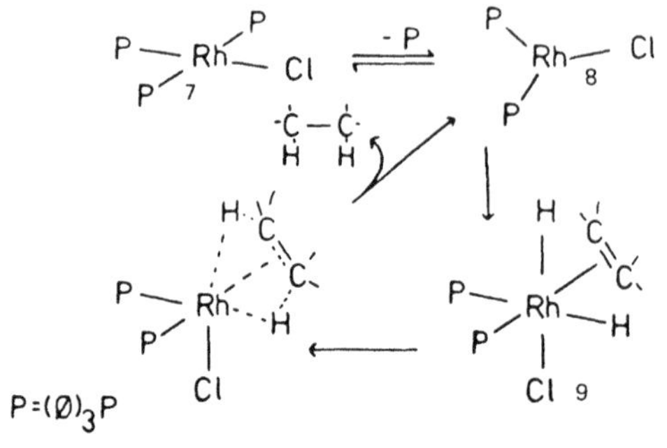

(a)

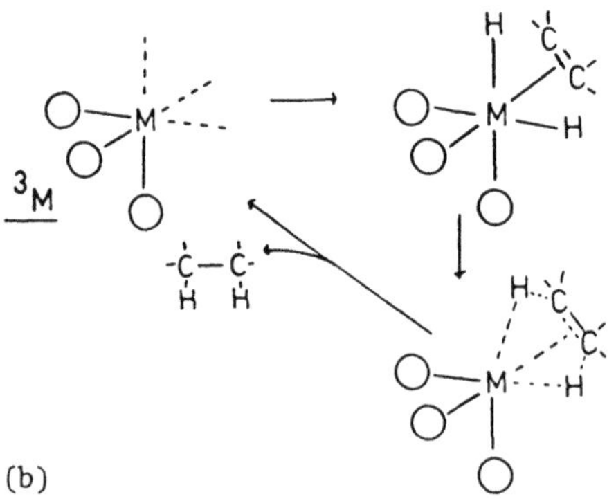

(b)

Fig. 6 Mechanism for alkene hydrogenation. (a) Over $(\Phi_3P)_3RhCl$.[9] (b) Over ^3M direct saturation sites.[6,8]

types of surface sites. One of the most obvious for the ^3M sites is the analogy with the mechanism for alkene hydrogenation known to occur with the Wilkinson catalyst, $(\Phi_3P)_3RhCl$, 7,[9] depicted in Fig. 6a. In this mechanism, loss of a phosphine generates the catalytically active tricoordinate species, 8, which then adds both a double bond and a dihydrogen molecule to give the dihydride 9. The nearly simultaneous transfer of the two hydrogen atoms to the double bond gives the alkane and regenerates 8 for further reaction. The similarity between 8 and the ^3M site is obvious as the proposed ^3M-promoted hydrogenation mechanism given in Fig. 6b shows. The primary problem with the acceptance of this analogy for the surface-promoted hydrogenation of a double bond is that this mechanism does not provide for the intermediacy of a half-hydrogenated state or metal alkyl and thus cannot be used to explain the formation of double-bond isomers or the production of multiple deuterium-exchanged products. However, if one considers the possibility that different types of sites can promote different reactions rather than having all of the reaction possibilities incorporated onto one active site, all of the vagaries of the double-bond hydrogenation can be explained.

Isomerization could take place on edge or ^{2}M sites by the procedure outlined in Fig. 7.[8] This mechanism is very similar to that proposed[6] utilizing the ^{3}MH sites (Fig. 4) in that a ^{2}MH site is the

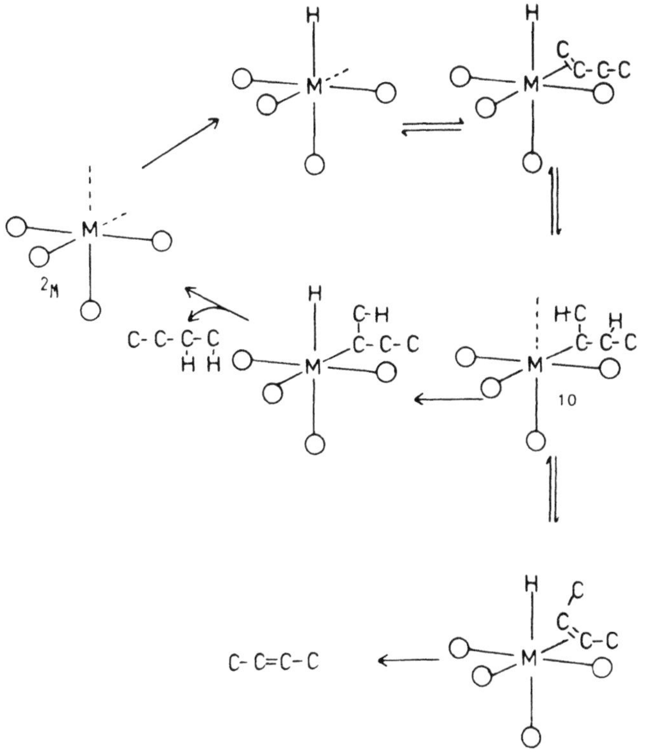

Fig. 7 Mechanism for double-bond isomerization over ^{2}M sites or edge atoms.[8]

active species. The primary difference is that there is no addition of dihydrogen to the metal alkyl, **10**, but instead only a single hydrogen atom is added, presumably by way of dissociative adsorption of dihydrogen on a two-atom system. Because of this there is no direct homogeneous catalyst analog for this mechanism. β Elimination on **10** would give either the starting olefin or its isomer. It is envisioned that the decrease in coordinative unsaturation of the ^{2}M sites results in a decrease in the stability of the metal alkyl, **10**, as compared to the more unsaturated metal alkyl, **6** (Fig. 5). This, and the difficulty anticipated in adding a single hydrogen atom to **10**, leads to the assumption that under standard hydrogenation conditions isomerization will take place on these ^{2}M sites but double-bond saturation will occur on both the ^{3}M and ^{3}MH sites.

III. HETEROGENEOUS CATALYST CHARACTERIZATION

Combining differing reactivities with the different types of sites should provide some means of producing a product site correlation that would give some idea of the nature of the active surface of the metal catalyst particles. Such a correlation, however, is not possible under normal reaction conditions because it is highly unlikely that every site will react at the same rate, and thus the product-site relationship would be skewed toward the more active sites. The single-turnover (STO) procedure,[8,10] does provide for such a correlation since with this technique the metal surface is treated as a reagent, so that each surface site reacts only once per turnover cycle. By use of this technique five different types of sites have been identified on dispersed metal surfaces. There are two types of direct hydrogenation sites presumably reacting as depicted in Fig. 6b. One of these, termed $^{3}M_{I}$, has the hydrogen adsorbed rather strongly and on the other, called the $^{3}M_{R}$, the hydrogen is more weakly held. There is also a hydrogenation site on which the hydrogens are added to the olefin in two distinct steps as depicted in Fig. 5. This site has been labeled ^{3}MH because of this mechanism. The isomerization sites were formally called $^{2}M_{C}$ but now are listed simply as ^{2}M. In

Table 1 Surface Site Reaction Characteristics for Alkene Hydrogenation

Type	Reaction	Mechanism
^3M_I	One-step hydrogenation	Fig. 6b
^3M_R	One-step hydrogenation	Fig. 6b
^3M_H	Two-step hydrogenation	Fig. 5
^2M	Isomerization	Fig. 7
^1M	H_2 Adsorption, no other reaction	

addition, there are sites which adsorb hydrogen but do not take part in the hydrogenation reaction. These sites are considered to be on crystal faces and are analogous to the ^1M site shown in Fig. 3. A listing of these sites and their reaction characteristics is given in Table 1.

With the correlation of the cyclohexane dehydrogenation activity of these hydrogenation sites, both direct and two-step, with that observed for kink atoms on Pt single-crystal catalysts, it was concluded that all of these alkene saturation sites are corner or kink atoms.[11] The principle of microscopic reversibility supports the premise that C-H bond breaking and making should take place on the same sites. Thus, it would appear that the site descriptions proposed by Siegel[6] and the mechanisms presented above are all reasonably valid. There is a problem with this supposition, though. In the first place, it is difficult to rationalize the presence of three different types of corner atoms using the octahedral site description described above. Secondly, and of greater importance, the presence of metal atoms in the octahedral orientation needed for these site descriptions is not possible with fcc crystals, the crystal orientation of the most common catalytically active metals.

IV. SURFACE SITES AND THEIR PROPOSED REACTIONS

Examination of models of fcc crystals shows that one can have 11 different types of atoms present on the surface. The description

Table 2 Surface Atom Descriptions for fcc Metal Crystallites

Type	Planes	No. near neighbors	No. next-near neighbors
Face	111	9	3
Face	100	8	5
Edge	111-111 (narrow)	6	1
Edge	111-111 (wide)	7	2
Edge	111-100 (narrow)	5	2
Edge	111-100 (wide)	7	3
Corner	111-111-100 (cubooctahedron corner)	6	2
Corner	111-111-100-100	5	2
Corner	111-111-111-111 (100 adatom-octahedron corner)	4	1
Corner	111-111-111 (111 adatom-tetrahedron corner)	3	1
Corner	111-111-100 (square pyramidal corner)	3	1

of these sites as related to the low Miller Index 111 and 100 planes is given in Table 2 along with a tabulation of the number of nearest and next-nearest metal atom neighbors associated with each one. The 110 plane is not included because it is, in reality, a highly stepped series of 111 planes. While the band theory is commonly used to assess the electronic nature of bulk metals, there is reasonably common agreement that the surface electron orbitals, especially on the smaller particles, are at least somewhat delocalized and can be treated as simple extensions of the atomic orbitals. In one of the first applications of this concept, Bond[12] showed that the surface orbitals on 111 and 100 faces could be depicted as shown in Figs. 8a and 9a, respectively. Figures 8b and 9b show models of these faces with the d orbitals projecting from the surface. With the 111 face there appears to be a rather general broad "smear" of electrons, both e_g and t_{2g}, over the surface. On the 100 face the single e_g orbital is perpendicular to the surface with the t_{2g} orbitals overlapping over the surface. If one considers the mode of double-bond

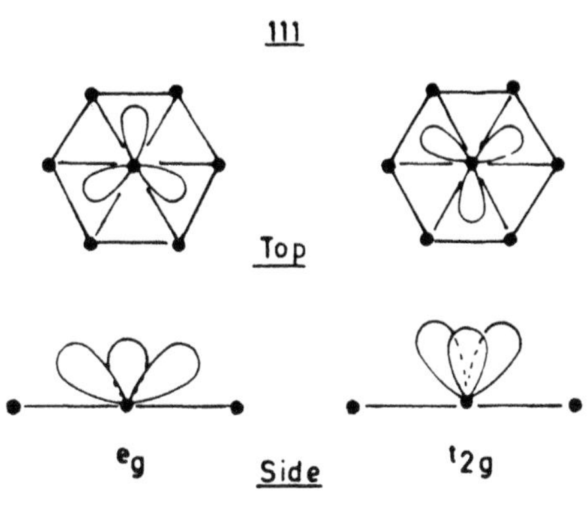

(a)

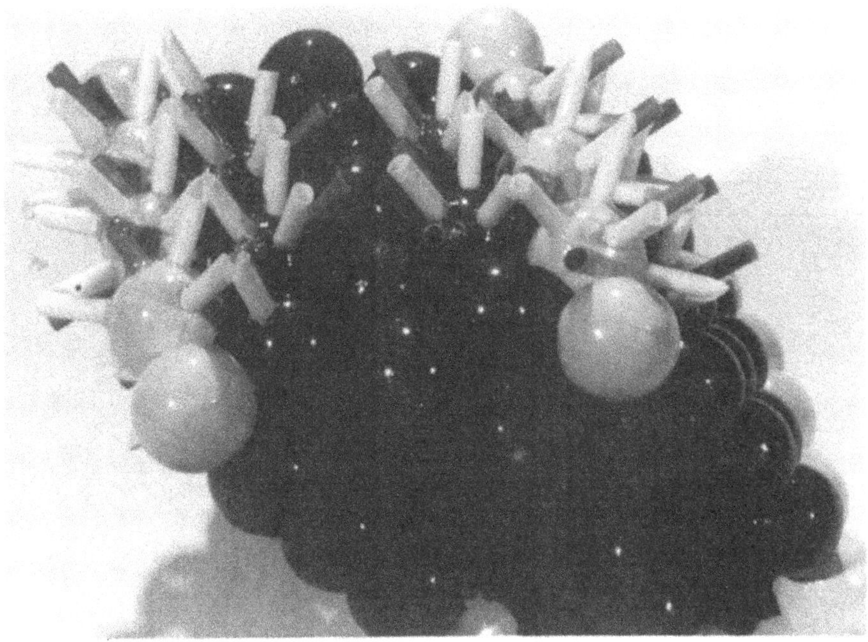

(b)

Fig. 8 Surface orbital description of atoms in a 111 plane. (a) Orbital diagram.[12] (b) Picture of model.

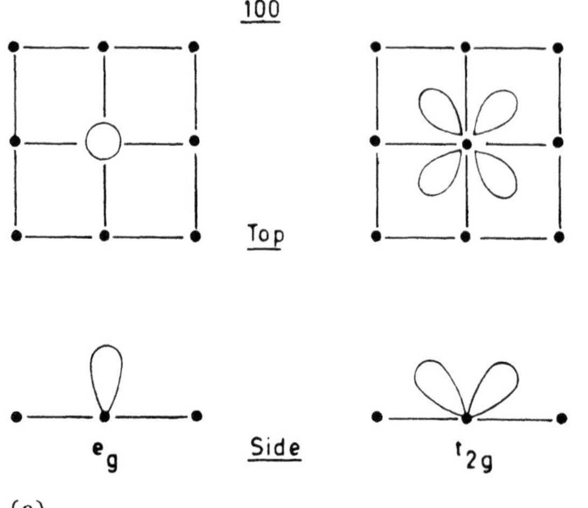

(a)

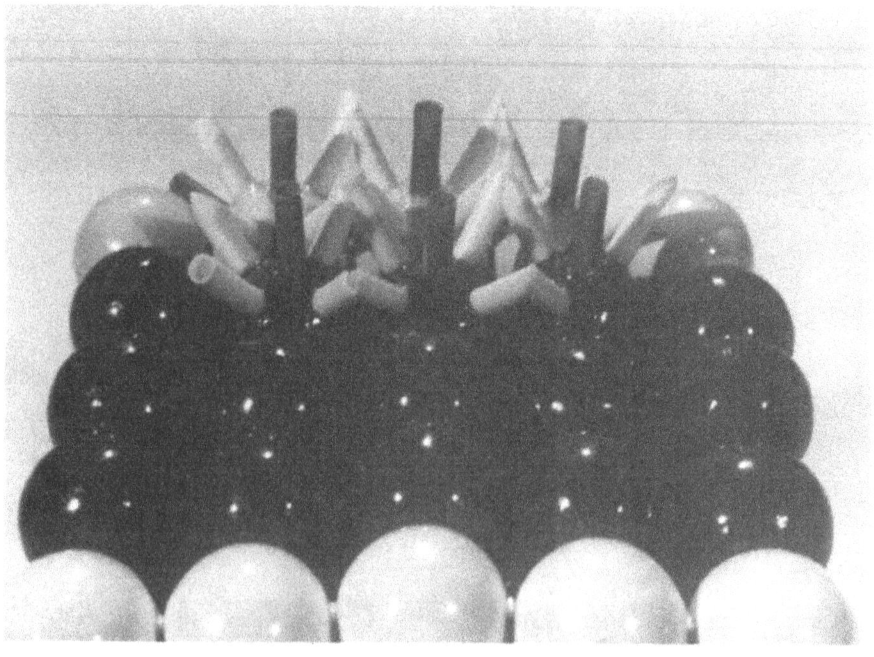

(b)

Fig. 9 Surface orbital description of atoms in a 100 plane. (a) Orbital diagram.[12] (b) Picture of model.

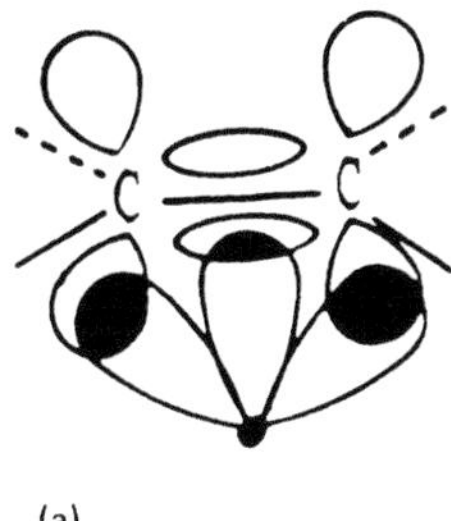

Fig. 10 Modes of surface adsorption. (a) Double-bond adsorption on a single surface atom. (b) Dissociative H_2 adsorption on a single surface atom. (c) Dissociative H_2 adsorption involving several surface atoms.

adsorption on a metal surface to be the same as that involved in the formation of metal-alkene complexes (Fig. 10a), then it might be considered possible to have a double bond adsorb on the 100-face atoms with π-electron donation into an empty perpendicular e_g orbital and back donation from the t_{2g} orbitals into the π^* orbitals of the alkene. This does not appear to be verified experimentally. Our preliminary angular overlap calculations for these sites indicate that the LUMO of the 100 site is not the $d_z{}^2$ as expected but rather a p orbital and, further, that the symmetry is not correct for double-bond interaction.

One mode of dissociative adsorption of dihydrogen, depicted in Fig. 10b, is similar to that proposed for the olefin adsorption with σ-electron donation and back bonding into the σ^* bonds of the dihydrogen molecule to produce two M-H bonds. Since hydrogen chemisorption is the common method of determining metal surface area in dispersed metal catalysts, one must consider that hydrogen has the capability

(a)

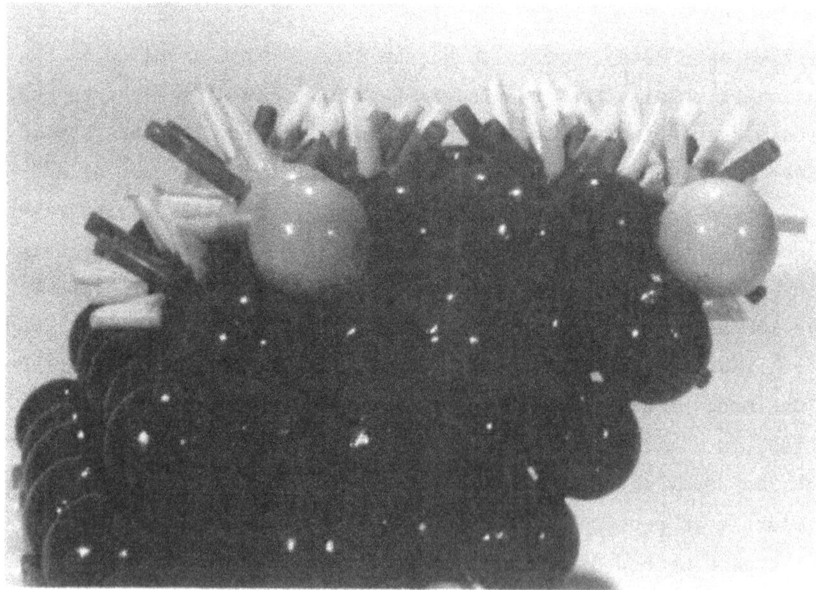

(b)

Fig. 11 Models of 111-100 edges. (a) Narrow angle between planes. (b) Wide angle between planes.

of adsorbing on all of the different types of surface atoms, so that other modes of dissociative adsorption, such as that shown in Fig. 10c, which involves more than one surface atom, must also be possible.

As described in Table 2, there are four different types of edges possible with the fcc crystal arrangement; two 111-111 and two 111-100 edges differing in the angle of intersection of the planes. Models showing both the narrow and wide 111-100 edges are pictured in Fig. 11. It is readily apparent from these pictures that the narrow edge (Fig. 11a) is considerably more coordinatively unsaturated than is the wide edge (Fig. 11b). A similar pattern exists for the narrow and wide 111-111 edges. With the overlap of the angularly protruding orbitals with those on adjacent atoms it is possible that the wide edges do not have sufficient coordinative unsaturation to promote hydrogen-olefin reactions so reactions such as isomerization could be taking place only on the narrow-type edges.

Pictures of models showing some of the five different corner atoms are shown in Fig. 12. The most coordinatively unsaturated of

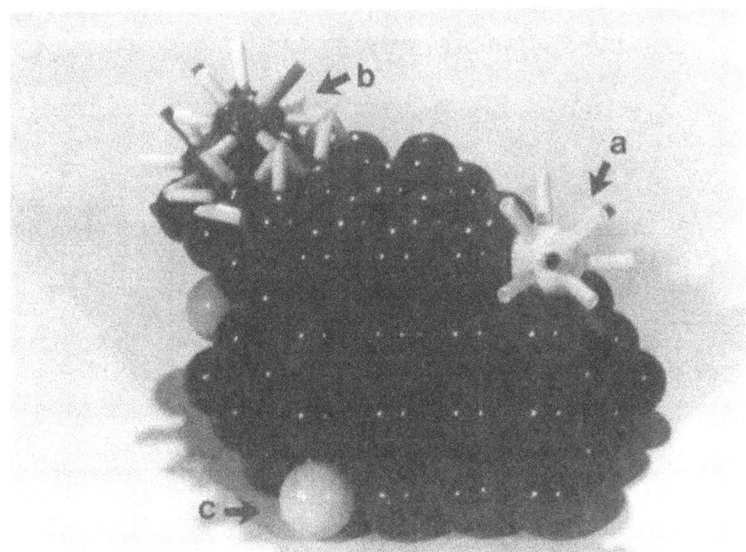

Fig. 12 Model of fcc metal crystallite showing several different corner atoms. (a) 111-111-100-100 corner. (b) 111-111-111 corner (111 adatom). (c) 111-111-100 corner.

these is the 111 adatom (Fig. 12b), which is a tetrahedral corner having only three nearest neighbors and one next-nearest neighbor. Looking at the orbital arrangement on these corners, it appears that they all have the capability of adsorbing both an olefin and a dihydrogen, as shown in Fig. 10a and b, and so should be able to react in the manner depicted in Fig. 6b. However, it is apparent from experimental data that at least one of these corner atoms must not be able to adsorb both atoms of the dihydrogen molecule but must depend on cooperation with neighboring atoms to produce the ^{3}MH species, although it is not apparent from an examination of the models as to which one this would be. Further, there is no obvious reason for the observed[8]

(a)

Fig. 13 Models showing direct, one-step alkene hydrogenation reaction sequence. (a) Active ^{3}M site. (b) ^{3}M site with H_2 adsorption. (c) ^{3}M site with H_2 and alkene adsorption.

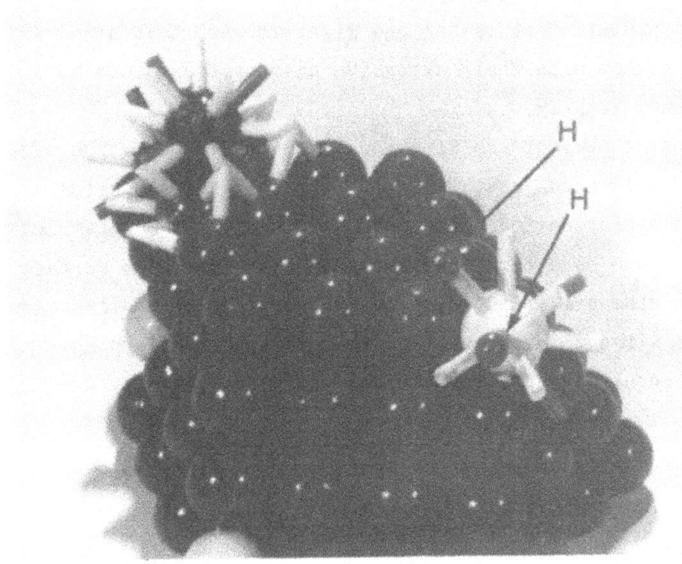

(b)

(c)

difference in the strength of hydrogen chemisorption between the two types of direct saturation sites but one might imagine that sites such as the 111 adatom with their extensive unsaturation would hold the hydrogen rather tightly.

One could use these surface site models to predict the type of reactions which could occur on the various types of surface sites during an alkene hydrogenation. The easiest to see is the possible reaction taking place on a corner as shown in Fig. 13a. The surface orbitals on this atom are arranged so that adsorption of a dihydrogen could take place without involvement of the orbitals overlapping those of the adjacent surface atoms. This would give the surface

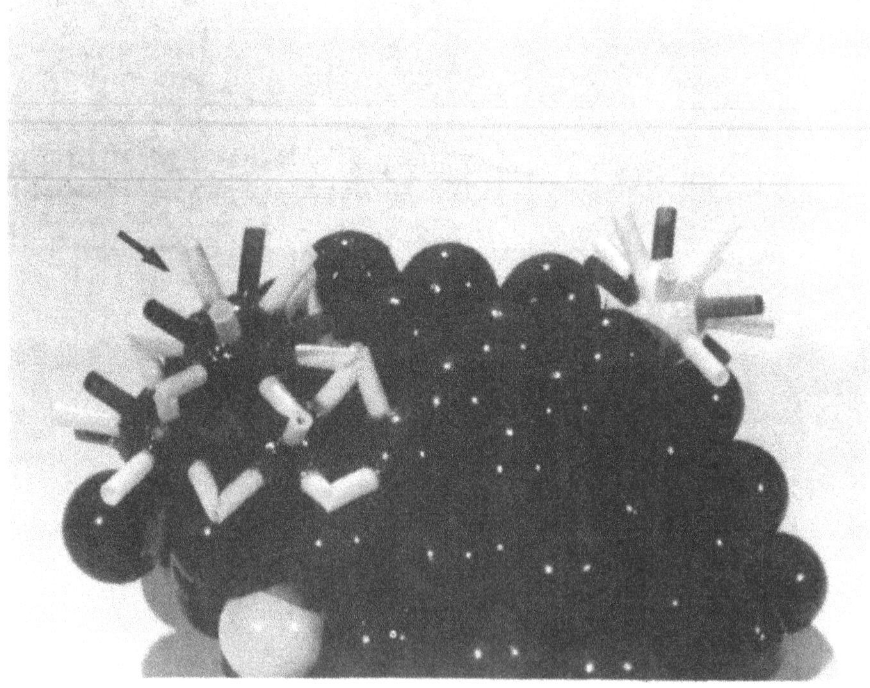

(a)

Fig. 14 Models showing two-step alkene hydrogenation reaction sequence. (a) Active corner site. (b) ^{3}MH site formation, dissociative adsorption of H_2 on two adjacent atoms. (c) ^{3}MH site with adsorbed alkene. (d) Intermediate surface metal alkyl on ^{3}MH site. (e) Dihydridometal alkyl species.

(b)

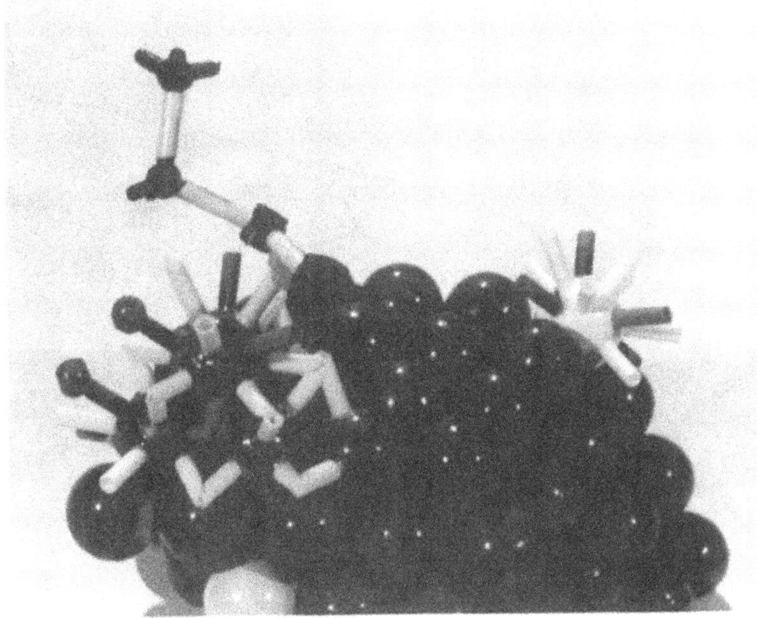

(c)

(d)

(e)

Fig. 14 (Continued)

dihydride shown in Fig. 13b. Adsorption of the olefin leads to the species pictured in Fig. 13c. Transfer of both hydrogens to the double bond as depicted in Fig. 6b gives the alkane and regenerates the empty surface site for further catalytic reactions.

A two-step hydrogenation sequence similar to that outlined in Fig. 5 is pictured in Fig. 14. Adsorption of hydrogen by two surface atoms one of which is on the corner as pictured in Fig. 14a gives the dihydride in Fig. 14b. Adsorption of the olefin (Fig. 14c) and hydrogen insertion gives the surface metal alkyl (Fig. 14d). Addition of dihydrogen (Fig. 14e) and reductive elimination leads to alkane formation and regeneration of the ^{3}MH active site (Fig. 14b).

A ^{2}M-type isomerization site is pictured in Fig. 15a. A two-atom dissociative adsorption of dihydrogen gives the surface hydride shown in Fig. 15b. Adsorption of the olefin (Fig. 15c) and hydrogen insertion produces the surface metal alkyl (Fig. 15d) but the addition of further hydrogen is inhibited so the reaction proceeds via β elimination to give the isomerized adsorbed olefin (Fig. 15e), which desorbs to regenerate the active site (Fig. 15b). Some reasons can be proposed to explain why these ^{2}M sites react by isomerization while the mechanistically similar ^{3}MH sites react to give the saturated product (Fig. 14). As discussed above, the difference in coordinative unsaturation between these two types of sites could result in a difference in metal alkyl stability. This premise is supported by STO findings,[8] which show that on the two-step hydrogenation site (^{3}MH) the intermediate metal alkyl is stable under STO reaction conditions for prolonged periods of time and is removed from the catalyst only by the addition of further hydrogen to complete the reduction. It is this factor which is used to distinguish between the two-step hydrogenation sites and the isomerization sites in the STO characterization procedure.

Another factor to consider is the difference in the ease of adsorption of the second hydrogen atom needed for saturation. This is facilitated on the ^{3}MH sites because the degree of unsaturation present on these sites permits the adsorption of a dihydrogen molecule on the reactive atom itself. With the ^2M sites this is not possible and addition of more hydrogen can take place only with the assistance of neighboring atoms which, under standard reaction conditions, are themselves probably saturated with hydrogen, so that further addition to the ^2M site is restricted.

V. CONCLUSIONS

The presence of a multiplicity of surface sites each with its own reactivity and selectivity is not contradicted by the data accumulated to support the Horiuti-Polanyi mechanism. The isomerization sites can promote not only isomerization but also multiple deuterium exchange on the desorbed olefin, as observed.[3] What is also noted in the deuteration of double bonds, especially over Pt catalysts, is that the bulk

(a)

Fig. 15 Models showing the double-bond isomerization reaction sequence. (a) ^2M sites. (b) ^2MH site formation, dissociative adsorption of H_2 on two adjacent atoms. (c) ^2MH site with adsorbed alkene. (d) Intermediate surface metal alkyl. (e) ^2MH site with adsorbed isomerized alkene.

(b)

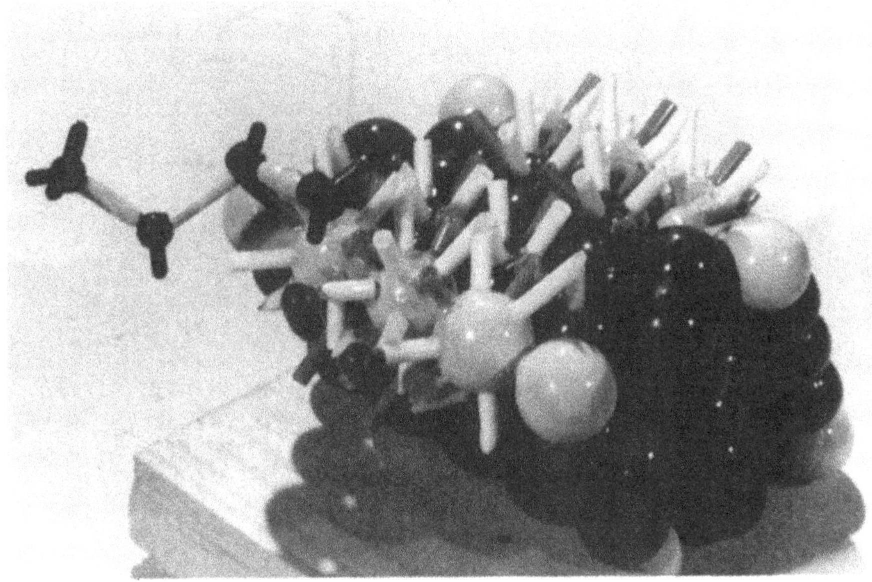

(c)

(d)

(e)

Fig. 15 (continued)

of the deuterated product is the d_2 species,[3] which would be expected from reaction on the direct saturation sites. The obtaining of H-D-containing products from mixtures of H_2 and D_2 can be explained by the presence of the ^{3}MH sites. Thus, all of the data supporting the Horiuti-Polanyi mechanism can be accounted for by the presence of these different types of sites on the surface of the metal catalysts.

Obviously, the picture presented here is somewhat simplified and based extensively on analogy and the reasonability of the mechanistic proposals. The fact remains, however, that the surface of a dispersed metal catalyst particle is composed of many different types of sites and it is perfectly reasonable to expect these different types of sites to have different reactivities and selectivities. Progress is presently being made in rendering these concepts more quantitative by employing angular overlap and other MO type of calculations on the different types of sites and then correlating the observed d and p electron pattern with the reactivities proposed above.

ACKNOWLEDGMENT

This research was supported by grant DE-FG02-84ER45120 from the U.S. Department of Energy, Office of Basic Energy Science.

REFERENCES

1. I. Horiuti and M. Polanyi, *Trans. Faraday Soc., 30,* 1164 (1964).
2. R. L. Burwell, Jr., *Catal. Revs., 7,* 25 (1972).
3. R. L. Augustine, F. Yaghmaie, and J. F. Van Peppen, *J. Org. Chem., 49,* 1865 (1984).
4. S. Siegel, *Adv. Catal., 16,* 123 (1966).
5. R. L. Augustine and J. F. Van Peppen, *Ann. N.Y. Acad. Sci., 172,* 244 (1970).
6. S. Siegel, J. Outlaw, Jr., and N. Garti, *J. Catal., 52,* 102 (1978).
7. C. O'Connor and G. Wilkinson, *J. Chem. Soc. A,* 2665 (1968).
8. R. L. Augustine and R. W. Warner, *J. Catal., 80,* 358 (1983).
9. J. A. Osborne, F. H. Jardine, J. F. Young, and G. Wilkinson, *J. Chem. Soc. A,* 1711 (1966).

10. R. L. Augustine and R. W. Warner, *J. Org. Chem.*, *46*, 2614 (1981).

11. R. L. Augustine and M. M. Thompson, *J. Org. Chem.*, *52*, 1911 (1987).

12. G. C. Bond, *Disc. Faraday Soc.*, *41*, 200 (1966); *Surf. Sci.*, *18*, 11 (1969).

9

CIR-FTIR Studies of Palladium-Catalyzed Carboalkoxylation Reactions

WILLIAM R. MOSER, ANDREW W. WANG, and NICHOLAS K. KILDAHL

Worcester Polytechnic Institute, Worcester, Massachusetts

ABSTRACT

Palladium-catalyzed carboalkoxylation reactions, in which an aryl halide and an alcohol react in the presence of carbon monoxide and a tertiary amine, are potentially important both industrially and in the laboratory as alternatives to conventional routes to ester production. In spite of this, the current understanding of the mechanism of this reaction is limited. A novel in situ infrared technique, cylindrical internal reflectance-Fourier transform infrared spectroscopy (CIR-FTIR), has been utilized in this study to obtain a clearer picture of the principal steps involved in catalytic ester formation. Infrared spectra of the active reaction at high alcohol concentrations show that the dominant palladium complex does not contain a carbonyl group, consistent with a rate-limiting step involving oxidative addition of the aryl halide to a palladium-phosphine complex. At low methanol concentrations, the palladium-benzoyl ("acyl") complex predominates and the rate is a strong function of the basicity of the amine used. Stoichiometric studies have shown that the acyl complex is quite stable unless both alcohol and amine are present, suggesting that the amine activates the alcohol by deprotonating it and thus converting it to a more reactive alkoxide ion. These observations are consistent with the mechanism presented in Fig. 8.

I. PALLADIUM-CATALYZED CARBOALKOXYLATION REACTIONS: BACKGROUND

A. The Reaction and Its Applications

The palladium-catalyzed carboalkoxylation of organic halides with alcohols, reported by Heck[1,2] in the late 1970s, is of both commercial and laboratory significance as a route to ester production. From a commercial standpoint, it potentially represents a process by which haloalkenes or aryl halides, produced as byproducts of major hydrocarbon processes, can be converted into commercially important chemicals. In the laboratory, carboalkoxylation may be useful as a means of synthesizing esters when more conventional preparations, such as acylation of alcohols by acyl halides or carboxylic acids, are either cumbersome or unavailable.

Catalytic carboalkoxylation involves reaction of an organic halide with an alcohol and carbon monoxide under mild pressures and temperatures of 35-200 psi and 85-100°C. A stoichiometric amount of a tertiary amine must be added to take up the hydrogen halide produced. The homogeneous catalyst consists of tertiary phosphine-modified palladium complexes, usually introduced as tetrakis(phosphine)palladium(0) or palladium(II) acetate with added phosphine. Typically, triphenylphosphine or a similar tertiary phosphine is employed. The overall reaction may be written as:

$$\mathrm{RX + R'OH + R''_3N + CO \xrightarrow{\text{"Pd"},\ L_3P} RCOOR' + R''_3NH^+X^-}$$

B. Rates and Scope

Heck found that this carboalkoxylation reaction may be carried out with a variety of aryl, vinyl, and benzyl halides.[1,2] The reaction is most facile for organic iodides, while rates using chlorides may become prohibitively slow. Intermediate rates are observed for bromides. The reaction is also very sensitive to the organic group.

Kudo and coworkers[3] reported the following order of reactivity: allyl > benzyl > phenyl = methyl > vinyl > propyl > ethyl. This correlates with the dissociation energy of the R-X bond, except in cases of unsaturated compounds, which have the advantage of being able to form π-complexes first. When aryl halides are used as the substrate, electron-withdrawing substituents on the aryl ring increase the reaction rate.[2] Selectivities to the ester product are generally over 90%. The use of high pressures of carbon monoxide in conjunction with phosphine-to-palladium mole ratios of less than 4 is to be avoided, as it tends to lead to catalyst decomposition via formation of palladium carbonyl clusters.

Carboalkoxylation is just one member in the large family of homogeneous palladium-catalyzed organic reactions. Related reactions include the double carbonylation of organic halides with alcohols or secondary amines to produce α-keto esters[4] or amides respectively,[5-7] the coupling of organometallics (applicable to a host of substrates),[8-21] and the arylation of allyl alcohols to prepare aldehydes and ketones,[22-28] to name a few prominent examples.

II. PREVIOUS MECHANISTIC WORK

A. Mechanisms Involving Acyl Intermediates

The original mechanism proposed by Heck[1] for the carboalkoxylation reaction is shown in Fig. 1. The salient points of the mechanism include oxidative addition of the organic halide to a palladium(0) complex forming a palladium(II) acyl complex, alcoholysis of this complex yielding ester and a halo(hydrido)palladium(II) complex, and elimination of HX to regenerate the original catalyst. It is important to note that the amine does not directly enter into this mechanism at all and merely serves to react with the HX produced. The oxidative addition of organic halides to palladium(0) complexes has been examined by a number of researchers. The work of Fitton and coworkers[29-31] in particular has shown that oxidative addition of aryl halides to tetrakis(triphenylphosphine)palladium(0) is facile for iodides and bromides, while chlorides may not add at all. This

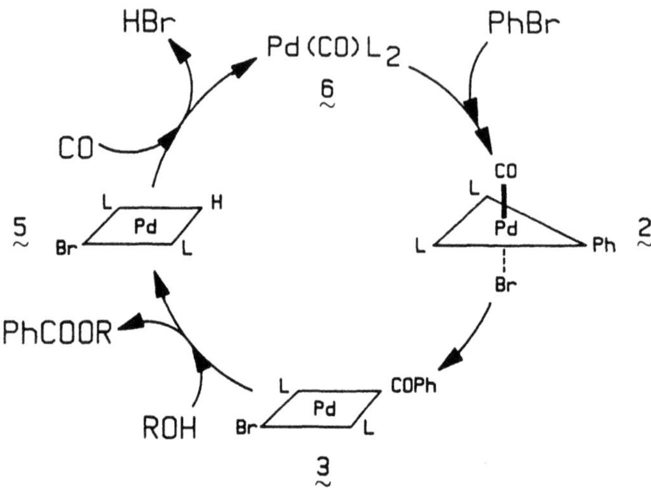

Fig. 1. Carboalkoxylation mechanism originally published by Heck.[1] L = tertiary phosphine ligand, Ph = phenyl.

is consistent with oxidative addition of the organic halide to a palladium(0) complex being the rate-limiting step for the carbo-alkoxylation reaction under typical catalytic conditions. A second mechanism[1] proposed by Heck involves formation of the acyl complex followed by reaction with the alcohol and the amine producing an organo(carboalkoxy)palladium(II) intermediate. Heck[1] also suggested a third possibility, in which the acyl complex (3) reductively eliminates the corresponding acyl halide via reaction with carbon monoxide, regenerating the original catalyst. The acyl halide would then react rapidly with the tertiary amine and the alcohol to form ester and trialkylammonium halide. The preparation of esters from acyl halides and alcohols in the presence of a tertiary amine is a standard laboratory reaction.[32]

Experiments carried out by Kudo and coworkers[3] showed that bromobenzene does not oxidatively add to $Pd(P(C_6H_5)_3)_3(CO)$, implying that this complex is too electron-poor to undergo oxidative addition due to the presence of the carbonyl ligand. This suggests that the catalytic carboalkoxylation mechanism may proceed via addition of the organic halide to bis(phosphine)palladium(0) rather than a zero-valent palladium carbonyl complex. Garrou and Heck[33] examined the carbonylation of haloorganobis(phosphine)nickel, -palladium, and -platinum complexes in detail, using a wide variety of organic halides and phosphines. In all cases, they observed the corresponding acyl complex as the final product, and reductive elimination of the acyl halide was never observed. In general they found that the use of excess phosphine ligand depressed the rate of carbonylation and that the carbonylation rate approached a constant value with continued addition of phosphine. From this they concluded that carbonylation proceeded via two pathways: the more facile route, involving prior dissociation of a phosphine ligand, was suppressed by the addition of excess phosphine, while the slower direct carbonylation route was independent of phosphine concentration. Later work has shown that this direct route is far more sensitive to changes in the electronic nature of the migrating group than the dissociative pathway. In most cases, Garrou and Heck found that the rate of production of the carbonyl intermediate (**2**) was much slower than that of the subsequent insertion step, and this intermediate was not observed. However, when strongly electron-donating phosphines were used in conjunction with iodoplatinum complexes, they were able (via ^{31}P NMR and infrared spectroscopy) to observe initial formation of the intermediate carbonyl complex (**2**), followed by gradual conversion to the acyl complex (**3**). By analogy, one can assume that all carbonylation reactions of haloorganobis(phosphine)nickel, -palladium, and -platinum complexes proceed via a haloorganobis(phosphine)metal carbonyl intermediate.

The double-carbonylation reaction to produce α-keto amides from organic halides and secondary amines has been studied extensively by Ozawa and coworkers.[5-7] Significantly, they found that their data were consistent with a mechanism for double carbonylation which proceeds via oxidative addition of the organic halide to a palladium(0) complex, followed by carbonylation and insertion forming the acyl intermediate. This conclusion is in complete agreement with the first part of the Heck mechanism for the carboalkoxylation (single-carbonylation) reaction.

B. Mechanism Involving a Carbomethoxy Intermediate

The preparation of chloro(carbomethoxy)bis(triphenylphosphine)palladium(II) was described by Hidai and coworkers in 1973.[34] They found that a solution of $(PPh_3)_2PdCl_2$ in methanol would react with carbon monoxide to form the carbomethoxy complex in the presence of a tertiary amine and that this complex would react with methyl iodide to form methyl acetate. Building on this work, Stille and Wong[35] reported that when the carbomethoxylation of iodobenzene was carried out with $(PPh_3)_2PdCl_2$ as the catalyst, the palladium was recovered as the halo(carbomethoxy)bis(phosphine) complex. Furthermore, a methanol solution of the carbomethoxy complex in the presence of sodium acetate reacted with an equimolar amount of iodobenzene to form methyl benzoate. In this study, a 25% yield was obtained after 12 hr at 60°C under 200 psi of carbon monoxide. Based on this observation, Stille and Wong suggested that, in some cases, a catalytic mechanism involving reaction of organic halide with a carbomethoxy intermediate to form an ester may be operative (Fig. 2).

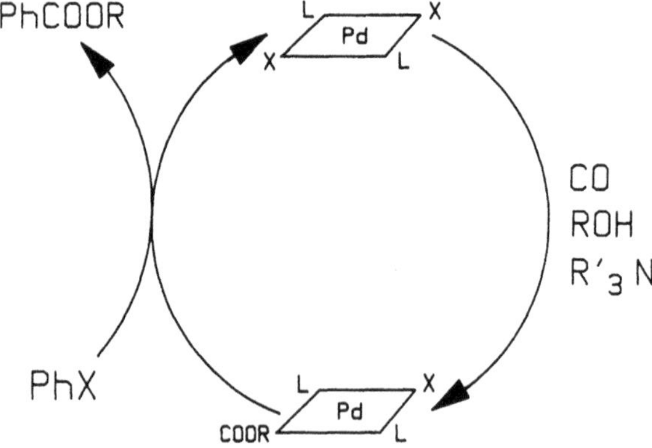

Fig. 2 Alternate carboalkoxylation mechanism suggested by Stille and Wong.[35] L = tertiary phosphine ligand, Ph = phenyl, X = halide ligand.

III. CURRENT MECHANISTIC STUDIES

A. CIR-FTIR In Situ Reactors

Current mechanistic studies of this reaction in our laboratory have been centered around closely examining the mechanistic steps suggested in the literature to determine their feasibility on a stoichiometric basis and their consistency with the data measured under catalytic conditions. For the purposes of this study, the carbomethoxylation of bromobenzene with methanol and triethylamine to produce methyl benzoate was chosen as a model system. In situ infrared spectra obtained via cylindrical internal reflectance-Fourier transform infrared spectroscopy (CIR-FTIR) have proven invaluable in monitoring the dominant catalytic intermediates and measuring the rates of catalytic processes.[36-40] In this technique, the incident infrared beam is directed onto the conical end of a cylindrical silicon rod. The reactor used in these studies is constructed in such a way that this silicon element passes directly through the reacting solution and out the other side (see Fig. 3). As the infrared beam travels through the infrared element, it undergoes a series of "reflections" off the interface between the silicon and its surroundings. At each reflection, the infrared radiation penetrates into the surrounding medium in accordance with the principles of attenuated total reflectance[41]; thus infrared absorbances of the surrounding medium are detected by the infrared beam. The transmitted beam from the end of the element is directed by mirrors to the infrared detector.

This technique offers several experimental advantages in the study of solution chemistry. Most significantly, it permits the acquisition of in situ infrared spectra of reacting solutions without seriously undermining the quality of the mixing or the gas-liquid contacting in the reactor. Thus, the portion of the solution being analyzed is truly representative of the solution as a whole. Furthermore, the pathlength in a CIR-FTIR experiment is much shorter (approximately 10 μm) and less sensitive to changes in temperature or pressure than that of a conventional infrared technique. These features enable one to analyze solutions in strongly absorbing solvents and greatly

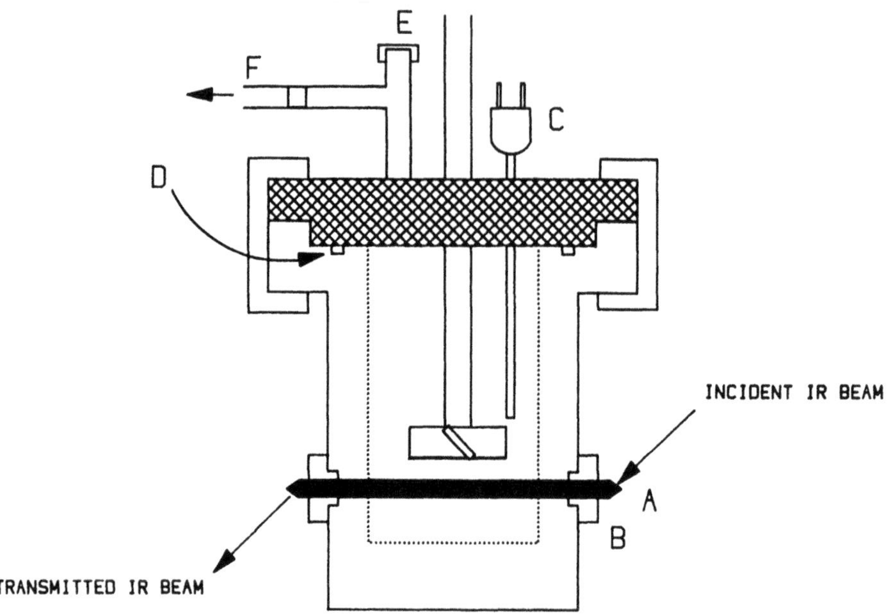

Fig. 3 Schematic of the CIR-FTIR reactor. The incident infrared beam is directed at an angle onto the CIR element (A, black) by mirrors. The beam then undergoes a series of reflections while passing through the element before exiting through the far end, where it is collected by mirrors and directed to the detector. Teflon O rings captured by stainless steel endplates (B) provide a seal around the CIR element good to 1500 psi. An additional O ring (D) seals the top of the reactor. A thermocouple (C) and pressure gage (not shown) enable the researcher to accurately monitor conditions in the reactor. The reactor is also equipped with a septum (E) for adding or removing liquids during operation, and a gas inlet (F).

facilitate the use of the infrared spectrometer in quantitative or kinetic studies. Reference 36 provides a more complete description of the CIR-FTIR technique and its applications.

Figure 4 shows a typical infrared spectrum of the carbomethoxylation of bromobenzene to form methyl benzoate, obtained at 90°C and 50 psi. Infrared absorbances due to methanol (3300, 2900-2700), bromobenzene (1578), methyl benzoate (1725), and triethylamine (2900-2700) are clearly seen. Note that this system is a particularly convenient one to study by infrared spectroscopy, since the carbonyl stretching region (2300-1600) is largely devoid of peaks.

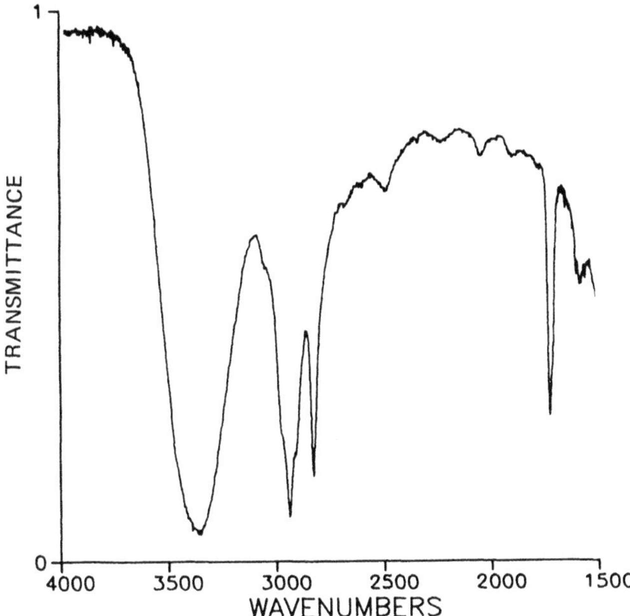

Fig. 4 A full-scale infrared spectrum of an active carboalkoxylation reaction: the carbomethoxylation of bromobenzene in the presence of triethylamine at a temperature of 90°C and a pressure of 40 psi.

B. Carbonylation of Haloorganobis(phosphine)palladium(II)

The reaction of bromo(phenyl)bis(triphenylphosphine)palladium(II) with carbon monoxide has been examined via CIR-FTIR.[42] Consistent with the work of Garrou and Heck,[33] the reaction proceeds smoothly to form bromo(benzoyl)bis(triphenylphosphine)palladium(II), and no reductive elimination of benzoyl bromide is observed. Moreover, when the carbomethoxylation of bromobenzenene is carried out catalytically and allowed to consume all of the available carbon monoxide, the appearance of infrared peaks attributed to $Pd(Br)(C_6H_5)(P(C_6H_5)_3)_2$ is observed in the CIR-FTIR spectrum of the reacting solution. This is consistent with a mechanism in which reaction of carbon monoxide with $Pd(Br)(C_6H_5)(P(C_6H_5)_3)_2$ becomes rate limiting under conditions of carbon monoxide depletion.

C. Reductive Elimination of Acyl Halide[42]

It was of interest to examine the possibility that ester formation results principally from reaction of the amine and the alcohol with an acyl halide reductively eliminated from the acyl complex.[3] Although Garrou and Heck were unable to observe any reductive elimination from the acyl complexes they prepared, it is possible that an unfavorable yet rapid equilibrium to the acyl halide exists. Such an equilibrium could constitute the dominant pathway to ester formation, since the subsequent reaction of the acyl halide with the amine and the alcohol are known to be very rapid.[32] This possibility was evaluated by examining the reactivity of bromo(phenyl)bis(triphenylphosphine)palladium(II) with triethylamine at reaction temperature. CIR-FTIR spectra of this system taken over a 20-min period revealed that this acyl complex is completely stable in the presence of triethylamine, indicating that reductive elimination of benzoyl bromide does not occur to any measurable extent under these conditions.

D. Reaction of Alcohols with Haloorganobis(phosphine)palladium(II) Complexes[42]

The reaction of bromo(benzoyl)bis(triphenylphosphine)palladium(II) with methanol was studied to determine whether direct alcoholysis of the acyl complex (**3**) could be the primary mechanism of ester formation in the carboalkoxylation reaction. A solution of bromo(benzoyl)bis(triphenylphosphine)palladium(II) in methanol was heated to reaction temperature in the CIR-FTIR reactor. Infrared spectra obtained at 4-min intervals over a 20-min period indicated only a slow rate of methyl benzoate formation. However, when triethylamine was added to the reacting solution, rapid formation of methyl benzoate was immediately observed. Similarly, addition of methanol to the previously described unreactive solution of (**3**) with triethylamine resulted in a rapid rate of esterification. From these data, we conclude that direct attack of alcohol on the acyl complex (**3**) is not the principal route to ester but that both amine and alcohol are necessary for facile esterification.

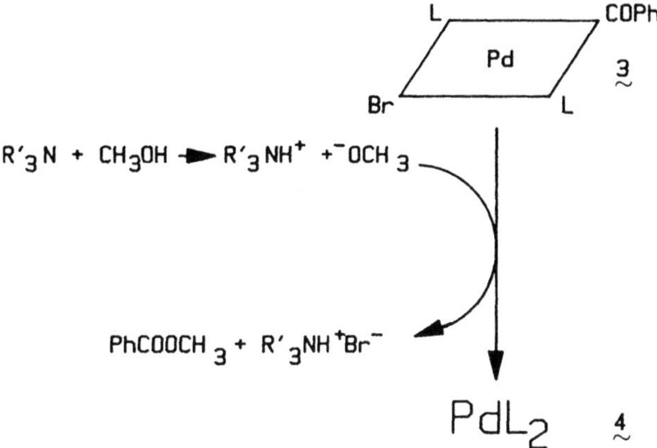

Fig. 5 Proposed mechanistic step in which the amine activates the alcohol by deprotonating it in a preequilibrium step. The resulting alkoxide ion then attacks the acyl complex (3), producing ester product and the bis(phosphine)palladium(0) complex. L = tertiary phosphine ligand.

E. Role of Amines in the Carboalkoxylation Mechanism

It is apparent that triethylamine in some way activates methanol with respect to its reactivity with bromo(benzoyl)bis(triphenylphosphine)palladium(II). The highly basic nature of tertiary amines suggests that this activation may occur via an equilibrium deprotonation of the methanol by the amine, forming methoxide and triethylammonium ions.[43] Methoxide ion is a far superior nucleophile to methanol, and therefore its reaction with the acyl complex (3) should be markedly faster (see Fig. 5).

This hypothesis was tested in a series of stoichiometric experiments in which the rate of reaction of bromo(benzoyl)bis(triphenylphosphine)palladium(II) with methanol was measured in the presence of a variety of amines of varying basicity.[42] Consistent with our expectations, the observed rate of methyl benzoate formation (and acyl disappearance) was found to be a linear function of the amine basicity and therefore of the concentration of methoxide ion present in the reactor. This provides further evidence that it is methoxide ion, rather than methanol, which reacts with the acyl complex to produce ester.

This amine basicity effect was also explored in the catalytic reaction. The carbomethoxylation of bromobenzene was carried out catalytically using the same amines as were examined noncatalytically. A low methanol concentration was used in these experiments to magnify the amine effect by ensuring that the methoxide-involving step would be rate determining. The rates of ester formation measured catalytically correlate well with those observed in the noncatalytic experiments. From this, we conclude that this reaction, in which methoxide ion produced from equilibrium deprotonation of methanol by triethylamine reacts with the acyl complex (3), is a part of the overall catalytic mechanism.

F. Effect of Changing Methanol Concentration

In the previous section, it was assumed that below some threshold methanol concentration, the reaction of methoxide ion with the acyl complex becomes rate determining. Under these conditions, the overall catalytic rate of methyl benzoate formation should be very sensitive to methanol concentration and the acyl complex should be the dominant catalytic intermediate in the reacting solution. CIR-FTIR studies of the catalytic reaction carried out over a range of initial methanol concentrations confirmed these hypotheses. At methanol concentrations of roughly 6 M or less, the rate of methyl benzoate formation was found to be a strong function of methanol concentration; at higher concentrations, the rate increased only gradually with increasing methanol concentration. Furthermore, CIR-FTIR spectra taken in the low-methanol regime display a peak at 1648 cm^{-1}, indicative of the presence of the acyl complex (3); in spectra of high-methanol concentration reactions, this peak is either much weaker or is absent.

G. Regeneration of $Pd(Br)(C_6H_5)(P(C_6H_5)_3)_2$ from $Pd(Br)(COC_6H_5)(P(C_6H_5)_3)_2$

The reaction of methoxide ion with the acyl complex should produce bis(triphenylphosphine)palladium(0). In the overall catalytic mechanism, one would suspect that this complex would immediately undergo oxidative addition of bromobenzene, regenerating $Pd(Br)(C_6H_5)(P(C_6H_5)_3)_2$. When this reaction is carried out in the CIR-FTIR reactor, the disappearance of the acyl infrared peak at 1648 cm^{-1} is accompanied by the

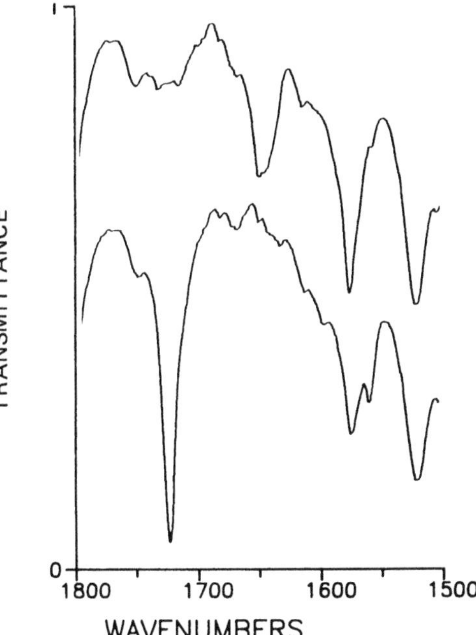

Fig. 6 Regeneration of the "alkyl" complex ($Pd(C_6H_5)(Br)(P(C_6H_5)_3)_2$) from the "acyl" complex ($Pd(COC_6H_5)(Br)(P(C_6H_5)_3)_2$) via reaction with methanol, triethylamine, and bromobenzene. The top spectrum contains the acyl complex (1648 cm^{-1}) and bromobenzene (1578 cm^{-1}), before addition of methanol and triethylamine. The lower spectrum was obtained 30 min after addition of methanol and triethylamine, and shows that the acyl complex has been converted to the alkyl complex (1562 cm^{-1}) and methyl benzoate (1725 cm^{-1}).

appearance of peaks for methyl benzoate and bromo(phenyl)bis(triphenylphosphine)palladium(II)[42] (Fig. 6). This complex has been isolated and its identity verified by ^1H NMR and infrared spectroscopy. This confirms the hypothesis that the palladium-containing product of the methoxide reaction can undergo oxidative addition of bromobenzene to regenerate the active catalyst and thus complete a catalytic cycle.

The bis(phosphine) intermediate is highly reactive. In the absence of sufficient quantities of phosphine, it readily scavenges dissolved carbon monoxide resulting in the irreversible formation of

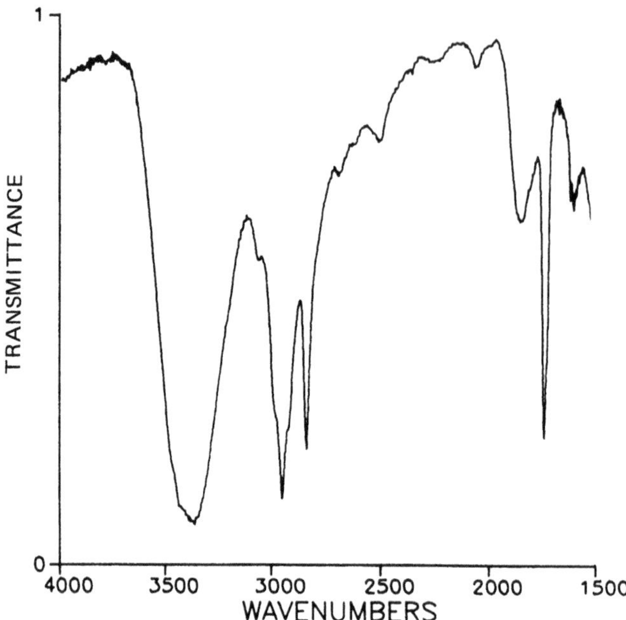

Fig. 7 Full-scale infrared spectrum of a catalytic system in which the catalyst has partially deactivated via formation of palladium homopolymers or complexes on the CIR element (broad band at 1850-1900 cm^{-1}).

palladium-carbonyl clusters or homopolymers. This mode of catalyst decomposition is easily observed spectroscopically because these compounds tend to precipitate directly onto the CIR element, causing very intense absorbances to be observed in the CIR-FTIR spectra (see Fig. 7). The use of excess phosphine stabilized the catalyst by shifting the equilibrium from the bis(phosphine) to the tetrakis(phosphine) complex. It is not clear whether oxidative addition of aryl halide occurs primarily to the two-coordinate or the four-coordinate compound.

H. Proposed Catalytic Cycle for Carboalkoxylation

As a result of this work, we proposed a new catalytic mechanism for the carbomethoxylation of bromobenzene and for palladium-catalyzed carboalkoxylation reactions of aryl halides in general[42] (Fig. 8).

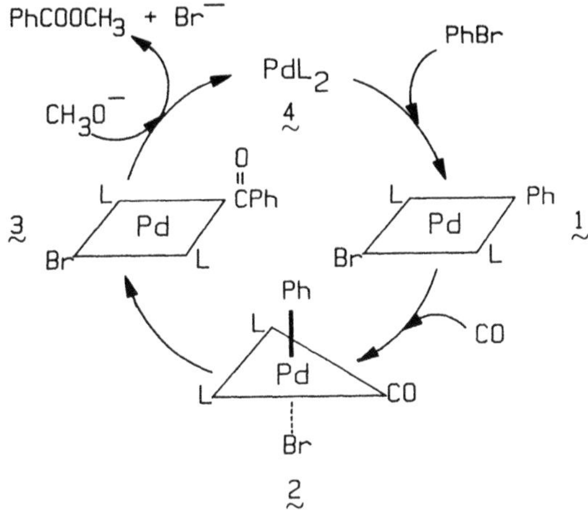

Fig. 8 Novel carboalkoxylation mechanism proposed in this study.[42] L = tertiary phosphine ligand, Ph = phenyl.

This scheme consists of four principal steps: oxidative addition of aryl halide to bis(phosphine)palladium(0) forming a halo(aryl)bis(phosphine)palladium(II) complex (1), carbonylation of this complex producing a carbonyl complex (2), insertion of the carbonyl ligand into the palladium-aryl bond yielding the acyl complex (3), and reaction of this complex with alkoxide ion producing ester and regenerating bis(phosphine)palladium(0). The alkoxide ion is produced in a pre-equilibrium deprotonation of the alcohol by the amine.

I. Hammett Studies of the Catalytic Reaction

Since infrared spectra of the active catalytic system at high methanol concentrations reveal no evidence of either the acyl complex (3) or the "alkyl" complex (1), one must conclude that if the mechanism suggested herein is correct, oxidative addition of bromobenzene to bis(triphenylphosphine)palladium(0) must be the rate-limiting step under these conditions. If this is the case, then facilitating the process of oxidative addition by using more electron-donating phosphines should markedly increase the overall catalytic rate of methyl benzoate formation. The rate of methyl benzoate production utilizing a variety of para-substituted triarylphosphines was monitored via

CIR-FTIR, and it was confirmed that electron-donating phosphines, such as tri(*p*-tolyl)phosphine, substantially increased reaction rate relative to triphenylphosphine, while electron-withdrawing phosphines suppressed it. This observation was consistent with our expectations based on the proposed mechanism.

J. Reactions Involving Carbomethoxy Intermediates

It is important to consider the alternate catalytic scheme outlined by Stille and Wong[36] in which a palladium-carbomethoxy complex, $Pd(X)(CH_3OCO)(P(C_6H_5)_3)_2$, reacts with the organic halide to produce ester. To evaluate the applicability of this mechanism to the bromobenzene-triethylamine system, a set of experiments were conducted in which solutions of the carbomethoxy intermediate were stirred at reaction temperature in the presence of bromobenzene and triethylamine.[42] A variety of solvent mixtures were examined; in some cases methanol was omitted altogether. The reaction was attempted both in the presence and absence of carbon monoxide. However, in none of the experiments were more than minute traces of methyl benzoate produced. From this we conclude that the principal mechanism of catalytic methyl benzoate formation does not involve reaction of a carbomethoxy intermediate with bromobenzene. It is possible that in catalytic systems involving more reactive organic halides, carboalkoxy intermediates may play a role in ester formation.

IV. SUMMARY AND CONCLUSIONS

Application of CIR-FTIR to the study of the carboalkoxylation of aryl halides has provided new insight into the mechanism by which palladium complexes serve as homogeneous catalysts for this reaction. It is apparent that earlier models involving formation of ester via direct attack of alcohol on a palladium-acyl complex do not adequately describe the primary route to this product. Instead, nucleophilic attack of alkoxide ion on this complex is so much more rapid that it constitutes the main mechanism of esterification, in spite of the fact that the alkoxide concentration is much lower than that of the parent alcohol. The rate of ester formation in both the catalytic and noncatalytic studies was found to be a function of the basicity of the amine used, providing support for the hypothesis that it is the amine which deprotonates the alcohol to form alkoxide ion.

Other experiments demonstrated that neither reductive elimination of acyl halide nor participation of a carboalkoxy intermediate, mechanistic steps suggested elsewhere in the literature, are significant paths to methyl benzoate formation in the bromobenzene-methanol system used as a model reaction in this study. A new mechanism, which is consistent with all the existing data on this reaction, is presented in Fig. 8.

REFERENCES

1. R. F. Heck, *Advances in Catalysis, 26,* 323 (1977).
2. A. Schoenberg, I. Bartoletti, and R. F. Heck, *J. Org. Chem., 23,* 3318 (1974).
3. K. Kudo, M. Sato, M. Hidai, and Y. Uchida, *Bull. Chem. Soc. Jap., 46,* 2820 (1973).
4. M. Tanaka, T. Kobayashi, T. Sakakura, H. Itatani, S. Danno, and K. Zushi, *J. Mol. Catal., 32,* 115 (1985).
5. F. Ozawa, H. Soyama, H. Yanagihara, I. Aoyama, H. Takino, K. Izawa, Y. Yamamoto, and A. Yamamoto, *J. Am. Chem. Soc., 107,* 3235 (1985).
6. F. Ozawa, H. Soyama, T. Yamamoto, and A. Yamamoto, *Tet. Lett., 23,* 3383 (1982).
7. F. Ozawa, T. Sugimoto, Y. Yuasa, M. Santra, Y. Yamamoto, and A. Yamamoto, *Organometallics, 3,* 683 (1984).
8. D. Milstein and J. K. Stille, *J. Am. Chem. Soc., 101,* 4992 (1979).
9. W. J. Scott, G. T. Crisp, and J. K. Stille, *J. Am. Chem. Soc., 106,* 4630 (1984).
10. J. W. Labadie and J. K. Stille, *J. Am. Chem. Soc., 105,* 6129 (1983).
11. A. O. King, N. Okukado, and E. Negishi, *J. Chem. Soc. Chem. Commun.,* 683 (1977).
12. E. Negishi, T. Takahashi, S. Baba, D. Van Horn, and N. Okukado, *J. Am. Chem. Soc., 109,* 2393 (1987).
13. J. K. Stille and J. H. Simpson, *J. Am. Chem. Soc., 109,* 2138 (1987).

14. J. W. Labadie and J. K. Stille, *J. Am. Chem. Soc., 105,* 669 (1983).

15. V. P. Baillargeon and J. K. Stille, *J. Am. Chem. Soc., 105,* 7175 (1983).

16. F. K. Sheffy and J. K. Stille, *J. Am. Chem. Soc., 105,* 7173 (1983).

17. I. Pri-Bar, P. S. Pearlman, and J. K. Stille, *J. Org. Chem., 48,* 4629 (1983).

18. D. E. Van Horn and E. Negishi, *J. Am. Chem. Soc., 100,* 2252 (1978).

19. D. E. Van Horn and E. Negishi, *J. Am. Chem. Soc., 99,* 3168 (1977).

20. S. Baba and E. Negishi, *J. Am. Chem. Soc., 98,* 6729 (1976).

21. D. Milstein and J. K. Stille, *J. Am. Chem. Soc., 101,* 4981 (1979).

22. R. F. Heck, *J. Am. Chem. Soc., 90,* 5518 (1968).

23. R. F. Heck, *J. Am. Chem. Soc., 90,* 5526 (1968).

24. R. F. Heck, *J. Am. Chem. Soc., 90,* 5531 (1968).

25. R. F. Heck, *J. Am. Chem. Soc., 90,* 5535 (1968).

26. R. F. Heck, *J. Am. Chem. Soc., 90,* 5538 (1968).

27. R. F. Heck, *J. Am. Chem. Soc., 90,* 5542 (1968).

28. R. F. Heck, *J. Am. Chem. Soc., 90,* 5546 (1968).

29. P. Fitton and E. A. Rick, *J. Organometallic Chem., 28,* 287 (1971).

30. P. Fitton and J. E. McKeon, *J. Chem. Soc. Chem. Commun.,* 4 (1968).

31. P. Fitton, M. P. Johnson, and J. E. McKeon, *J. Chem. Soc. Chem. Commun.,* 6 (1968).

32. A. Streitwieser and C. H. Heathcock, *Introduction to Organic Chemistry,* 2nd ed., Macmillan, New York, 1981.

33. P. E. Garrou and R. F. Heck, *J. Am. Chem. Soc., 98,* 4115 (1976).

34. M. Hidai, M. Kokura, and Y. Uchida, *J. Organometallic Chem., 52,* 431 (1973).

35. J. K. Stille and P. K. Wong, *J. Org. Chem., 40,* 532 (1975).

36. W. R. Moser, J. E. Cnossen, A. W. Wang, and S. A. Krouse, *J. Catal., 95,* 21 (1985).

37. W. R. Moser, C. C. Chiang, J. E. Cnossen, and R. A. Condrate (eds.), A New FT-IR Technique for the in situ Study of the Mechanism of Solid State Materials Preparations, *Advances in Materials Characterization,* Vol. 25, Plenum Publishing, New York, 1985.

38. W. R. Moser, C. J. Papile, D. A. Brannon, R. A. Duwell, and S. J. Weininger, *J. Mol. Catal.*, *41*, 271 (1987).

39. W. R. Moser, C. J. Papile, and W. J. Weininger, *J. Mol. Catal.*, *41*, 293 (1987).

40. P. Tooley, C. Ovalles, S. C. Kao, D. J. Darensbourg, and M. Y. Darensbourg, *J. Am. Chem. Soc.*, *108*, 5465 (1986).

41. Standard Practices for Internal Reflection Spectroscopy, *Annual Book of ASTM Standards*, American Society for Testing and Materials, Designation E573-81 (1981).

42. W. R. Moser, A. W. Wang, and N. K. Kildahl, *J. Am. Chem. Soc.*, *110*, 2816 (1988).

43. M. Hidai, T. Hikita, Y. Wada, Y. Fujikura, and Y. Uchida, *Bull. Chem. Soc. Jap.*, *48*, 2075 (1975).

10

Mass Transfer Limitations During Hydrogenation Revealed by CS_2 Titration of Active Sites

GERARD V. SMITH and DANIEL J. OSTGARD

Southern Illinois University, Carbondale, Illinois

FERENC NOTHEISZ, ÁGNES G. ZSIGMOND, ISTVÁN PÁLINKÓ, and MIHÁLY BARTÓK

Attila József University, Szeged, Hungary

ABSTRACT

We suggest a new method to test for mass transfer effects in catalytic reactions which accomplishes the same results as the Madon-Boudart test but does not require the preparation of two catalysts with different loadings and the same percent dispersion. This method works for liquid phase hydrogenations and depends on the linear poisoning of addition by CS_2. CS_2 can be used to titrate active sites on Pt and Pd on which the decrease in turnover frequency for addition is linear with CS_2 molecules unless mass transfer occurs. This is true for the hydrogenation of (-)-apopinene and cyclohexene. During mass transfer limitations the decrease in turnover frequency as a function of CS_2 molecules is slower during diffusion control and faster after the rate has been reduced below the diffusion limitation. Extrapolation back to zero CS_2 of the faster decrease (non-diffusion-limited) reveals the true rate. Moreover, if it is assumed that one CS_2 poisons one active site, the true turnover frequency can be determined from the true rate and the end point for the titration. The method allows a simple test of all mass transfer effects for the liquid phase hydrogenation reaction. It may be adaptable to other experimental situations.

I. INTRODUCTION

The Koros-Novak test for mass transfer effects in catalytic reactions[1] consists of diluting a supported catalyst with inert support material. Its limitation is the inability to detect diffusion limitations across the stagnant layer surrounding each catalyst particle. Consequently, this test was modified by Madon and Boudart[2] to use a constant particle size (percent exposed atoms or percent dispersion) with different loadings as an identifier of mass transfer effects. If the turnover frequency does not depend on loading, no mass transfer effects are occurring. Experimentally, this test requires the preparation of at least two catalysts with different loadings but the same percent dispersion, and it requires a chemisorption apparatus.

We describe a new method which accomplishes the same results as the Madon-Boudart test but does not require the preparation of two catalysts with different loadings and the same percent dispersion. This method works for liquid phase hydrogenations and depends on the linear poisoning of addition by CS_2.

II. EXPERIMENTAL

Hydrogenations were conducted in the liquid phase at ambient temperatures (298-303°K) under 1 atm of hydrogen or deuterium in an apparatus composed of a reactor which is shaken in a vortex manner at 2000 rpm and a gas-handling system for monitoring volume of gas absorbed at constant pressure. This apparatus and these procedures have been described in our previous publications.[3-5] The (+)-apopinene (6,6-dimethyl-1*R*,5*R*-bicyclo[3.1.1]hept-2-ene) and its (-) isomer were prepared and purified as previously described.[4-6] The gases were all high purity. The catalysts have all been described elsewhere.

The relative rates of hydrogenation (addition) were determined for each catalyst over the range of no poisoning to complete poisoning by incrementally adding pentane solutions of CS_2. The catalyst (20 mg) was soaked in deuterium overnight at room temperature in the reaction vessel. Then 200 μl of apopinene was injected onto the catalyst and the initial hydrogenation rate measured. Next, a definite amount of CS_2 (usually a 0.1% *n*-pentane solution) was injected into the reaction mixture. After each injection of the poison, followed by a 30-sec equilibration period, the rate of hydrogenation was measured. This procedure was repeated until hydrogenation ceased. This experimental procedure determined the change in absolute rate of hydrogenation as a function of the amount of poison.

III. RESULTS

Typically, CS_2 poisons both Pd and Pt linearly such that the rate of hydrogenation of (+)-apopinene decreases monotonically until hydrogenation ceases as shown in Figs. 1 and 2.[7,8] That is, each CS_2 molecule destroys the same amount of hydrogenation activity. A similar result occurs for the hydrogenation of cyclohexene (Fig. 3), although a different end point is reached. However, if too much catalyst is used, the decrease in rate is not linear over the entire range

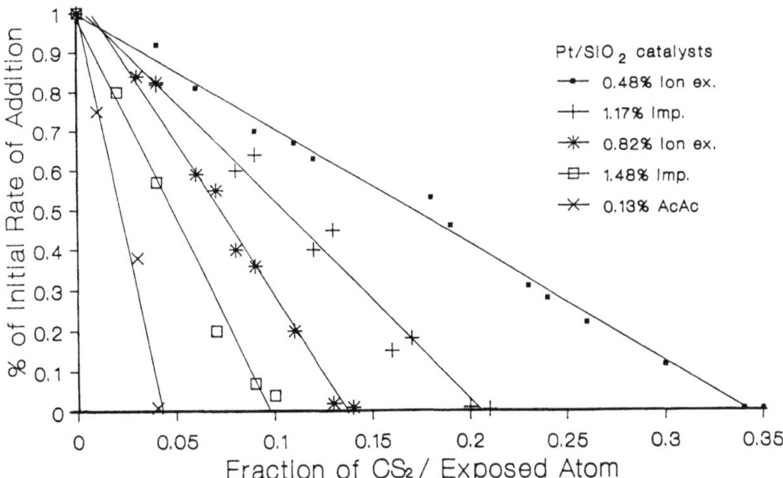

Fig. 1 Percent addition rate vs. CS_2/exp. atom for the (-)-apopinene reaction on Pt/SiO_2.

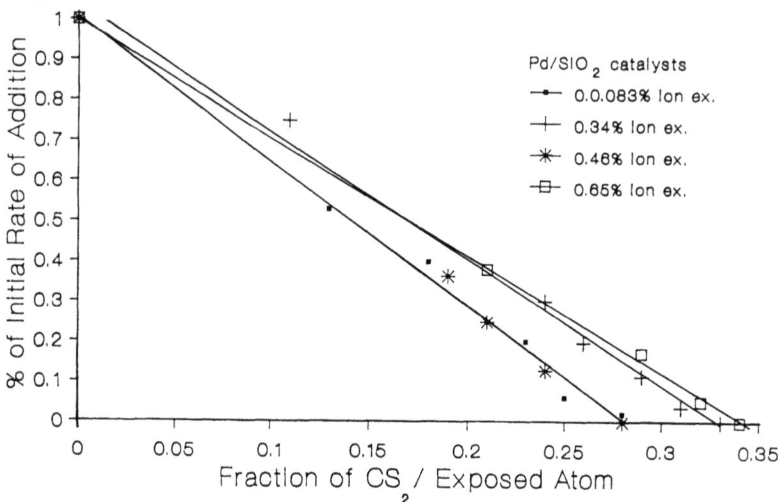

Fig. 2 Percent addition rate vs. CS_2/exp. atom for the (-)-apopinene reaction on Pd/SiO_2.

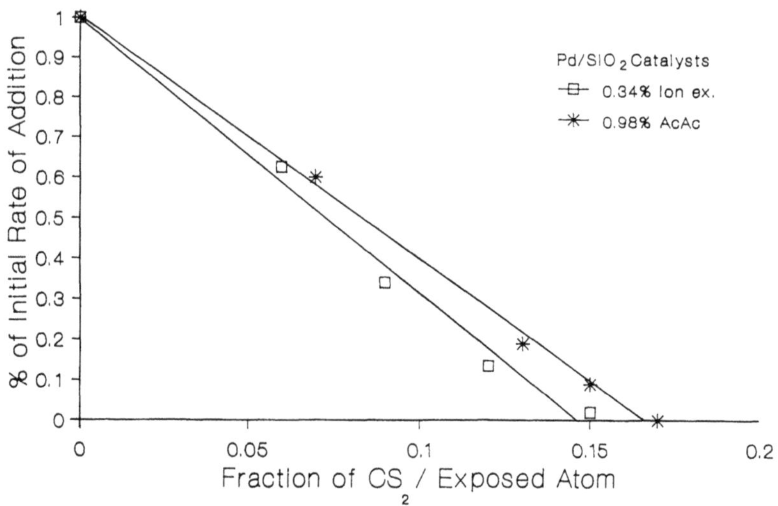

Fig. 3 Percent addition rate vs. CS_2/exp. atom for cyclohexene on Pd/SiO_2.

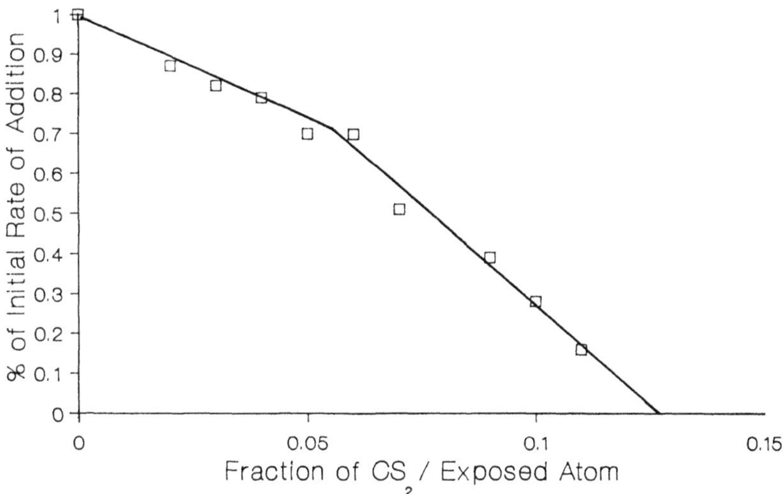

Fig. 4 Percent addition rate vs. CS_2/exp. atom for (-)-apopinene on 0.69% Pd/SiO_2 AcAc.

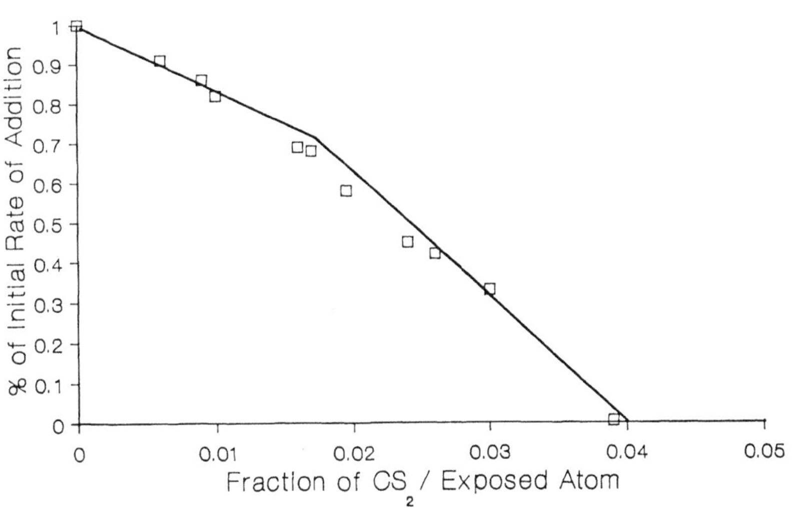

Fig. 5 Percent addition rate vs. CS_2/exp. atom for (-)-apopinene on 0.15% Pt/Al_2O_3 AcAc.

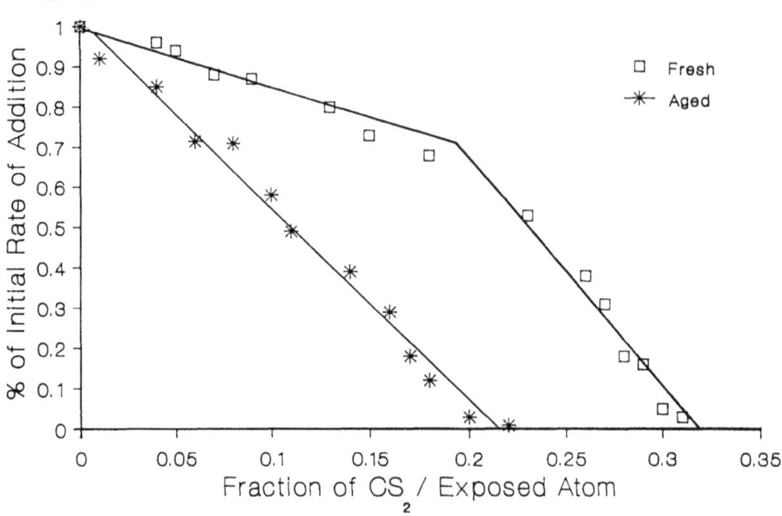

Fig. 6 Percent addition rate vs. CS_2/exp. atom for (-)-apopinene on 0.46% Pd/SiO_2 ion ex.

of titration. In fact, the straight-line portion of the last part of the titration does not extrapolate back to the starting rate at zero CS_2 (Figs. 4 and 5). This occurs because the original starting rate was slowed by mass transfer problems.

Finally, Fig. 6 shows results from two experiments with the same catalyst before and after aging. The "fresh" catalysts shows definite mass transfer problems while the "aged" catalyst, which exhibited approximately one-half of the activity of the fresh, shows none.

IV. DISCUSSION

CS_2 can be used to titrate active sites on Pt and Pd on which the decrease in turnover frequency for addition is linear with CS_2 molecules unless mass transfer occurs. This is true for the hydrogenation of apopinene and cyclohexene. During mass transfer limitations the decrease in turnover frequency as a function of CS_2 molecules is slower during diffusion control and faster after the rate has been reduced below the diffusion limitation (Figs. 4-6). Extrapolation back to zero CS_2 of the faster decrease (non-diffusion-limited) reveals the true rate as shown in Table 1. Moreover, if it is assumed that one CS_2 molecule poisons one active site, the true turnover frequency can be determined from the true rate and the end point for the titration. For example, in other work in press we have shown that the average turnover frequency per CS_2 site (TOF_{CS2}) is 17.0 ± 3.4 sec^{-1} $site^{-1}$ and have suggested that this number is actually the turnover frequency for hydrogen dissociation on edges.[9] Therefore, simply dividing by 17 the true rate (in turnovers per second, as determined by extrapolating the titration back to zero CS_2) gives the fraction of active sites available for hydrogenation (dissociation of hydrogen).

$$\frac{\text{True TOF}}{TOF_{CS2}} = P$$

TABLE 1 Experimental and Extrapolated Turnover Frequencies

Catalyst	Preparation	% D[a]	TOF[b] measured	TOF extrapolated
1.17% Pd/SiO_2	$PdCl_2$[c]	41.2	2.56	2.9
0.69% Pd/SiO_2	$Pd(AcAc)_2$[d]	37.4	1.53	1.82
0.46% Pd/SiO_2	Ion Ex.[e] (fresh)	61.6	7.1	12.2
0.12% Pd/SiO_2	$Pd(AcAc)_2$ (fresh)	93.7	1.26	1.44
0.15% Pt/Al_2O_3	$Pt(AcAc)_2$[f]	98.6	1.16	1.4

[a]These dispersions were obtained by the H_2 isotherm technique at 70°C.
[b]TOF: turnover frequency (the number of reactions per surface site per second).
[c]This catalyst was prepared by impregnating (Imp.) the silica with $PdCl_2$.
[d]These catalysts were prepared by ligand-exchanging $Pd(AcAc)_2$ onto the silica.
[e]This catalyst was prepared by ion-exchanging $Pd(NH_3)_2(Cl)_2$ onto the silica.
[f]This catalyst was prepared by ligand-exchanging $Pt(AcAc)_2$ onto the silica.

where P is the fraction of active sites not poisoned by another poison. This number, P, will be useful for evaluating the actual loss in activity of a spent or partially spent catalyst.

Additionally, in the manufacture of supported catalysts the efficiency E of the metal can be evaluated by dividing the number of CS_2 sites by the total number of metal atoms, i.e.,

$$\frac{\text{No. } CS_2 \text{ sites}}{\text{Total no. metal atoms}} = E$$

This number will be useful for determining the fraction of metal atoms which are actually useful for hydrogenation (dissociation of hydrogen).

Previously, we reported a correlation between the fraction

$$\frac{\text{Number of } CS_2}{\text{Fraction exposed (D)}}$$

and the fraction of edge sites on the metal crystallites of supported Pt and Pd.[9] If the number of CS_2 molecules is in fact the number of edge sites, then it is possible to estimate the average particle size and therefore the % D by assuming a shape for the metal crystallites.

However, at this time the number of CS_2 molecules required to poison one metal atom (presumably edge atom) is not confirmed. For example, Gonzalez-Tejuca et al.[10] assumed that each molecule of CS_2 poisoned two surface atoms, yet our results seem to be more consistent by assuming each molecule of CS_2 poisons one edge atom. Moreover, the CS_2 titration of Chen et al.[7,8] suggests that edge sites are poisoned first and then plane sites. If CS_2 dissociates on edges into CS and S fragments, then two vacancies on one edge site or one vacancy on an edge site and one vacancy on an adjacent plane site could be poisoned. Since hydrogen dissociation will be poisoned by occupation of either one or two of the unsaturated vacancies on an edge site, either process will explain how one CS_2 molecule can poison only one surface (edge) atom even if it dissociates into two fragments. Of course, the effect of CS_2 poisoning will be different for reactions occurring on plane sites because these sites will be poisoned after edge site poisoning.[5]

The method allows a simple test of all mass transfer effects for the liquid phase hydrogenation reaction. It may be adaptable to other experimental situations.

ACKNOWLEDGMENTS

We gratefully acknowledge support from the National Science Foundation through grant INT-8403357 and from the Hungarian Academy of Sciences through grant 319/82/1.3. We thank the Materials Technology Center at SIU for financial aid as well as critical support services.

REFERENCES

1. R. M. Koros and E. J. Novak, *Chem. Eng. Sci., 22,* 470 (1967).
2. R. J. Madon and M. Boudart, *Ind. Eng. Chem. Fundam., 21*(2), 438 (1982).
3. G. V. Smith, O. Zaharaa, Á. Molnár, M. Khan, B. Rihter, and W. E. Brower, *J. Catal., 83,* 238 (1983).
4. G. V. Smith, Á. Molnár, M. Khan, D. Ostgard, and N. Yoshida, *J. Catal., 98,* 502 (1986).
5. F. Notheisz, M. Bartók, D. Ostgard, and G. V. Smith, *J. Catal., 101,* 212 (1986).
6. G. V. Smith and D. S. Desai, *Ann. N.Y. Acad. Sci., 214,* 30 (1973).
7. (a) G. V. Smith, M. Bartók, F. Notheisz, Á. G. Zsigmond, and I. Pálinkó, *J. Catal., 110,* 203 (1988); (b) B. J. McCoy and J. M. Smith, *J. Catal., 110,* 206 (1988).
8. S. Y. Chen, J. M. Smith, and B. J. McCoy, *J. Catal., 102,* 365 (1986).
9. G. V. Smith, F. Notheisz, A. G. Zsigmond, D. Ostgard, T. Nishizawa, and M. Bartók, *Proceedings 9th International Congress on Catalysis,* Calgary, 1988 (in press).
10. L. Gonzalez-Teujca, K. Aika, S. Namba, and J. Turkevich, *J. Phys. Chem., 81*(14) (1977).

Part IV

Topics in Catalysis

11

Alkene and Alkyne Reactions with Cyclopalladated Organopalladium Complexes

RICHARD F. HECK, GUANGZHONG WU, WEIJING TAO, and ARNOLD L. RHEINGOLD

Department of Chemistry, University of Delaware, Newark, Delaware

ABSTRACT

Cyclopalladated azobenzene tetrafluoroborate catalyzes the dimerization of styrene to 2,3-diphenyl-1-butene and the isomerization of allylbenzene. The complex reacts stoichiometrically with various disubstituted alkynes to form 2-phenylcinnolinium tetrafluoroborate derivatives, usually in good yields. Similar reactions occur with cyclopalladated N-substituted benzalimine salts. Disubstituted alkynes, for example, react to form N-substituted isoquinolinium salts. Other cyclopalladated complexes form naphthalene or pentasubstituted cyclopentadiene derivatives. 1,2,3,4-Tetraphenylnaphthalene derivatives have been obtained in modest yields catalytically from iodobenzene derivatives and diphenylacetylene. Vinylic bromides and diphenylacetylene with a palladium catalyst form fulvene derivatives in low yields.

I. INTRODUCTION

Reactions of transition metal compounds with alkynes have been known for many years. Perhaps the most studied of these reactions is the catalytic cyclotrimerization of alkynes to benzene derivatives.[1]

Reactions of palladium chloride and its derivatives with alkynes were extensively studied by Maitlis in the 1960s and 1970s.[2] Most recently, Pfeffer[3] prepared a variety of unusual complexes from alkynes and cyclopalladated complexes.

We became involved in this area of chemistry unexpectedly while attempting to cause the very unreactive azobenzene-palladium chloride cyclopalladation product[4] to undergo reactions with alkenes or alkynes in the hope of ultimately being able to form heterocyclic products. We observed that the complex failed to react with styrene even at 100°C[5] while in the absence of the chelating azo group reaction occurred below room temperature to form stilbene in high yield.[6] We reasoned that the inertness of the cyclopalladated complex toward alkenes might be due to the lack of dissociation of ligands from the very stable complex and thus the probably essential formation of intermediary π-alkene complexes was prevented from occurring. Therefore, we removed the chloro ligands by reacting the complex with silver tetrafluoroborate to form weakly solvated tetrafluoroborate salts. These complexes proved to be very much more reactive than the chloro complexes and a study of their chemistry has led to the discovery of a variety of new reactions.[7]

II. ALKENE REACTIONS

Treatment of the cyclopalladated azobenzene chloro dimer with silver tetrafluoroborate in acetonitrile solution leads to the formation of the stable crystalline, yellow diacetonitrile solvate of the cyclopalladated complex in high yield:

$$[(\text{PhN=NC}_6\text{H}_4)\text{PdCl}]_2 + 2\text{AgBF}_4 + 4\text{CH}_3\text{CN} \longrightarrow 2\ (\text{PhN=NC}_6\text{H}_4)\text{Pd(CH}_3\text{CN)}_2{}^+\text{BF}_4{}^- + 2\text{AgCl}$$

The same reaction in nitromethane solution yields a viscous liquid product which is probably analogous to the acetonitrile complex. The nitromethane solvate, however, is much more reactive, presumably because the solvent is much less strongly bound to the metal.

The most reactive form of the complex is formed in the presence of the reacting alkene without an additional solvent. Under these conditions, at 25°C, styrene is catalytically dimerized in 90% yield overnight.[8] None of the expected stilbene-type product is formed. Interestingly, the styrene dimer that is formed has turned out to have a different structure than the one made with related palladium catalysts having different ligands. A catalyst, for example, prepared from π-allylpalladium chloride-tri-*n*-butylphosphine and silver tetrafluoroborate dimerized styrene at room temperature, also in 90% yield, to 1,3-diphenyl-1-butene[9] while our catalyst gave only 2,3-diphenyl-1-butene.

$$PhCH{=}CH_2 \xrightarrow[AgBF_4,\ 25^\circ,]{(\pi\text{-}C_3H_5PdCl)_2,\ P\text{-}n\text{-}Bu_3} 90\%$$

Ph, Ph

$$PhCH{=}CH_2 \xrightarrow[25^\circ]{(AzobzPdCl)_2,\ AgBF_4} 90\%$$

Ph, Ph

The reason for the formation of different products from these otherwise similar reactions is probably the difference in the size of the ligands on the catalyst. It has been suggested that the actual catalyst in the dimerization is a hydride although the source of this compound is not known.[9] Presumably, the metal hydride adds to styrene to form the α-phenethylpalladium derivative no matter which ligands are present. The ligands play a role in the second step in which the α-phenethyl derivative adds to a second styrene molecule. Addition may occur in two possible directions to the styrene double bond. With the smaller ligands employed previously, the addition is electronically controlled and the 1,3-diphenylbutylpalladium intermediate is formed. A final metal hydride elimination yields the 1,3-diphenyl-1-butene. When a large ligand, e.g., azobenzene, is present, addition apparently is sterically controlled and the smaller alkyl group, the

2,3-diphenylbutyl group, is formed. Elimination of metal hydride from this intermediate would give 2,3-diphenyl-1-butene.

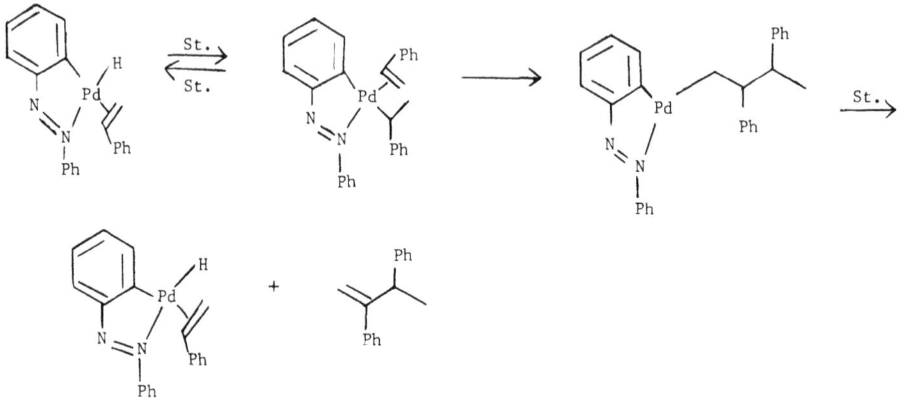

Terminal alkenes are isomerized by these catalysts to internal alkenes if neighboring methylene groups are present.

III. ALKYNE REACTIONS

A. Dimers, Tetramers, and Oligomers

We then attempted a reaction of the cyclopalladated azobenzene tetrafluoroborate with phenylacetylene hoping to replace the palladium group with a phenylethynyl substituent as occurs with nonchelated arylpalladium compounds.[10] However, an exothermic polymerization occurred instead of substitution. Reducing the reactivity of the catalyst gave lower molecular weight products. The cyclopalladated azobenzene triphenylphosphine palladium acetate at 55°C gave a mixture of oligomers with phenylacetylene from which 43% of a tetramer, 1-phenyl-(2,4,6-triphenylphenyl)ethene, was isolated.

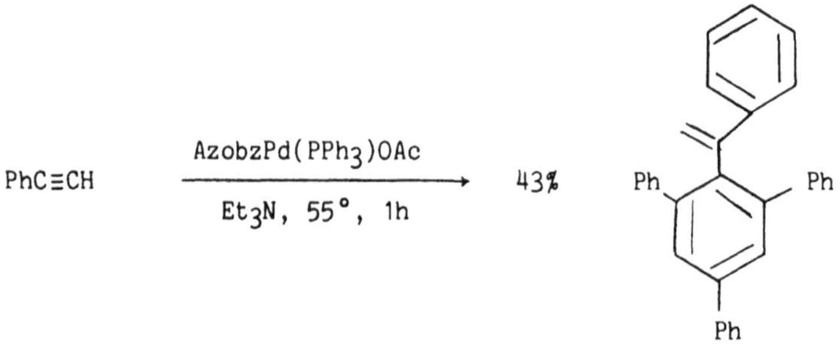

PhC≡CH $\xrightarrow[\text{Et}_3\text{N, 55°, 1h}]{\text{AzobzPd(PPh}_3\text{)OAc}}$ 43%

A similar reaction at 25°C with the azobenzene-palladium acetate dimer as catalyst produced mainly (54%) 1,4-diphenylbutenyne while an azobenzene-tri-*o*-tolylphosphine palladium chloride catalyst formed mainly the isomeric 1,3-diphenyl-3-buten-1-yne (51%) along with 9% of the 1,4 isomer. Thus, steric factors appear to control the direction of phenylacetylene dimerization as they did in the styrene dimerization.

$$\text{PhC≡CH} \xrightarrow[\text{CH}_2\text{Cl}_2,\ 25°\text{C},\ 21\ \text{hr}]{\text{AzobzPd(PPh}_3\text{)OAc}} \text{PhC≡CCH=CHPh}$$

$$\text{PhC≡CH} \xrightarrow[\text{Et}_3\text{N},\ 25°\text{C},\ 19\ \text{hr}]{\text{AzobzPd[P(}o\text{-tol)}_3\text{]Cl}} \text{PhC≡C(Ph)C=CH}_2$$

The use of triethylamine in these reactions often leads to improved yields of the lower molecular weight products.

B. Cinnolinium Salt Formation

We next carried out similar reactions with diphenylacetylene expecting that we might be able to isolate stable alkyne π complexes. With the azobenzenepalladium tetrafluoroborate complex, however, we found that palladium was lost even at room temperature and a new organic product was formed. The product proved to be 2,3,4-triphenylcinnolinium tetrafluoroborate as determined by X-ray crystallography. This reaction was found to be quite general for disubstituted alkynes. Yields of heterocycles were only in the range of 25-35%, however, when all reactants were initially mixed. The other products formed were alkyne oligomers. We observed that slow addition of the alkyne to the palladium complex at elevated temperatures (60-100°C) led to much higher yields. Some examples of the reaction are shown below[7]:

$$\text{(AzobzPdCl)}_2 + 2\text{AgBF}_4 + \text{RC≡CR} \xrightarrow{\text{CH}_3\text{NO}_2} \text{3,4-R}_2\text{-2-Ph-cinnolinium BF}_4^- + 2\text{AgCl} + \underline{\text{Pd}}$$

R	Alkyne addition method (yields) (%)	
	All at once at 25°C	Over 1-4 hr at 100°C
Ph^-	35	87
$C_2H_5^-$	33	83
$CH_3O_2C^-$	28	80
$(C_2H_5O)_2CH^-$	—	46
$HOC(CH_3)_2^-$	—	39

The mechanism of this cyclization presumably involves an insertion of the alkyne into the palladium-carbon bond to give a seven-membered ring palladacycle. A reductive elimination of the palladium then would give the cinnolinium product. A model for the intermediate is the cyclopalladated *N,N*-dimethylbenzylamine chlorodimer reaction product with hexafluorobutyne reported by the Pfeffer group.[11]

CH_3 CH_3 N $PdCl)_2$ + $CF_3C{\equiv}CCF_3$ $\xrightarrow[12h]{60°}$ CH_3 CH_3 $N\text{-}PdCl)_2$ CF_3 CF_3

Unsymmetrically disubstituted alkynes have been added to the cyclopalladated azobenzene tetrafluoroborate also. Phenylpropynal-diethylacetal gave exclusively the 3-phenyl-4-diethoxymethyl isomer. The structure of the product was established by X-ray crystallography.

$PdL_2^+BF_4^-$ N N Ph + $PhC{\equiv}CCH(OC_2H_5)_2$ ⟶ $CH(OC_2H_5)_2$ Ph N N^+ Ph BF_4^- + 2L + Pd

L = CH_3NO_2

On the other hand, the addition of methyl phenylpropiolate to the azobenzene complex under the same conditions led to the exclusive formation of the 4-phenyl-3-carbomethoxy isomer.

$$\text{(azobenzene-}PdL_2^{+}BF_4^{-}\text{ complex)} + PhC{\equiv}CCO_2CH_3 \longrightarrow \text{(4-Ph-3-}CO_2CH_3\text{-2-phenylcinnolinium } BF_4^{-}) + 2L + \underline{Pd}$$

It appears that the difference in behavior of the above two alkynes in the addition is due to a steric effect such that the larger of the two substituents prefers to be in the 4 position.

Now that the ring closure reaction was well documented, we returned to study the cyclopalladated chlorodimer to determine if it was necessary to employ the tetrafluoroborate salt in the reaction. We found that at about 140°C in DMF solution that the chlorodimer would react with diphenylacetylene to form the cinnolinium salt but only in about 40% yield.[8]

$$\text{(azobenzene-PdCl)}_2 + 2PhC{\equiv}CPh \xrightarrow[\text{DMF}]{140^\circ} 2\ \text{(3,4-diphenyl-2-phenylcinnolinium } Cl^{-}) + 2\ \underline{Pd}$$

While 2-phenylcinnolinium salts have not previously been reported, there is no reason to expect these compounds to be of anything but academic interest. We therefore turned our attention to analogous reactions of cyclopalladated benzalimines with alkynes where therapeutically interesting isoquinolinium salts might be expected as products.

C. Isoquinolinium Salt Formation

Indeed, we found that the reaction did proceed analogously with various N-substituted benzalimine-palladium complexes. Yields were

R	Alkyne addition method (yield) (%)	
	All at once at 25°C	Over 1-4 hr at 100°C
Ph^-	0	80
$PhCH_2^-$	5	42
$o\text{-}CH_3C_6H_4^-$	5	60
CH_3^-	0	0

also sensitive to the rate of alkyne addition. For example, 3-hexyne reacted as indicated with the complexes listed below.[12]

$$\text{(benzaldimine N-R PdCl)}_2 + AgBF_4 + C_2H_5C{\equiv}CC_2H_5 \xrightarrow{CH_3NO_2} \text{N-R-3,4-diethylisoquinolinium } BF_4^- + 2\ AgCl + \underline{Pd}$$

The failure of the *N*-methyl derivative to give any product at 100°C is due to the instability of the tetrafluoroborate at 100°C. Even in the absence of alkyne, this salt fairly rapidly deposited palladium metal at 100°C and it did so slowly even at 0°C. The above data also show that competing oligomerization of the alkyne is quite serious at 25°C when excess alkyne is present.

The reaction of the *N*-methylbenzalimine complex with diphenylacetylene at 100°C with addition in 1 hr did give 30% of the isoquinolinium salt. The success with this alkyne probably means that it reacts more rapidly than 3-hexyne does in the ring closure step and therefore the cyclization competes more successfully with alkyne oligomerization.

$$\text{(N-}CH_3\text{ benzaldimine)}PdL_2^+\ BF_4^- + PhC{\equiv}CPh \longrightarrow \text{N-}CH_3\text{-3,4-diphenylisoquinolinium } BF_4^- + 2L + \underline{Pd}$$

Use of the large *N*-*t*-butyl substituent in the benzalimine complex not only favors alkyne oligomerization more than smaller substituents do, but this group is partly lost during the reaction because of the serious crowding of the three adjacent aromatic substituents in the initial product.

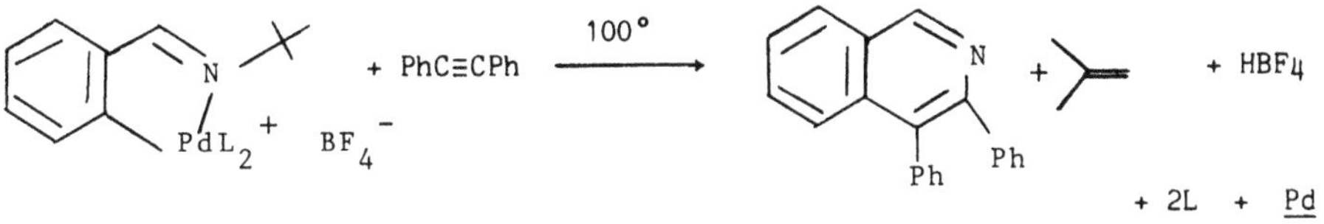

The cyclopalladated *N*-methylbenzalimine tetrafluoroborates are more stable when methoxyl substituents are present. The 2,5-dimethoxy derivative, for example, yields 48% of the isoquinolinium salt with diphenylacetylene with a 1-hr addition of the alkyne. The yield is improved to 66% if the reaction is carried out at 70°C. Apparently, even this *N*-methyl salt is decomposing slowly thermally at 100°C and perhaps even at 70°C.

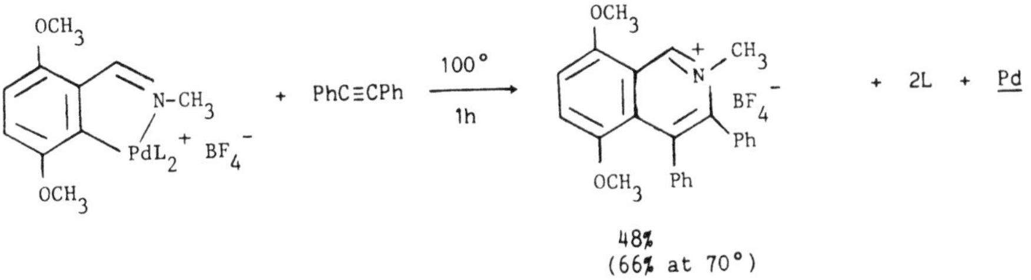

48%
(66% at 70°)

An attempt to simplify the isoquinolinium salt synthesis by employing a cyclopalladated benzalimine chlorodimer instead of the tetrafluoroborate led to a low yield of the product with alkyne oligomerization being the major side reaction. The *N*-phenylbenzalimine palladium chloride dimer at 150°C in DMF with 3-hexyne gave the expected isoquinolinium chloride, but only in 29% yield.[12]

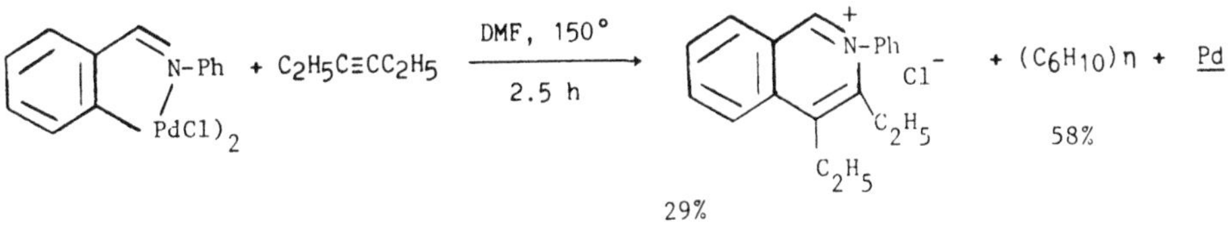

29% 58%

Efforts to make these reactions catalytic in palladium by using 2-halobenzalimines as reactants in the presence of triethylamine or sodium acetate as bases at 150°C have been totally unsuccessful. The reason for the failure is not obvious. Perhaps the imine inhibits the catalyst by strongly coordinating with the cyclopalladated intermediates.

D. Naphthalene Derivatives

We attempted to prepare 3,4-diethylisoquinoline by the reaction of cyclopalladated *N*-*t*-butylbenzalimine tetrafluoroborate with 3-hexyne at 25°C similarly to the diphenylacetylene reaction at 100°C mentioned above. Instead of getting an isoquinoline product, however, a naphthalene derivative and a minor amount of a curious 3:1 alkyne/imine adduct were formed. Structures of the products were determined by NMR spectroscopy. The major product (22%) was *N*-*t*-butyl-5,6,7,8-tetraethyl-1-naphthalimine and the minor one (5%) was the pentaethylcyclopentadienyl derivative shown below.

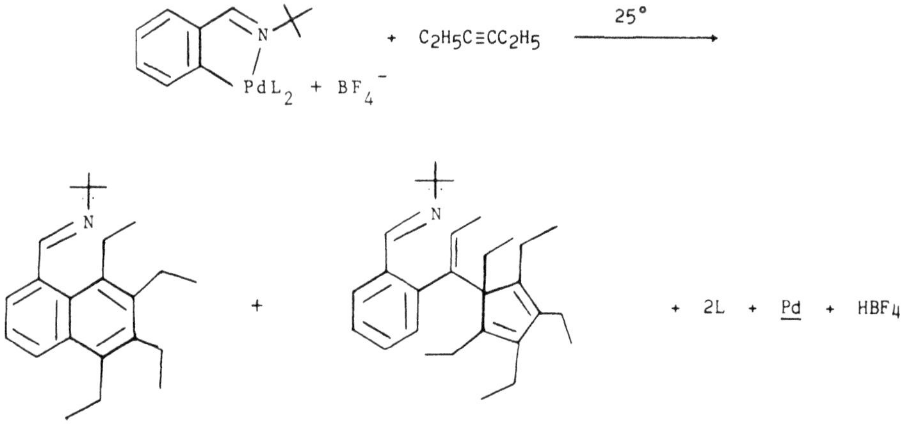

PdL_2 + BF_4^- + $C_2H_5C{\equiv}CC_2H_5$ $\xrightarrow{25°}$ + 2L + Pd + HBF_4

Naphthalene formation also has been observed in the reaction of cyclopalladated *N*,*N*-dimethylbenzylamine tetrafluoroborate with 3-hexyne in yields up to 55% along with smaller amounts of the 3:1 adduct.

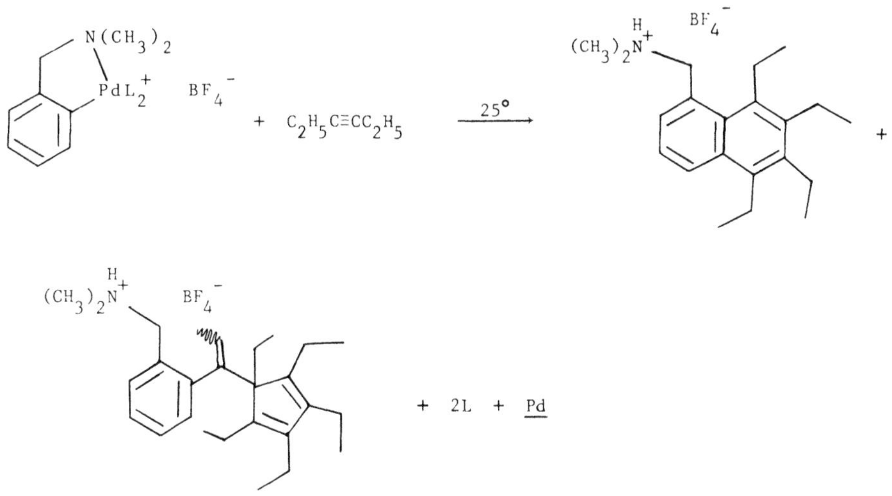

$N(CH_3)_2$ PdL_2^+ BF_4^- + $C_2H_5C{\equiv}CC_2H_5$ $\xrightarrow{25°}$ $(CH_3)_2\overset{H}{N}^+$ BF_4^- +

$(CH_3)_2\overset{H}{N}^+$ BF_4^- + 2L + Pd

While the reason for the difference between the 3-hexyne and the diphenylacetylene reactions is not clear, the mechanisms of formation of the new products may be rationalized on the basis of known complexes.

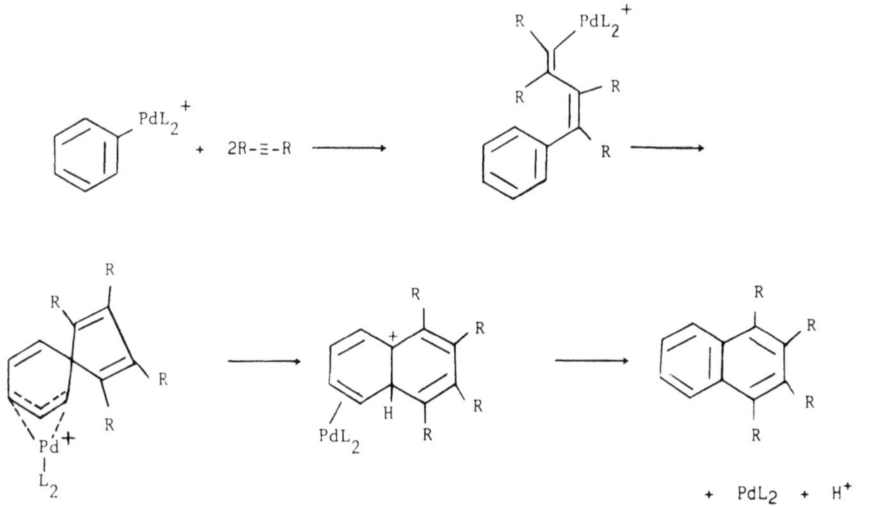

PdL_2^+ + 2R-≡-R ⟶ R PdL_2^+ R R R ⟶

R R R R Pd+ L_2 ⟶ R R + R H PdL_2 R ⟶ R R R R

+ PdL_2 + H^+

In this mechanism, chelation of the palladium group is not essential and indeed we found that naphthalene derivatives could be made catalytically from simple aryl iodides, diphenylacetylene, triethylamine, and a catalytic amount (2%) of $Pd(OAc)_2 + 2PPh_3$ in 30-50% yields.[13]

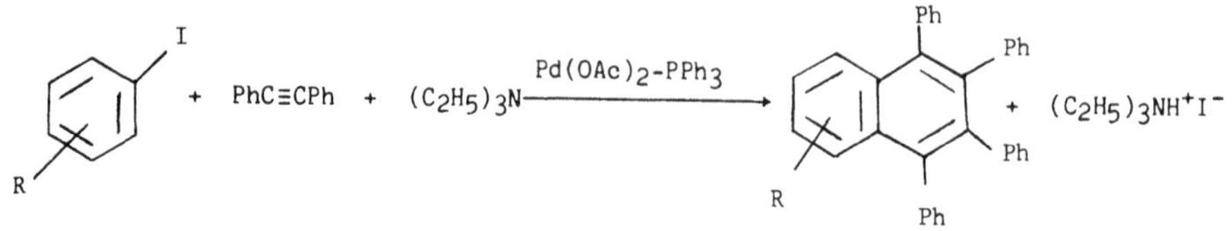

R	% Yield of naphthalene
H	47
4-CH_3	36
4-CH_3O	30
3-CH_3O	21% *p* + 13% *o*

E. 3:1 and 4:2 Alkyne-to-Aryl Halide Products

Curiously, in the absence of the chelating side chain, 3-hexyne does not yield naphthalene products with aryl iodides but instead gives mainly oligomers, 3:1 adducts, and, very unusual, 4:2 alkyne-to-aryl-halide products.[13] For example, iodobenzene with 3-hexyne under the reaction conditions employed for the naphthalene synthesis gave 10% (isolated) of the 3:1 adduct and 1% of the 4:2 adduct along with hexyne oligomers. The structure of the 4:2 adduct was established by X-ray crystallography.

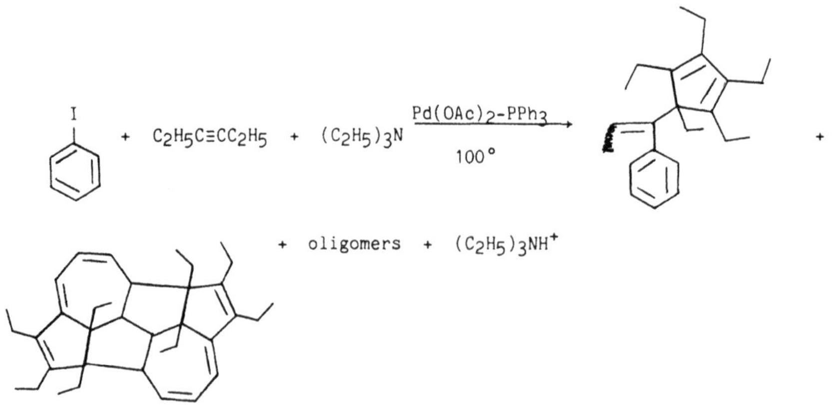

p-Iodoanisole and 3-hexyne produce analogous products in comparable yields under the same conditions. We have attempted to cleave the 3:1 adducts to the pentaethylcyclopentadienide ion with potassium tri-*sec*-butylborohydride but so far without success.

The 4:2 products are of interest for another reason. The X-ray structures of the parent and the dimethoxy derivatives show that the molecules have bowllike shapes and that the cavity may be large enough to accommodate some small molecules. Perhaps molecular complexes can be formed, particularly if 4:2 complexes with other functional groups can be prepared. We are currently attempting to increase the yields of these products.

We speculate that the mechanism of formation of the 4:2 complexes may be by way of the same spiro[4.5]deca-1,3-diene intermediate as is formed with diphenylacetylene but that the intermediate rearranges to a bicyclo[5.3.0]decadiene rather than to the bicyclo[6.6.0]decapentaene (naphthalene) system. Dimerization of the bicyclic palladium complex with loss of palladium followed by an electrocyclic process could explain the formation of the observed 4:2 products.

The formation of 3:1 alkyne-to-aryl-iodide products is easily explained as a series of three alkyne insertions into the Pd-C bond followed by a cyclization by addition of the terminal Pd-C bond to the remaining double bond from the alkyne unit inserted first. The adduct finally loses palladium hydride to give the observed 3:1 product. These reactions are rather similar to the 2-butyne cyclizations reported by Maitlis.[2] We isolated a model for the cyclized palladium complex in the mechanism from the reaction of cyclopalladated 8-methylquinoline tetrafluoroborate with 3-hexyne.[7]

PdL_2^+ BF_4^- + 3 $C_2H_5C{\equiv}CC_2H_5$ ⟶ Pd BF_4^-

22%

It occurred to us that if cyclization of the 2:1 and 3:1 adducts is so favorable, that it might be possible to cyclize similar complexes prepared by other routes. We carried out a few reactions to test this idea.

The addition of diphenylacetylene to 2-iodobiphenyl, for example, catalyzed by palladium should yield 9,10-diphenyphenanthrene if the cyclization occurs. Indeed this reaction does take place, although the yield is only 22%.

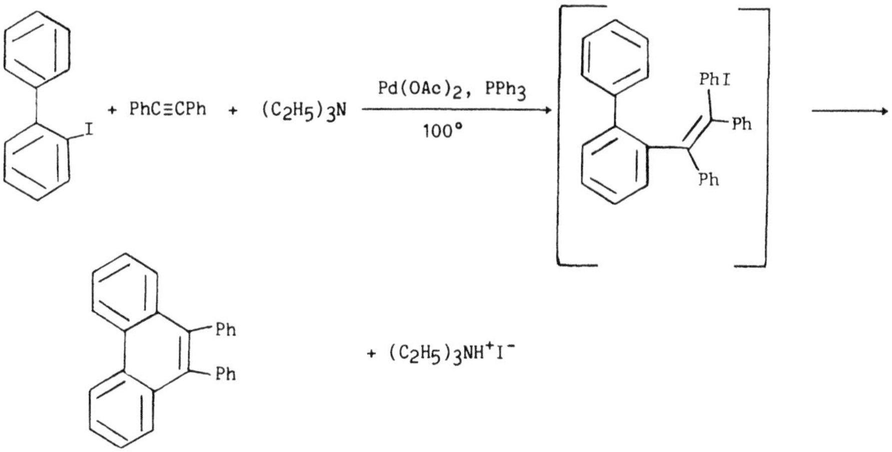

+ PhC≡CPh + $(C_2H_5)_3N$ $\xrightarrow[100°]{Pd(OAc)_2,\ PPh_3}$ [PhI, Ph, Ph] ⟶

+ $(C_2H_5)_3NH^+I^-$

F. Fulvene Synthesis

An attempted variation of this reaction, the addition of diphenylacetylene to *cis*-β-bromostyrene, which was expected to produce 2,3-diphenylnaphthalene by analogy with the above reaction, produced instead pentaphenylfulvene in 27% yield (45% based on the β-bromostyrene which actually reacted).[14]

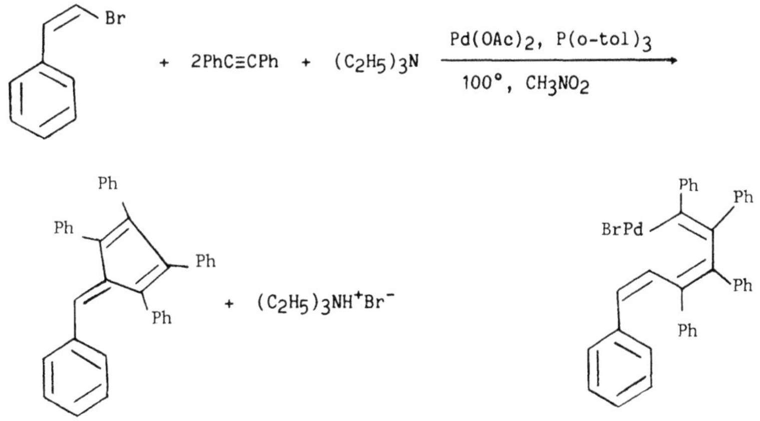

Br + 2PhC≡CPh + $(C_2H_5)_3N$ $\xrightarrow[100°,\ CH_3NO_2]{Pd(OAc)_2,\ P(o\text{-}tol)_3}$

+ $(C_2H_5)_3NH^+Br^-$

BrPd, Ph, Ph, Ph, Ph

intermediate

This fulvene synthesis seems to be rather general when a hydrogen substituent is present on the halogen-bearing carbon of the vinylic halide. Thus, E-methyl 3-bromo-2-methylpropenoate and diphenylacetylene give the expected fulvene in 11% yield.

$$CH_3O_2C(CH_3)C{=}CHBr + 2PhC{\equiv}CPh + (C_2H_5)_3NH \xrightarrow{Pd(OAc)_2,\ P(o\text{-}tol)_3} \text{fulvene } (=C(CH_3)CO_2CH_3;\ Ph,\ Ph,\ Ph,\ Ph)$$

The last reaction was carried out in triethylamine solution. In this case if we use nitromethane as solvent, a new product is formed, a nitromethane adduct of the fulvene.

$$CH_3O_2C(CH_3)C{=}CHBr + 2PhC{\equiv}CPh + (C_2H_5)_3N + CH_3NO_2 \xrightarrow[100^\circ]{Pd(OAc)_2,\ P(o\text{-}tol)_3}$$

$$\text{tetraphenylcyclopentadiene (H, } CH(CO_2CH_3)CH_2CH_2NO_2) + (C_2H_5)_3NH^+Br^-$$

35%

The formation of fulvenes also occurs with 3-hexyne in place of diphenylacetylene. E-*p*-Methoxy-β-bromostyrene and 3-hexyne, for example, appear to form mainly the expected fulvene at 100°C, but after isolation the product undergoes isomerization on warming or even standing in chloroform solution. The isomerized product, obtained in 83% yield by GLC, is believed to be the conjugated triene in which one double bond has moved from the ring to a side chain.[14]

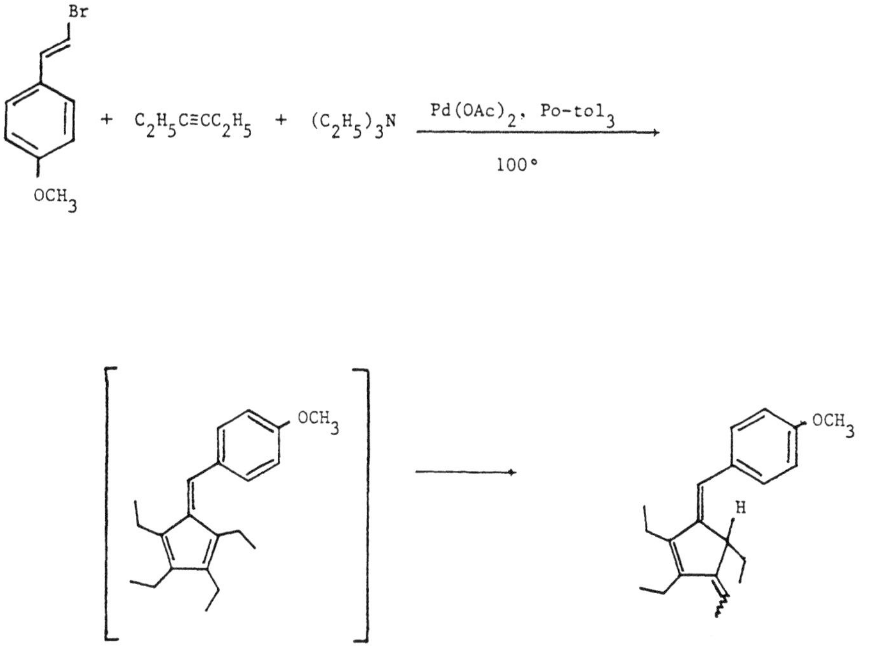

Br

OCH$_3$ + $C_2H_5C{\equiv}CC_2H_5$ + $(C_2H_5)_3N$ $\xrightarrow[100°]{Pd(OAc)_2,\ Po\text{-}tol_3}$

OCH$_3$ ⟶ OCH$_3$ H

83%

Similar results were obtained when 1-bromo-1-propene was reacted with 3-hexyne. The fulvene appeared to be a little more stable in this case although it also isomerized (to a mixture of five compounds) on heating.

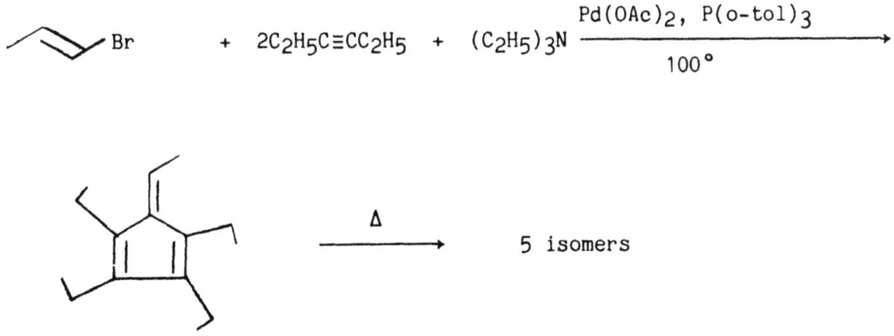

Br + $2C_2H_5C{\equiv}CC_2H_5$ + $(C_2H_5)_3N$ $\xrightarrow[100°]{Pd(OAc)_2,\ P(o\text{-}tol)_3}$

$\xrightarrow{\Delta}$ 5 isomers

Another attempt was made to cyclize an aromatic compound with a four-carbon side chain. *o*-Bromobenzyl phenyl ether was subjected to our usual reaction conditions; triethylamine, palladium acetate, and triphenylphosphine at 100°C, but only the debrominated benzylphenylether was formed. At this time, however, we discovered that the cyclization of this compound and several similar ones with palladium had already been achieved.[15] It was apparently necessary to use sodium acetate as the base rather than an amine to better control the reduction of the bromide.

CH_2 — O, Br + NaOAc $\xrightarrow[\text{2h, 160°, DMA}]{Pd(PPh_3)_2Cl_2}$ O + O-CH_2 + NaBr + HOAc

36% 35%

IV. SUMMARY

This study of alkyne reactions with organopalladium complexes has led to the discovery of several new and useful reactions. The results may be summarized in a single scheme since all of the products arise from insertions of one, two, three, or more alkyne units into a Pd-C bond of an organopalladium complex.

$RPdL_2^+$ + $R^1C{\equiv}CR^1$ ⟶ $R^1(R)C{=}C(R^1)PdL_2^+$ ⟶ Cinnolinium Salts, Isoquinolinium Salts, Phenanthrenes, etc

$R^1C{\equiv}CR^1$ ↓

$R(R^1)C{=}C(R^1)\text{–}C(R^1){=}C(R^1)PdL_2^+$ ⟶ Naphthalenes, Fulvenes, etc

$R^1C{\equiv}CR^1$ ↓

$R(R^1)C{=}C(R^1)\text{–}C(R^1){=}C(R^1)\text{–}C(R^1){=}C(R^1)PdL_2^+$ ⟶ Cyclopentadienes

RC≡CR ↓

Oligomers

REFERENCES

1. C. W. Bird, *Transition Metal Intermediates in Organic Synthesis,* Academic Press, New York, Chap. 1, 1967.
2. P. M. Maitlis, *J. Organometal. Chem., 200,* 161 (1980).
3. F. Maassarani, M. Pfeffer, and G. Leborgne, *Organometallics, 6,* 2043 (1987) and references therein.
4. A. C. Cope and R. W. Siekman, *J. Am. Chem. Soc., 87,* 3272 (1965).
5. D. W. Hart, R. Bau, C. H. Chao, and R. F. Heck, *J. Organometal. Chem., 179,* 301 (1979).
6. R. F. Heck, *J. Am. Chem. Soc., 90,* 5518 (1968).
7. G. Wu, A. L. Rheingold, and R. F. Heck, *Organometallics, 5,* 1922 (1986).

8. G. Wu, A. L. Rheingold, and R. F. Heck, *Organometallics, 6,* 2386 (1987).

9. P. Grenouillet, D. Neibecker, and I. Thatchenco, *Organometallics, 3,* 1130 (1984).

10. H. A. Dieck and R. F. Heck, *J. Organometal. Chem., 93,* 259 (1975).

11. A. Bahsoun, J. Dehand, M. Pfeffer, M. Zinsius, S. Bauaoud, and G. LeBorgne, *J. Chem. Soc., Dalton (1979)*, 547.

12. G. Wu, S. J. Geib, A. L. Rheingold, and R. F. Heck, *J. Org. Chem., 53,* 3238 (1988).

13. G. Wu, A. L. Rheingold, and R. F. Heck, *Organometallics, 6,* 1941 (1987).

14. L. J. Silverberg, G. Wu, A. L. Rheingold, and R. F. Heck, *Tetrahedron Lett.*, submitted.

15. D. E. Ames and A. O. Palko, *Tetrahedron, 40,* 1919 (1984).

12

Reduction of Sulfonyl Chlorides to Thiols

VICTOR L. MYLROIE AND JOSEPH K. DOLES

Eastman Kodak Company, Rochester, New York

ABSTRACT

Reduction of sulfonyl chlorides to thiols can be achieved by the use of lithium aluminum hydride, or zinc or with hydroiodic acid generated in situ from iodine and red phosphorus. However, little is known or reported regarding the catalytic reduction of sulfonyl chlorides to yield the corresponding thiols.

We report here the reduction of aromatic sulfonyl chlorides, such as *p*-toluenesulfonyl chloride, naphthalenesulfonyl chloride, or *p*-(chlorosulfonyl)benzoic acid, to yield the aromatic thiols. This reduction was accomplished by use of a palladium catalyst under a moderate pressure of hydrogen. The reduction was carried out in the presence of a mild base to neutralize the strong acid formed during the reduction. When using a basic Amberlite resin to neutralize the acid, disulfide compounds are generated. The disulfides can be reduced to the thiols by use of Raney cobalt catalyst.

I. INTRODUCTION

We have been interested in the preparation of aromatic thiols as intermediates for photographically useful compounds. Aromatic thiols are also useful as intermediates in various fields including pharmaceuticals.

Traditionally, thiols from sulfonyl chlorides have been prepared by chemical reduction with zinc[1,2] or lithium aluminum hydride[3,4] or with hydroiodic acid prepared from an in situ reaction of red phosphorus and iodine.[5] Dr. Milos Hudlicky[6] of Virginia Polytechnic Institute and State University presented a brief summary of these reduction techniques.

Allied Chemical Corporation[7] reported the catalytic reduction of sulfonyl chlorides to aromatic thiols treated with hydrogen sulfide in the presence of a presulfided cobalt-molybdenum catalyst. They reported yields of 78-79% by reaction at 110°C at 60 psi for 17 hr.

We found in some early work we attempted that 2,5-diisopropylbenzenesulfonyl chloride, for example, was rather easily reduced with a palladium-on-carbon catalyst to 2,5-diisopropylbenzenethiol with fairly good activity shown by the catalyst. We also found that by-products from the reaction created a very corrosive environment which quickly etched any stainless steel exposed to the reaction medium. We concluded that for this reaction to be practical we would have to use a scavenger to neutralize the effect of strong acid. We tried basic Amberlite resins (Rohm and Haas Co.) obtained from Aldrich Chemical. The ion exchange resin Amberlite GC-400 is a strong-base ion exchange resin. This appeared to work well, but with some of the compounds small to moderate amounts of the disulfide were generated during the reduction. We also investigated the use of *N*,*N*-dimethylaniline and *N*,*N*-dimethylacetamide. The *N*,*N*-dimethylacetamide was later shown to be the preferred choice.

II. EXPERIMENTAL PROCEDURES

A. Materials

Most chemicals were of ACS reagent or Eastman Grade and used as supplied from the Kodak Laboratory and Research Products catalog. The 2,5-diisopropylbenzenesulfonyl chloride was prepared by David Vogel of Eastman Kodak Company. Catalysts were purchased from

Engelhard Industries and W. R. Grace. Most solvents were Eastman Grade.

B. Apparatus

Hydrogenations were carried out on Parr shakers using both 500-ml glass vessels and 1700-ml stainless steel vessels. Some reactions were carried out in a 500-ml stainless steel Zipperclave. This is a quick disconnect autoclave manufactured by Autoclave Engineers. All samples of catalyst were mixed with solvent under an inert atmosphere prior to charging the vessel. All catalysts used are pyrophoric and should be handled with care to prevent fires. Some typical examples of the catalytic reductions are shown below.

C. Procedure A—2,5-Diisopropylbenzenethiol

SO_2Cl $\xrightarrow[\text{Pd/CS}]{3\,H_2}$ SH (1)

A mixture of 20 g (0.0767 mole) 2,5-diisopropylbenzenesulfonylchloride, 5.0 g of 5% palladium sulfided on carbon catalyst (50% wet weight), and 250 ml of tetrahydrofuran (THF) was added to a 500-ml capacity Parr bottle. The Parr bottle and hydrogen line were purged twice with hydrogen gas and then charged with hydrogen to a pressure of 60 psi. The bottle and contents were heated to 40°C and shaken for about 18 hr. The reaction appeared to be complete after about 12 hr. The Parr shaker was stopped and vented of excess hydrogen gas. The reaction mixture was filtered through a pad of filter aid to remove the catalyst. The solvent was removed from the filtrate by evaporation on a rotating evaporator. Contact of this compound with air may cause oxidation of the thiol to disulfide. The red liquid can be used directly or distilled to a clear liquid. The crude liquid weighed 15 g (95% yield).

Analysis: The boiling point was 137-139°C at 7 mm Hg. The IR is consistent with proposed structure [IR (S): 2960, 1560, 1500, 1490, 1100, 875, and 640 cm^{-1}]. The NMR spectra showed: singlet 1.15 ppm, 12 H; multiplet 2.7-3.8 ppm, 3 H; multiplet 6.9-7.2 ppm, 3 H. The thin-layer chromatograph eluted on silica gel plates in ligroine/ethyl acetate 95:05 showed no spots corresponding to the starting material and one major spot corresponding to the mercaptan and a trace spot corresponding to the disulfide compound. The liquid chromatograph showed 80.0% 2,5-diisopropylphenylmercaptan and 10.5% disulfide.

D. Procedure B—1-Naphthalenethiol

$$\text{1-}C_{10}H_7SO_2Cl \xrightarrow[\text{Pd / C}]{3\ H_2} \text{1-}C_{10}H_7SH \tag{2}$$

A mixture of 5.0 g (0.0221 mole) 1-naphthalenesulfonyl chloride, 1.0 g 5% palladium on carbon catalyst, 250 ml THF and 3.0 g basic Amberlite GC-400 (100-200 mesh)[8] was added to a 500-ml capacity Parr bottle. The Parr bottle and hydrogen line were purged twice with hydrogen gas and then charged with hydrogen to a pressure of 60 psi. The bottle and contents were heated to 45-50°C and shaken for about 18 hr (for convenience), at which time the theoretical amount of hydrogen had been consumed. The reaction mixture was filtered through a fiberglass filter and the filtrate evaporated to dryness. The liquid was taken up in 50 ml of ethyl acetate and the solution washed twice with 25-ml portions of 5% hydrochloric acid solution followed by a wash with distilled water. The organic layer was dried over magnesium sulfate. The solvent was removed by distillation on a rotating evaporator, and the yield of colorless liquid was 3.2 g.

Analysis: In a mass spectrometer the sample gave a parent ion/base peak at 160 amu for $C_{10}H_8S$. The mass spectral data is consistent with the structure of 1-naphthalenethiol. A small amount of disulfide was detected ($C_{20}H_{14}S_2$). The infrared showed absorptions at 3020, 2550, 1675, 1620, 1500, 1375, 1340, 1250, 1205, 1025, 980, 790, and 768 cm^{-1}. The NMR spectra ($CDCl_3$) showed: (S 3.25, δ 1 H; -SH); multiplet 6.8-7.55, 6 H; aromatic); 7.85-8.0 δ (M, 1 H; aromatic).

E. Procedure C—2-Naphthalenethiol

S-S $\xrightarrow[\text{RaCo}]{H_2}$ SH (3)

A mixture of 2.5 g (0.00785 mole) of the disulfide of the 2-naphthalenethiol, 250 ml of THF, and 1.0 g of (wet weight) washed Raney cobalt catalyst was added to a 500-ml Zipperclave (stirred, stainless steel autoclave) and charged with hydrogen to a pressure of 500 psi. The temperature was raised to 63°C and stirring continued for about 7.5 hr. The sample was cooled to 27°C and vented of excess hydrogen. The reaction mixture was filtered through a fiberglass filter to remove the catalyst. The same care and caution in filtering the Raney cobalt catalyst should be used as when filtering a noble metal catalyst. Raney cobalt is very pyrophoric. The sample was stripped to dryness (2.5 g) and recrystallized from diethyl ether to yield 2.2 g of white solid with a melting point of 80-81°C.

Analysis: The infrared showed peaks: 3.35, 3.4, 3.9, 6.2, 6.35, 6.5, 8.9, 9.2, 10.4, 10.6, 11.5, 11.8, 12.6, and 22 μm. The NMR (DMSO) showed 3.38δ (S, 1H; -SH); 7.4-7.9δ (M, 7H; aromatic).

III. RESULTS AND DISCUSSION (see Table 1)

The catalytic reduction of aromatic sulfonyl chlorides to thiols appears to be a general reaction. However, our work suggests that several factors are important for a successful reduction. First, it appears to be necessary to have a mild base present to neutralize the strong hydrochloric acid and allow the reaction to go to completion and to prevent the corrosive effects of the acid on a metal

Table 1 Summary of Reductions of Aromatic Sulfonyl Chlorides to Thiols

Substrate	Product	% Yield	Note	% Purity
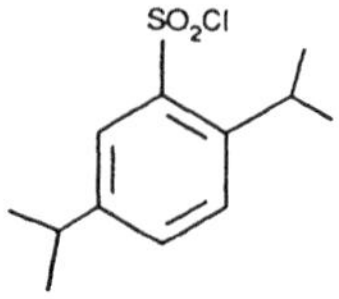 SO_2Cl	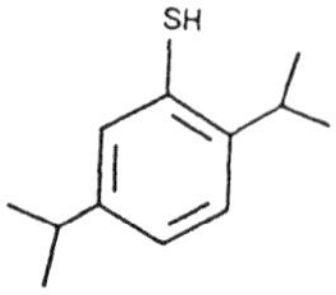 SH	95	a–c	80.0 (LC) 10.0 disulfide
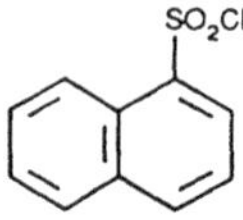 SO_2Cl	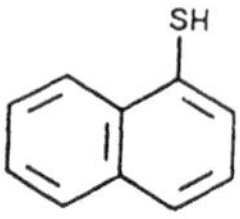 SH	90.5	a–d	94.7 (GC)
 SO_2Cl CH_3	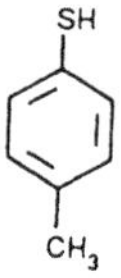 SH CH_3	99	b–e	50.0 (GC) 48.0 disulfide
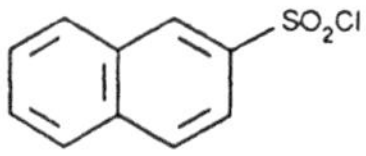 SO_2Cl	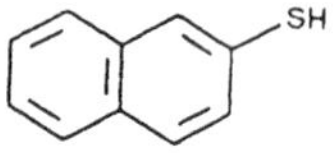 SH	87	b–e	98.6 (GC)
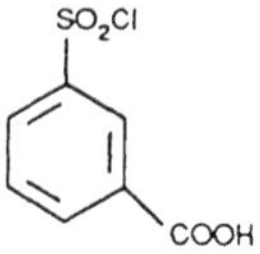 SO_2Cl COOH	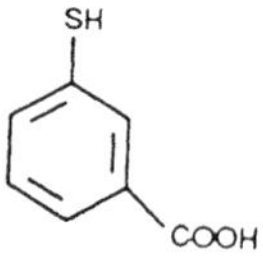 SH COOH	81.8	b–e	91.4 (GC)
 SO_2Cl COOH	 SH COOH	91.9	b–d	94.1 (GC)
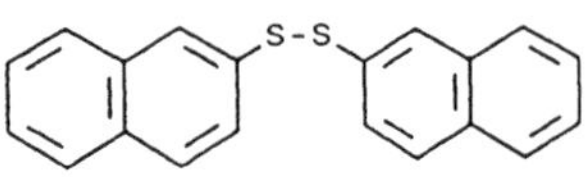 S-S	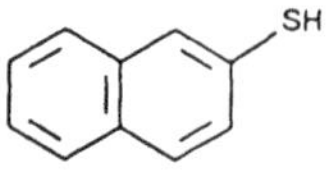 SH	87.4	b	98.6 (GC)

[a]Run with Amberlite resin (basic).
[b]IR and NMR are consistent with the proposed structure.
[c]Disulfide present in the product mixture.
[d]Run with *N*,*N*-dimethylacetamide.
[e]The sample was isolated by GC and then the structure confirmed by mass spectral data.

reactor. A number of bases were investigated. *N,N*-Dimethylaniline did not work well, little product was detected and in most cases the sample contained large quantities of the sulfonyl chloride. We found that a strongly basic Amberlite resin GC-400 worked well. However, in some cases fairly large amounts of disulfide were produced. The preferred base is *N,N*-dimethylacetamide.

Second, at least on our scale, the catalyst-to-substrate loading is fairly high.

Third, it appears to be necessary to heat the reaction to about 40-70°C to promote the hydrogenation of the last half of the reaction.

It is interesting that we were not successful in reducing disulfides using palladium on carbon catalyst but found that Raney cobalt catalyst was effective for the reduction and gave thiol in good yield. The Raney cobalt catalyst did not appear to be successful as a catalyst in reducing the sulfonyl chloride. We think the acid condition generated during the start of the reduction inhibits the catalyst and prevents further reduction.

Some preliminary work was initiated in the reduction of alkylsulfonyl chlorides. We found the desired product in the reaction mixture but many other components as well. It was not investigated further.

IV. CONCLUSION

We found the reduction of aromatic sulfonyl chlorides to thiols can be accomplished using a palladium-on-carbon catalyst at pressures of 60 psi or less in the presence of a solvent such as tetrahydrofuran (THF) containing a mild base such as *N,N*-dimethylacetamide. In some cases large amounts of the disulfide are generated when a basic Amberlite resin is used or when no base is used in the reaction medium. The disulfides can be reduced to the thiol by the use of Raney cobalt catalyst, at hydrogen pressures of 500 psi in a stirred autoclave.

ACKNOWLEDGMENTS

Appreciation is expressed to Mr. Michael Spitulnik of Eastman Kodak Company for helpful discussion and direction during the course of this work. We thank the Analytical Technology Laboratory Staff for help with the analyses.

REFERENCES

1. C. S. Marvel, *Org. Syn. Coll.*, *1*, 504 (1932).
2. F. C. Whitmore and F. H. Hamilton, *Org. Syn. Coll.*, *1*, 47 (1932).
3. L. Field and F. A. Grunwald, *J. Org. Chem.*, *16*, 946 (1951).
4. J. Strating and H. J. Backer, *Rec. Trav. Chim. Pays-Bas*, *69*, 638 (1950).
5. A. W. Wagner, *Chem. Ber.*, *99*, 375 (1966).
6. M. Hudlicky, *Reductions in Organic Chemistry*, John Wiley and Sons, New York, 1984.
7. C. T. Ratcliffe, U.S. Patent 4,128,586, Allied Chemical Corp. (Dec. 5, 1978); Appl. or P. 881952 (Feb. 27, 1978).
8. When the *N*,*N*-dimethylacetamide was used, an excess of four equivalents to one was used. No work was done to see if this ratio could be reduced.

13

Selective Hydrogenolysis of Benzyl and Carbobenzyloxy Protecting Groups for Hydroxyl and Amino Functions

LOUIS S. SEIF, KENNETH M. PARTYKA, and JOHN E. HENGEVELD

Abbott Laboratories, Abbott Park, Illinois

ABSTRACT

In synthesizing complex molecules, protected functional groups must be deblocked. Benzyl and carbobenzyloxy groups attached to oxygen or nitrogen are important protecting groups in many complex syntheses. The competitive catalytic debenzylation of N versus O and the hydrogenolysis of carbobenzyloxy groups in the presence of benzyloxime will be discussed.

Selective debenzylation studies will focus on the following examples: *N*-benzylcyclohexylamine, benzylcyclohexyl ether, dibenzyl-*p*-aminophenol, and 6-*O*-methyl-2'-*O*-(benzyl)-3'-*N*-(benzyl)erythromycin A 9-*O*-(benzyl)oxime ammoniol iodide. In addition, carbobenzyloxy group hydrogenolysis in the presence of benzyloxime protecting groups will be illustrated utilizing the following example: 3'-de(dimethylamino)-6-*O*-methyl-3'-[methyl(benzyloxycarbonyl)amino]-2'-*O*-(benzyloxycarbonyl)erythromycin A 9-[*O*-(2-chlorobenzyl)oxime].

I. INTRODUCTION

Synthetic organic chemistry abounds with examples of hydrogenolysis reactions. In the synthesis of multifunctional compounds, reactive sites must be temporarily blocked. Benzyl ethers, benzylamines, benzyl carbamates, and benzyl carbonates are several protecting groups which are commonly used.

The hydrogenolysis reaction can be explained by the reductive cleavage of a C-O or C-N bond (see Eqs. 1-4). These reactions may be considered as heterogeneous catalytic displacement reactions with hydrogen as the attacking species.[1]

$$C_6H_5{-}CH_2{-}OR + 2H \longrightarrow C_6H_5{-}CH_3 + HOR \tag{1}$$

$$C_6H_5{-}CH_2{-}NRR' + 2H \longrightarrow C_6H_5{-}CH_3 + H{-}NRR' \tag{2}$$

$$C_6H_5{-}CH_2{-}O{-}C(=O){-}NRR' + 2H \longrightarrow C_6H_5{-}CH_3 + CO_2 + H{-}NRR' \tag{3}$$

$$C_6H_5{-}CH_2{-}O{-}C(=O){-}OR + 2H \longrightarrow C_6H_5{-}CH_3 + CO_2 + HOR \tag{4}$$

Although hydrogenolysis is the method of choice for the removal of all of these protecting groups, examples from the literature show that cleavage of one group from a compound containing several benzyl groups is possible. The work presented here follows up on the theme of selective removal of benzyl groups by use of appropriate hydrogenolysis conditions. It is generally assumed that the hydrogenolysis of *O*-benzyl-type groups (Eq. 1) takes place with greater ease than *N*-benzyl-type groups (Eq. 2).[2,3] However, one cannot assume that an *O*-benzyl group will be selectively cleaved in the presence of an *N*-benzyl group. The selectivity may change if the *N*-protecting group is a more readily reducible one such as the carbobenzyloxy (Cbz) group.[2] An advantage to using a carbobenzyloxy radical as a protecting group is that it can be removed easily by hydrogenolysis under the mildest conditions (Eqs. 3 and 4),[4] even in the presence of other sensitive protecting groups, including benzyl,[5] if the molecule is not too sterically hindered.[6]

II. BACKGROUND

The nature of the alcohol blocked as a benzyl ether has a profound influence on its ease of hydrogenolysis relative to an *N*-benzylamine. The few available examples of reductions of substrates containing both a benzyl ether and a primary or secondary *N*-benzylamine show that the *O*-benzyl group is removed preferentially.[7] 1-Phenyl-2-[(4-benzyloxy-3-methoxy)phenyl]ethylamine was reduced in methyl alcohol with palladium on carbon (Pd/C) at 55°C and 1-4 atm of hydrogen to yield 1-phenyl-2-[(4-hydroxy-3-methoxy)phenyl]ethylamine (see Eq. 5).[8] The *O*-benzyl groups in *N*-[(3,4-dioxymethylene)benzyl]-2-[(3,4-dibenzyloxy)phenyl]ethylamine were selectively removed by reduction over 5% palladium on barium sulfate (see Eq. 6).[9] *N*-[(4-Benzyloxy-3-methoxy)benzyl]-2-[(4-benzyloxy-3-methoxy)phenyl]ethylamine on hydrogenolysis with Pd/C gave *N*-[(4-hydroxy-3-methoxy)benzyl]-2-[(4-hydroxy-3-methoxy)phenyl]ethylamine (see Eq. 7).[10]

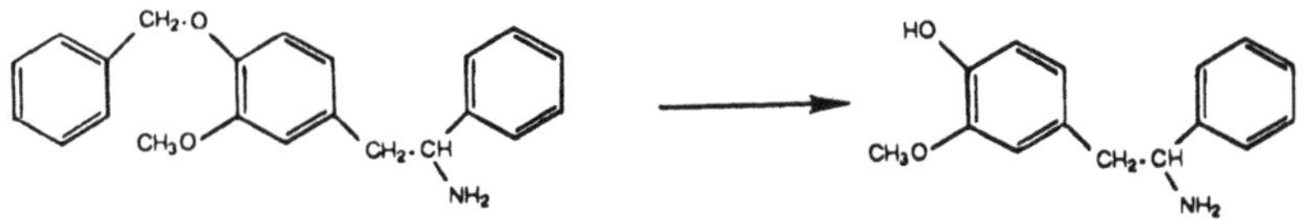

$CH_2{\cdot}O$ CH_3O $CH_2{\cdot}CH$ NH_2 → HO CH_3O $CH_2{\cdot}CH$ NH_2 (5)

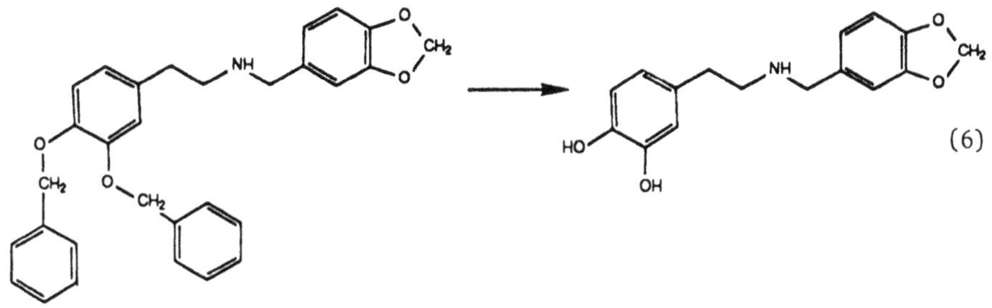

O CH_2 O CH_2 NH O CH_2 O → HO OH NH O CH_2 O (6)

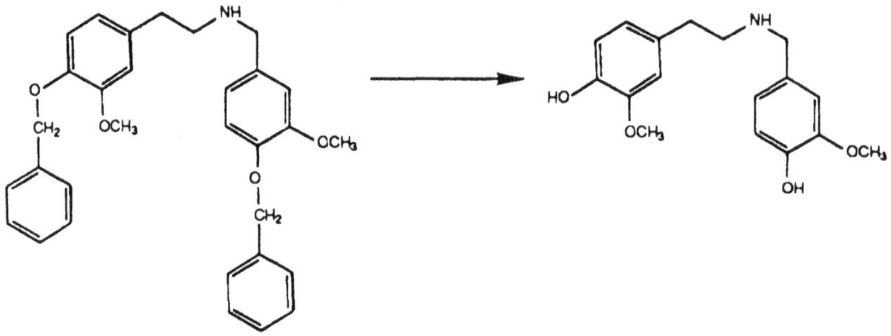

(7)

A recent investigation reported the opposite reactivity in the cases where alkylbenzyl ethers (as compared to phenolic benzyl ethers) were hydrogenolyzed in the presence of *N*-benzylamines.[11] When a solution of *N*,*O*-dibenzyl-*N*-ethyl-1-amino-8-octanol in 95% ethanol was shaken with Pd/C catalyst under 45 psi of hydrogen at room temperature for 24 hr, only the corresponding *N*-debenzylation product *N*-ethyl-*O*-benzyl-1-amino-8-octanol was isolated (see Eq. 8).[11] This result indicates that the presence of an amino function was inhibiting *O*-debenzylation. To further evaluate the inhibition of *O*-debenzylation by amines, a solution of benzyl *n*-nonyl ether in 95% ethanol was

(8)

shaken with Pd/C catalyst under 45 psi of hydrogen at room temperature for 24 hr; complete *O*-debenzylation was observed (see Eq. 9a). When the reaction was conducted under the same conditions but in the presence of 5 mol % of *n*-butylamine or *N*-benzylethylamine, benzyl *n*-nonyl ether was recovered (see Eq. 9b).[11] Further investigation confirmed that amines did not inhibit the cleavage of phenolic benzyl ethers. For example, in the presence of 5 mol % of *n*-butylamine, the hydrogenolysis of benzylphenyl ether proceeded smoothly to give a quanti-

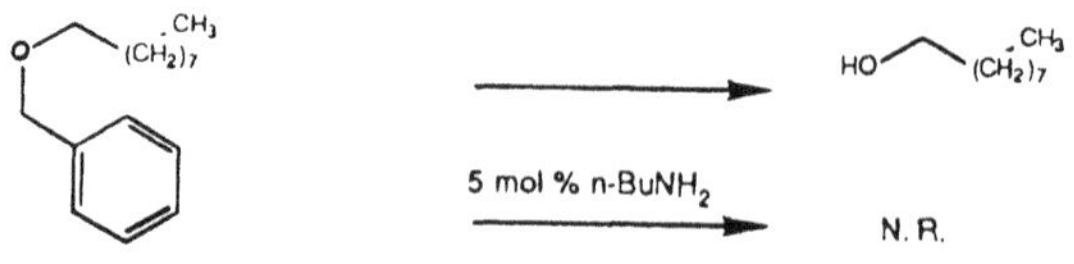

5 mol % n-$BuNH_2$

N. R.

(9a)

(9b)

$$C_6H_5CH_2OC_6H_5 \xrightarrow{\text{5 mol \% n-BuNH}_2} C_6H_5CH_3 + HOC_6H_5 \quad (10)$$

$$2,4\text{-}(CH_3)_2C_6H_3CH_2O\text{-}C_6H_4\text{-}NH_2 \longrightarrow 2,4\text{-}(CH_3)_2C_6H_3CH_3 + HO\text{-}C_6H_4\text{-}NH_2 \quad (11)$$

tative yield of phenol (see Eq. 10).[11] At 4 atm of hydrogen, the hydrogenolysis of 2,4-dimethylbenzyl-4-aminophenyl ether was complete after 15 hr (see Eq. 11).[12]

A more in-depth investigation was needed to focus on the selective removal of different protecting groups (benzyl ethers, *N*-benzylamines, benzyl carbonates, and benzyl carbamates) from multiprotected compounds. Through the use of information obtained from previous examples along with personally observed results, we have found that selective hydrogenolysis of protecting groups can be effected successfully.

III. DISCUSSION

A. Hydrogenolysis of *N*-Benzylcyclohexylamine and Benzylcyclohexyl Ether

To delve into the question of selective removal of benzyl groups from hydroxyl and amino functions, *N*-benzylcyclohexylamine and benzylcyclohexyl ether were chosen as substrates (see Eqs. 12 and 13). The reaction variables examined in this study were the quantity of catalyst, the solvent, and the presence or absence of acid.

Table 1 shows the effects of varying the hydrogenolysis conditions in the case of *N*-benzylcyclohexylamine. Complete hydrogenolysis

$$C_6H_5CH_2NH\text{-}C_6H_{11} \longrightarrow C_6H_5CH_3 + H_2N\text{-}C_6H_{11} \quad (12)$$

$$C_6H_5CH_2O\text{-}C_6H_{11} \longrightarrow C_6H_5CH_3 + HO\text{-}C_6H_{11} \quad (13)$$

Table 1 Hydrogenolysis of *N*-Benzylcyclohexylamine[a]

	Reaction conditions					GLC analysis (area %)		
Run	*N*-Benzyl-cyclohexylamine	5% Pd/C	Solvent	Acid	Time (hr)	*N*-Benzyl-cyclohexylamine	Cyclohexylamine	Toluene
1	2.5 g (0.013 M)	0.125 g	MeOH 100 ml	-	24	34.5	32.5	32.5
2	2.5 g	0.50 g	MeOH 100 ml	-	24	-	50.0	50.0
3	2.5 g	0.125 g	EtOAc 100 ml	-	18	95.9	2.0	2.0
4	2.5 g	0.50 g	EtOAc 100 ml	-	24	42.2	28.9	28.9
5	2.5 g	0.125 g	MeOH 99 ml	HOAc 1 ml (0.013 M)	24	23.5	38.3	38.3
6	2.5 g	0.50 g	MeOH 99 ml	HOAc 1 ml (0.013 M)	21	-	50.0	50.0
7	2.5 g	0.50 g	EtOAc 99 ml	HOAc 1 ml (0.013 M)	18	53.0	23.5	23.5
8	2.5 g	0.125 g	MeOH 99 ml	HCl 1.12 ml (0.013 M)	18	92.3	3.8	3.8

[a]Run on a Parr shaker, 50 psig, 20°C.

of benzylcyclohexylamine occurred after 24 hr using a 20% w/w load of 5% palladium on carbon in methanol as the solvent. Ethyl acetate was a poorer solvent than methanol for hydrogenolysis. Addition of acetic acid did not appear to be a factor in the rate of hydrogenolysis. However, adding hydrochloric acid to the reaction inhibited the hydrogenolysis reaction.

Table 2 shows the effects of varying the reaction parameters for the hydrogenolysis of benzylcyclohexyl ether. Once again methanol proved to be a better hydrogenolysis solvent than ethyl acetate. In methanol, in the absence of acid, the reaction required a 20% w/w load of 5% Pd/C to go to completion, while in ethyl acetate, the reaction did not proceed to an appreciable extent under any of the conditions tried. Addition of one equivalent of acetic acid to the methanol runs using a high catalyst concentration reduced the rate of benzylcyclohexyl ether cleavage, whereas using a low catalyst load increased the rate of hydrogenolysis. Hydrochloric acid greatly facilitated the hydrogenolysis with the reaction going to completion in 2 hr at a 5% w/w catalyst load (5% Pd/C).

Judging from the results shown in Tables 1 and 2, one would conclude that it is possible to hydrogenolyze an *O*-benzyl group in the presence of an *N*-benzyl group and, to a lesser extent, an *N*-benzyl group in the presence of an *O*-benzyl group. This was borne out experimentally as shown in Table 3. Equal molar amounts of *N*-benzylcyclohexylamine and benzylcyclohexyl ether were combined with a 2.5% w/w load of 5% Pd/C in methanol and shaken for 24 hr; no hydrogenolysis was observed. Repeating the above experiment but increasing the catalyst load to 10% w/w, only *N*-benzylcyclohexylamine reduced. When two equivalents of hydrochloric acid were added to the reaction mixture, only benzylcyclohexyl ether was hydrogenolyzed. Such uses of acid, amine, and catalyst load can promote the selective hydrogenolysis of either an *N*-alkylbenzylamine or an alkylbenzyl ether and has interesting possibilities in the synthesis of complex molecules.

B. Hydrogenolysis of *N*-,*O*-Dibenzyl-*p*-Aminophenol

Recent synthetic endeavors required a protecting group for phenols in the presence of *N*-benzylamines. Thus, hydrogenolysis parameters

Table 2 Hydrogenolysis of Benzylcyclohexyl Ether[a]

Run	Reaction conditions					GLC analysis (area %)		
	Benzyl-cyclohexyl ether	5% Pd/C	Solvent	Acid	Time (hr)	Benzyl-cyclohexyl ether	Cyclohexanol	Toluene
9	2.5 g (0.013 M)	0.125 g	MeOH 100 ml	-	24	95.0	2.0	2.0
10	2.5 g	0.50 g	MeOH 100 ml	-	24	-	50.0	50.0
11	2.5 g	0.50 g	EtOAc 100 ml	-	25	98.0	0.8	0.8
12	2.5 g	0.125 g	MeOH 99 ml	HOAc 1 ml (0.013 M)	24	77.0	10.6	10.6
13	2.5 g	0.50 g	MeOH 99 ml	HOAc 1 ml (0.013 M)	21	59.8	18.0	18.0
14	2.5 g	0.125 g	EtOAc 99 ml	HOAc 1 ml (0.013 M)	18	99.0	-	-
15	2.5 g	0.50 g	EtOAc 99 ml	HOAc 1 ml (0.013 M)	18	93.0	3.0	3.0
16	2.5 g	0.125 g	MeOH 99 ml	HCl 1.12 ml (0.013 M)	2	-	50.0	50.0

[a]Run on a Parr shaker, 50 psig, 20°C.

Table 3 Hydrogenolysis of *N*-Benzylcyclohexylamine + Benzylcyclohexyl Ether[a]

Run	Reaction conditions						GLC analysis (area %)				
	Benzyl-cyclohexyl ether	*N*-Benzyl-cyclohexyl amine	5% Pd/C	Solvent	Acid	Time (hr)	Benzyl-cyclohexyl ether	*N*-Benzyl-cyclohexyl amine	CyOH[b]	$CyNH_2$[c]	Toluene
17	2.5 g (0.013 M)	2.5 g (0.013 M)	0.125 g	MeOH 100 ml	-	24	50.9	49.1	-	-	-
18	2.5 g	2.5 g	0.125 g	MeOH 98 ml	HCl (0.026 M)	24	-	50.2	24.1	-	24.1
19	2.5 g	2.5 g	0.50 g	MeOH 100 ml	-	24	48.2	2.4	-	24.6	24.6

[a]Run on a Parr shaker, 50 psig, 20°C.
[b]CyOH = cyclohexanol.
[c]$CyNH_2$ = cyclohexylamine.

Table 4 Hydrogenolysis of *N*,*O*-Dibenzyl-*p*-aminophenol (1)

Run	Reaction conditions						GLC analysis (area %)[a]			
	N,*O*-Dibenzyl-*p*-aminophenol (1)	Catalyst weight (type)	Solvent	Acid	Base	Time	1	2	3	4[b]
1	1 g	0.1 g (10% Pd/C)	MeOH 100 ml	—	—	5 min	3	5	35	51
						15 min	—	95	—	5
						30 min	—	94	3	—
2	1 g	0.1 g (10% Pd/C)	MeOH 100 ml	HCl 0.3 ml	—	5 min	2	5	—	92
						15 min	—	94	—	5
3	1 g	0.1 g (10% Pd/C)	MeOH 90 ml	HOAc 10 ml	—	5 min	1	18	2	75
						15 min	—	90	2	5
4	0.5 g	0.05 g (10% $Pd(OH)_2/C$)	MeOH 50 ml	—	—	5 min	10	10	—	81
						15 min	6	87	—	3
						30 min	4	96	—	—
5	0.5 g	0.005 g (10% Pd/C)	MeOH 50 ml	—	—	5 min	95	—	—	5
						15 min	90	—	—	10
						60 min	90	—	—	10
						24 hr	78	—	—	20
						48 hr	36	7	2	55

6	0.5 g	0.05 g (10% Pd/C)	EtOAc 50 ml	—	—	5 min	96	—	—	4
						15 min	82	—	3	15
						30 min	60	2	8	30
						60 min	36	11	14	39
						4 hr	3	92	2	2
7	1 g	0.1 g (10% Pd/C)	MeOH 100 ml	—	*n*-Butylamine (5 mol %)	5 min	8	29	63	—
						15 min	—	45	54	—
						30 min	—	61	39	—
						60 min	—	70	23	—
						2 hr	—	95	2	—
8	0.5 g	0.5 g (RaNi #28)	MeOH 50 ml	—	—	30 min	78	2	17	—
						1 hr	32	10	50	4
						3 hr	21	24	55	1
						5 hr	5	36	50	—
						24 hr	—	80	16	1

[a]Run on a Parr shaker, 50 psig, 20°C.
[b]1, *N*,*O*-dibenzyl-*p*-aminophenol; 2, aminophenol; 3, *N*-benzyl-*p*-aminophenol; 4, *O*-benzyl-*p*-aminophenol.

for deprotecting *N*,*O*-dibenzyl-*p*-aminophenol were evaluated in order to develop conditions which would selectively remove either the *O*-benzyl or the *N*-benzyl group. Equation 14 depicts the catalytic hydrogenolysis scheme.

$$\underset{\underline{1}}{p\text{-}(\phi CH_2NH)C_6H_4OCH_2\phi} \longrightarrow \underset{\underline{2}}{p\text{-}H_2NC_6H_4OH} + \underset{\underline{3}}{p\text{-}(\phi CH_2NH)C_6H_4OH} + \underset{\underline{4}}{p\text{-}H_2NC_6H_4OCH_2\phi} \quad (14)$$

Table 4 shows the effects of varying the reaction parameters for the hydrogenolysis of *N*,*O*-dibenzyl-*p*-aminophenol. Our first attempts at hydrogenolysis of (1) led to complete debenzylation. Using a 10% w/w load of 10% Pd/C in methanol brought about *N*- and *O*-debenzylation in 15 min. However, with the addition of either hydrochloric acid or acetic acid to this reaction or by changing the catalyst to Pearlman's [10% $Pd(OH)_2/C$] in neutral methanol gave only *O*-benzyl-*p*-aminophenol (4) at reaction times less than 15 min. Continuing these three reactions for a longer time period led to complete *N*- and *O*-debenzylation. Reducing the catalyst load (10% Pd/C) by a factor of 10 slowed the rate of *O*-debenzylation. A 55% yield of *O*-benzyl-*p*-aminophenol (4) was observed after running the reaction over several days.

Using ethyl acetate instead of methanol and a 10% w/w load of 10% Pd/C decreased the reaction rate. Both *N*- and *O*-debenzylation occurred simultaneously to give the completely deprotected aminophenol (2).

Adding 5 mol % of *n*-butylamine to a 10% w/w load of 10% Pd/C in methanol had an interesting impact on the course of the reaction. Formation of *O*-benzyl-*p*-aminophenol (4) was never observed at any point in the reaction. Therefore, the addition of an amine enhanced selectivity by allowing rapid *O*-debenzylation to occur while slowing down the removal of the *N*-benzyl group to form a mixture of two products consisting mainly of *N*-benzyl-*p*-aminophenol (3) and, to a lesser extent, the completely debenzylated product (2) with reaction times less than 15 min. Allowing the reaction to proceed for a longer time period gave complete *N*- and *O*-debenzylation to form (2). The point to be emphasized here is that amines do not inhibit debenzylation of phenolic benzyl ethers, only alkylbenzyl ethers.

Substituting Raney nickel #28 for the palladium catalyst also resulted in faster *O*-debenzylation than *N*-debenzylation. When (1) was hydrogenolyzed over Raney nickel #28 in methanol at room temperature, *O*-debenzylation preferentially occurred to yield a mixture composed predominantly of (3) along with the final product (2) at shorter reaction times (less than 5 hr). However, at longer reaction times (24 hr), complete deprotection occurred to give predominantly (2).

Whenever nickel or palladium catalyst was used, the reactions had to be carefully monitored to optimize the yield of selectively debenzylated products. With increases in reaction times, the yields of selectively debenzylated products decreased (except at lower loads of Pd/C).

Different catalysts and catalyst concentrations will be evaluated in the near future to improve the selectivity of *N*- and *O*-debenzylation in this and analogous systems.

C. Selective Deprotection of Two Intermediates for the Preparation of 6-*O*-Methylerythromycin A: 6-*O*-Methyl-2'-*O*-(benzyl)-3'-*N*-(benzyl)erythromycin A 9-*O*-(benzyl)oxime ammoniol iodide and 3'-De(dimethylamino)-6-*O*-methyl-3'-[methyl(benzyloxycarbonyl)amino]-2'-*O*-(benzyloxycarbonyl)-erythromycin A 9-[*O*-(2-chlorobenzyl)oxime]

In the synthesis of a complex macrolide like 6-*O*-methylerythromycin A (Clarithromycin) (5), protecting group selection is important. 6-*O*-Methylerythromycin A was originally synthesized by Taisho Pharmaceutical Co.[13] and is currently being developed jointly by Abbott Laboratories and Taisho.

Compounds (6)[14] and (11)[15] are intermediates in two separate and independent processes for the synthesis of 6-*O*-methylerythromycin A (5).

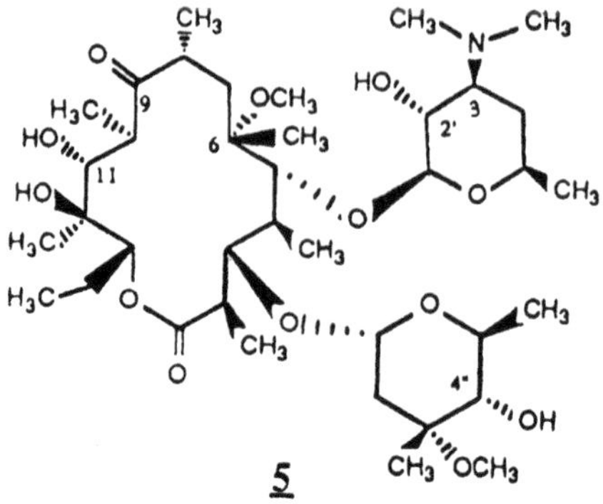

5

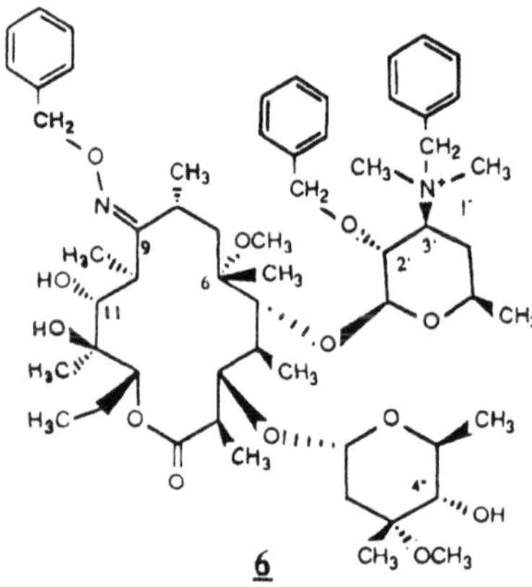

6

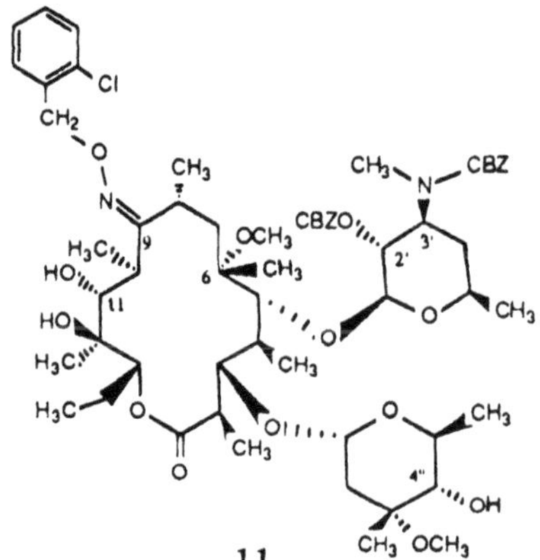

11

1. Hydrogenolysis of 6-*O*-Methyl-2'-*O*-(benzyl)-3'-*N*-(benzyl)erythromycin A 9-*O*-(benzyl)oxime ammoniol iodide, (6)[14]

The process to completely hydrogenolyze the three benzyl protecting groups on (6) required two steps (see Scheme 1). The catalyst used in both steps could be either palladium black or palladium on carbon (5, 10, or 20%), wet or dry. The first step in the process was the removal of the benzyl group from the quaternary amine (3' position) to yield (8). When methanol, ethyl acetate, or acetic acid were used as solvents, no noticeable hydrogenolysis occurred. However, with dimethylformamide as the solvent, using either a catalytic transfer hydrogenation method [10 g of (6), 100 ml of dimethylformamide, 10%

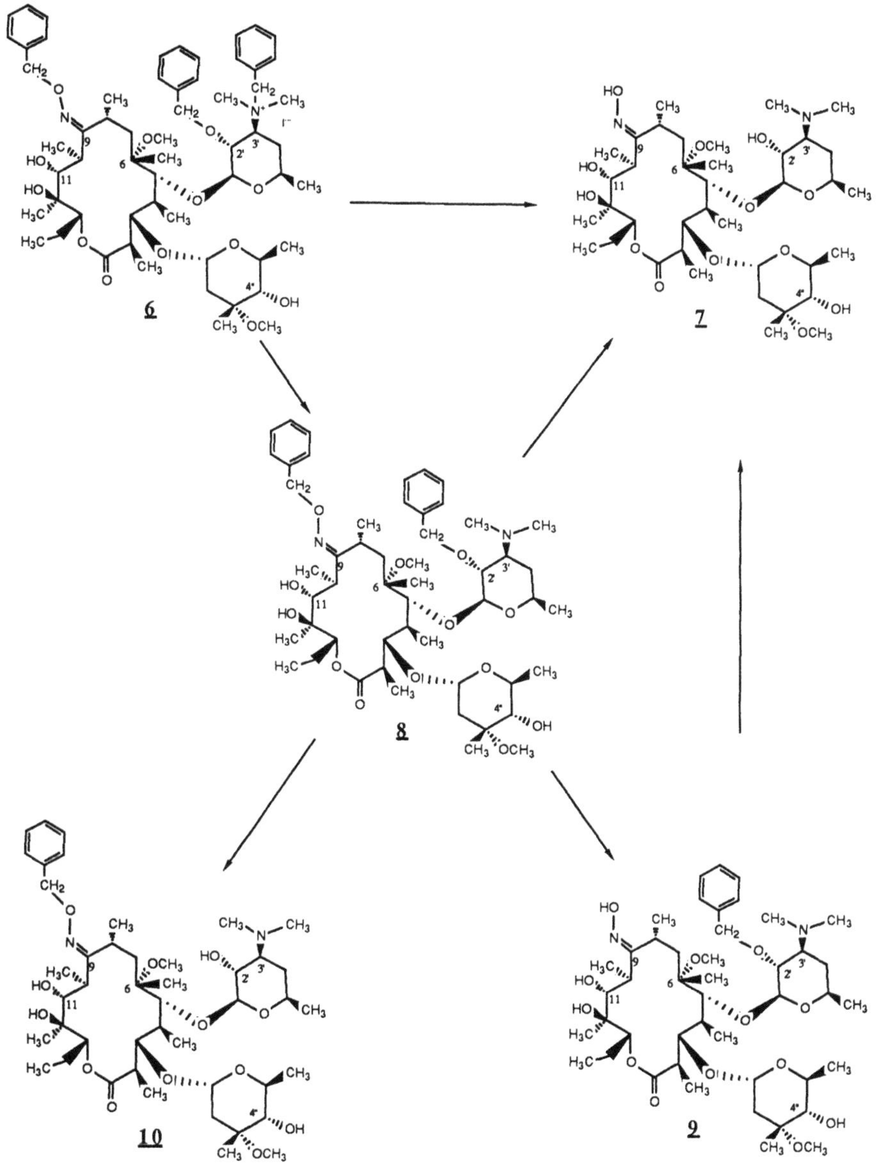

Scheme 1

w/w of 5% Pd/C and 9 g of ammonium formate at 85°C for 3 hr] or direct hydrogenolysis on a Parr shaker [10 g of (6), 100 ml of dimethylformamide, 10% w/w of 10% Pd/C and 50 psi of hydrogen at room temperature for 24 hr], the reaction proceeded to completion. Catalytic transfer hydrogenolysis gave more reproducible results than direct hydrogenolysis for the synthesis of (8).

The second step in the process for complete deprotection of (6) [conversion of (8) to (7)] was the catalytic hydrogenolysis of the benzyl ethers on both the 2'-hydroxyl and the 9-oxime. This reaction was also done using either catalytic transfer hydrogenolysis [40 g of (8), 700 ml of acetic acid, 22 g of ammonium formate, and 50% w/w of 10% Pd/C at room temperature for 24 hr] or direct hydrogenolysis [30 g of (8), 500 ml of acetic acid, 10% w/w of 10% Pd/C, and 50 psi of hydrogen at room temperature for 24 hr]. In this case, the direct hydrogenolysis method was more reproducible. Use of acetic acid as the solvent was crucial for complete debenyzlation. Methanol and ethyl acetate gave incomplete hydrogenolyses.

Several interesting observations were made in performing the last step of this deprotection sequence. First, by using the catalytic transfer hydrogenation method to cleave both the 2'-*O*-benzyl ether and the 9-*O*-benzyloxime [(8) to (7)], selectivity was observed. The benzyl groups were removed at different rates. When (8) was hydrogenolized using the catalytic transfer method [15 g of (8), 30% w/w of 5% Pd/C in 120 ml of methanol, and 9.3 ml of formic acid at reflux for 30 min], the benzyl group attached to the oxime (9 position) was cleaved first in less than one-half hour to give (9) in excellent yield. The reaction was repeated and continued for a longer time allowing the removal of the benzyl ether (2' position), thus forming the completely deprotected product (7) (Scheme 1).

When (8) was subjected to direct hydrogenolysis on a Parr shaker [0.2 g of (8), 12 ml of absolute ethanol, 1.0 g of Raney nickel #28, and 50 psi of hydrogen at room temperature for 24 hr], the benzyl ether (2' position) was hydrogenolyzed to form (10). Benzyloxime cleavage was undetected.

2. Hydrogenolysis of 3'-De(dimethylamino)-6-*O*-methyl-3'-[methyl(benzyloxycarbonyl)amino]-2'-*O*-(benzyloxycarbonyl)erythromycin A 9-[*O*-(2-chlorobenzyl)oxime], (11)[15]

Only one step was required to completely hydrogenolyze the three protecting groups in compound (11) [conversion of (11) to (12)] (see Scheme 2). The reaction could be accomplished by using either catalytic transfer hydrogenolysis [11 g of (11), 100 ml of methanol, 5 g of ammonium formate, 6 ml of formic acid, and 10% w/w of 10% Pd/C at reflux for 4 hr] or direct hydrogenolysis on a Parr shaker [150 g

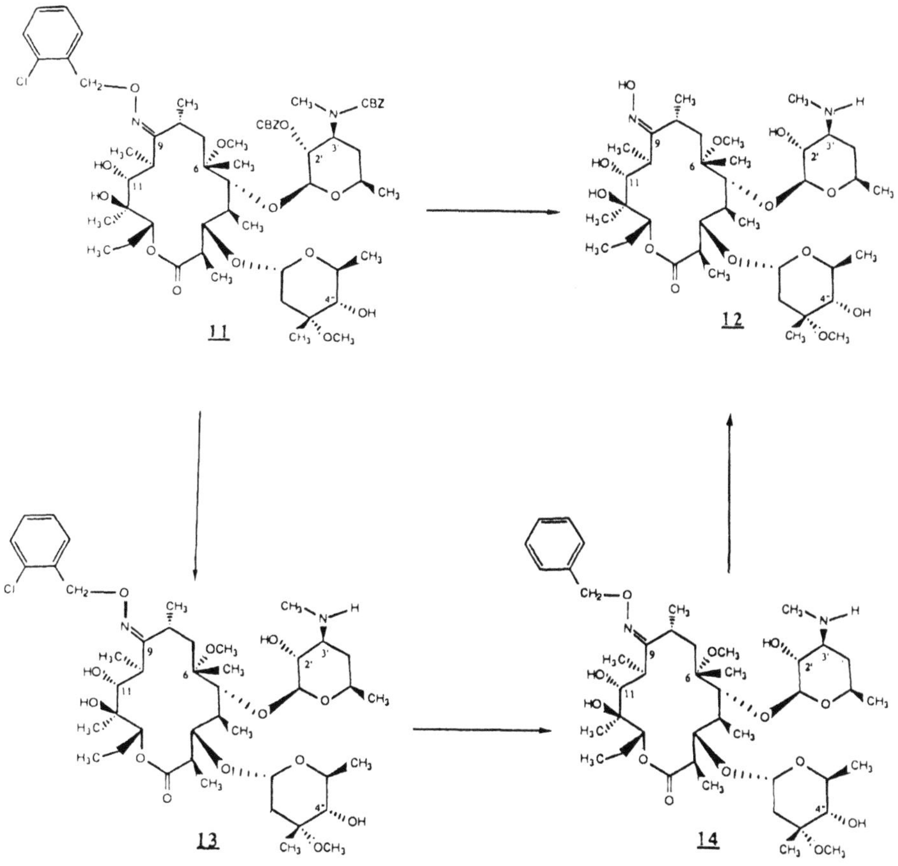

Scheme 2

of (11), 450 ml of methanol, 45 ml of water, 40 ml of acetic acid, 10% w/w of 10% Pd/C, and 50 psi of hydrogen at room temperature for 24 hr]. It was observed that by replacing methanol with ethanol in these direct hydrogenolysis reactions, the rate of deprotection [(11) to (12)] was enhanced and a cleaner product was isolated. In the absence of both acetic acid and water, hydrogenolysis did not occur. Thus, the acetic acid/water mixture may be nullifying the inhibiting effects of the amine substrate [(11), 3' position].[16] The direct hydrogenolysis method gave better results than the catalytic transfer method and was therefore used.

When (11) was hydrogenated on a Parr shaker [5 g of (8), 200 ml of methanol, 50 ml of ethyl acetate, 5.0 g of Raney nickel #28, and one equivalent of sodium acetate (used as a trap for HCl liberation) at 50 psi of hydrogen], complete removal of both carbobenzyloxy groups (carbonate, 2' position and carbamate, 3' position) resulted, yielding (13). However, the benzyloxime was not cleaved. Over prolonged time periods (24 hr), the compound underwent dehalogenation of the chlorobenzyl protecting group (9 position) to give (14) in excellent yield. Replacing the nickel catalyst with palladium allowed formation of the desired product (12) [conversion of (14) to (12)].

IV. CONCLUSION

Catalytic debenzylations using conditions similar to these can be found throughout the recent chemical literature. However, only conditions similar to those which we have found will succeed in the above-mentioned examples to

- Selectively cleave either *N*-benzyl or *O*-benzyl protecting groups on an alkyl- or aryl-substituted compound
- Selectively remove an *N*-benzyl group from a quaternary amine in the presence of both an *O*-benzyl ether and a benzyloxime
- Selectively cleave either an alkyl *O*-benzyl group or a benzyl group on an oxime
- Selectively remove carbobenzyloxy groups without cleaving a benzyloxime

Many other hydrogenolysis procedures were tried on these types of compounds and either failed or gave limited success.

ACKNOWLEDGMENT

We thank Dr. Steven M. Hannick and Dr. Richard J. Pariza for helpful discussions and technical assistance. Special appreciation must go to Abbott's structural chemistry group (IR, NMR, and MS). Mrs. Susan Clay has our special appreciation for her cheerful and skillful typing and related help. And, finally, our thanks to Dr. Bruce Surber for providing all of the graphics.

REFERENCES

1. A. P. G. Kieboom and F. van Rantwijk, *Hydrogenation and Hydrogenolysis in Synthetic Organic Chemistry,* Delft University Press, Rotterdam, 1977, p. 89.
2. M. Freifelder, *Practical Catalytic Hydrogenation,* John Wiley and Sons, New York, 1971, p. 430.
3. R. N. Rylander, *Catalytic Hydrogenation in Organic Synthesis,* Academic Press, New York, 1979, p. 280.
4. Ref. 3, p. 276.
5. S. S. Hall, D. Loebenberg, and D. P. Schumacher, *J. Med. Chem., 26,* 496 (1983).
6. B. H. Lee and M. J. Miller, *J. Org. Chem., 48,* 24 (1983).
7. Ref. 2, p. 431.
8. W. D. McPhee and E. S. Erickson, Jr., *J. Am. Chem. Soc., 68,* 624 (1946).
9. E. J. Forbes, *J. Chem. Soc.,* 3926 (1955).
10. G. W. Kirby and H. P. Tiwari, *J. Chem. Soc.,* 676 (1966).
11. B. P. Czech and R. A. Bartsch, *J. Org. Chem., 49,* 4076 (1984).
12. R. Davis and J. M. Muchowski, *Synthesis,* 987 (1982).
13. Y. Watanabe, S. Moriomoto, and S. Omura, U.S. Patent 4331803 (1982), Taisho Pharmaceutical Co.
14. S. Morimoto, T. Adachi, T. Asaka, M. Kashimura, Y. Watanabe, and K. Sota, Japanese Patent JP 53618 (March 1985), Taisho Pharmaceutical Co.

15. (a) Y. Watanabe, S. Morimoto, and S. Omura, Japanese Application JP 75258 (June 1980), Taisho Pharmaceutical Co. (b) Y. Watanabe, S. Morimoto, and S. Omura, Japanese Application JP 159128 (November 1980), Taisho Pharmaceutical Co. (c) S. Morimoto, T. Adachi, T. Asaka, Y. Watanabe, and K. Sota, Japanese Application JP 225543 (October 1984), Taisho Pharmaceutical Co.

16. R. L. Rylander in *Catalysis in Organic Syntheses* (W. Jones, ed.), Academic Press, New York, 1980, p. 171.

Part V

Selected Special Topics

14

Selective Catalytic Synthesis of Mixed Alkylamines and Polyfunctional Amines

MICHAEL E. FORD and THOMAS A. JOHNSON

Air Products and Chemicals, Inc., Allentown, Pennsylvania

ABSTRACT

Lower alkylamines are typically produced by reaction of an alcohol with ammonia or an amine in the presence of either acidic or supported metal catalysts. However, this technology is not suitable for the synthesis of unsymmetric (mixed) amines and polyfunctional amines. In the latter instances, selectivities to desired products are often low, owing to fast equilibration of the products of the initial alkylation reaction. To circumvent this problem, mixed amines and polyfunctional amines are currently prepared by alternate, more expensive routes, such as reductive amination of carbonyl compounds or amination of alkyl polyhalides. In this chapter, we describe a new approach to amine synthesis. A catalyst system has been developed which does not catalyze product equilibration under the conditions of product formation. Depending on the structure of the starting amine and alcohol components, group IIA and/or group IIIB metal acid phosphates selectivity catalyze amine formation. Specifically, selective syntheses of mixed alkyl amines have been achieved by amine-alcohol reactions over strontium hydrogen phosphate. In addition, strontium hydrogen phosphate catalyzes formation of more complex polyfunctional amines in high selectivity. Under appropriate conditions, strontium hydrogen phosphate, lanthanum hydrogen phosphate, or phosphoric acid on silica also catalyzes vapor phase production of noncyclic polyethylene amines from ethylenediamine, ethanolamine, and, optionally, ammonia. In all instances, products appear to be generated via initial formation of a phosphate ester on the catalyst surface with subsequent nucleophilic displacement of phosphate by the reactant amine.

I. INTRODUCTION

A. Synthesis of Alkylamines

Many methods have been developed for the preparation of aliphatic amines.[1,2] The general classes of reactions for amine synthesis are:

1. Reduction processes. These include reductive amination of aldehydes and ketones, both catalytically and by the Leuckart-Wallach reaction. Reduction of nitriles, nitro-organics, and amides also provides alkylamines.
2. Substitution processes. Stoichiometric reactions of amines with alkyl halides and catalytic reactions of amines with alcohols are used.
3. Amination of olefins. Catalytic addition of ammonia to olefins provides a direct route to alkylamines.

Each of these preparative methods has advantages and disadvantages. In conjunction with the catalytic reductive amination of aldehydes and ketones, reduction and reductive amination of nitriles provide routes to a wide variety of symmetric and unsymmetric amines (i.e., amines with mixed alkyl groups). The choice of catalyst is critical to obtain the desired selectivity, particularly in the reduction of nitriles.[3] Thus, cobalt and nickel catalysts convert nitriles to primary amines with good selectivities. Addition of ammonia to these reductions effectively suppresses secondary amine formation. Alternately, primary amines are obtained from nitriles in high yield with ruthenium catalysts. Under similar conditions, rhodium catalysts provide secondary amines, and platinum and palladium catalysts produce tertiary amines, both in essentially quantitative yield. These processes are usually operated at low to moderate pressures. However, many of the aldehyde, ketone, or nitrile starting materials are expensive, and costly precious metal catalysts such as rhodium or palladium are often required. Reduction of nitroalkanes provides a route to primary amines using a wide variety of catalysts under mild conditions. The high energy release potential of nitroalkanes is the major drawback to practice of this technology. The classical route to alkylamines is via substitution of halogen from an alkyl halide by ammonia or a simpler amine. Depending on the starting amine, a variety of mixed alkyl or arylalkyl amines can be obtained. However, when starting with ammonia or a primary amine, it is difficult to stop this process at an intermediate stage, e.g., at the corresponding primary or secondary amine. Generally, the symmetric tertiary amine is the major product. In some cases, formation of the tertiary amine is accompanied by generation of quaternary ammonium salts. A further limitation is the necessity of liberating the product amine from its hydrogen halide salt via neutralization with a stronger base. Disposal of the coproduct salt stream in an environmentally acceptable manner is a formidable barrier to practice of this technology. Moreover, from an operational standpoint, this technology requires either glass or exotic alloys as materials of construction to avoid the severe corrosion problems associated with its practice.

One of the most widely practiced routes to lower alkylamines is reaction of low-cost, readily available alcohols with ammonia over acid catalysts such as silica-alumina or hydrogen-activated base metal catalysts such as nickel or cobalt (on silica or alumina supports).[4] These processes are normally operated in either gas phase or trickle-bed modes. Product distributions are controlled by thermodynamics. While this route has the advantage that products for which no market exists can be recycled, recycle reduces reactor productivity. In addition, mixed alkylamines cannot be selectively produced.

B. Phosphate-Catalyzed Routes

Amine synthesis has been demonstrated with several phosphate catalysts. For example, a variety of unsymmetric *N*-alkylmorpholine derivatives has been prepared with aluminum phosphate.[5,6] However, selectivities and/or conversions are moderate at best. In contrast, metal phosphates are excellent catalysts for the preparation of diazabicyclo[2.2.2]octane (1) from *N*-(2-hydroxyethyl)piperazine (2)[7-10] (Eq. 1). While a number of phosphates are effective,[7,9]

$$\text{HN(CH}_2\text{CH}_2)_2\text{NCH}_2\text{CH}_2\text{OH} \xrightarrow{MPO_4} \text{N(CH}_2\text{CH}_2)_3\text{N} \qquad (1)$$

strontium hydrogen phosphate provides the highest yields.[10] Although these reactions illustrate amine synthesis with metal phosphate catalysts, they do not indicate whether metal phosphates are effective for preparation of unsymmetric amines. Only moderate yields of *N*-alkylmorpholines are obtained, while 1 is a symmetric, very stable product that could be formed over a catalyst which causes equilibration of unsymmetric amines.

II. RESULTS AND DISCUSSION

A. Mixed Alkylamines

To evaluate metal acid phosphates as catalysts for selective synthesis of unsymmetric amines, the reaction of ethylamine with methanol was studied as a model system.[11] If selective alkylation occurred, only two amines would be obtained (Scheme 1). However, if product equilibration were operative, up to eight products could be formed.

The object of the model study was to characterize amine-alcohol reactions over metal acid phosphates. Specifically, answers were sought to the following questions:

1. Does product equilibration occur?
2. Does the reaction follow a stepwise path?
3. How do process variables, such as temperature, feed ratio, and space velocity, influence the reaction?
4. How can the degree of alkylation be controlled?

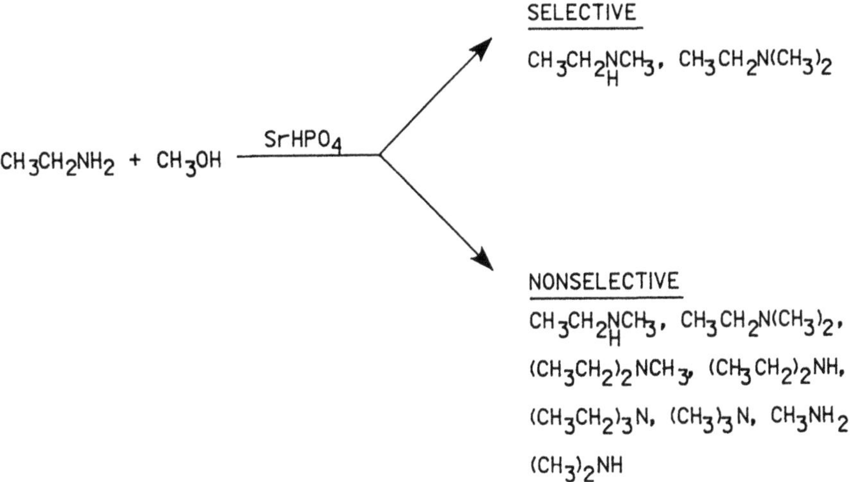

Scheme 1 Comparison of selective versus nonselective amine synthesis with $SrHPO_4$.

Reaction temperature, pressure, feed ratio (given as mole ratio of amine/alcohol), and space velocity were investigated for their effect on conversion and selectivity.

The influence of temperature was investigated over the range 150-375°C. Conversions of methanol and ethylamine increased at parallel rates, with methanol conversion exceeding ethylamine conversion at about 350°C (Fig. 1). This observation results from formation of dimethylethylamine at higher temperatures (conversions), a process which consumes 2 mol of methanol. Selectivity to the two expected products, methylethylamine and dimethylethylamine, was excellent (Fig. 2). The presence of only small amounts of byproducts ($\leq$10% selectivity) demonstrates that product formation is not followed by equilibration to the thermodynamic mix. At low temperature, conversion was low, and formation of methylethylamine predominated. At higher temperatures (conversions), selectivity to dimethylethylamine increased. This observation is consistent with a stepwise reaction path, in which ethylamine initially reacts with methanol to form methylethylamine. Alkylation of the latter by additional methanol provides dimethylethylamine.

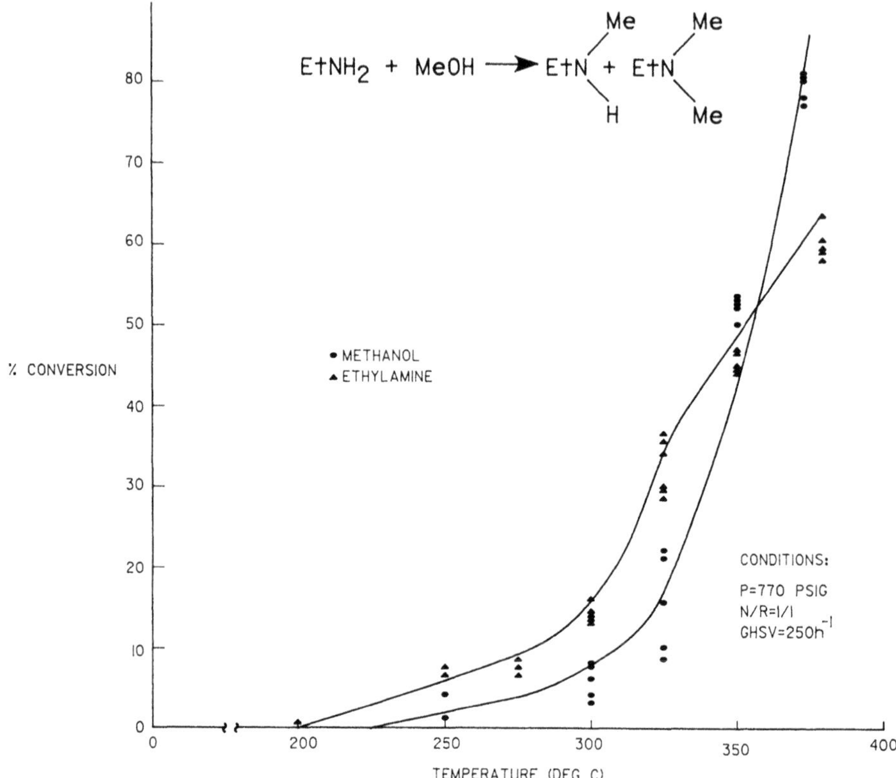

Fig. 1 Influence of temperature on conversion for ethylamine and methanol; reaction conditions: P = 785 psia; N/R = 1:1; GHSV = 250 hr^{-1}. ● Methanol, ▲ ethylamine.

The effects of pressure were evaluated between 10 and 785 psig. Conversion of both methanol and ethylamine increased smoothly with increasing pressure (Fig. 3). The increase in conversion probably results from an increasing concentration of the reactants at the catalyst surface. Formation of a liquid phase is not involved since 350°C is above the critical temperature of both reactants. Selectivity to the desired products was excellent throughout the pressure range (Fig. 4). The selectivity data are consistent with a kinetically controlled stepwise reaction in which methylethylamine is

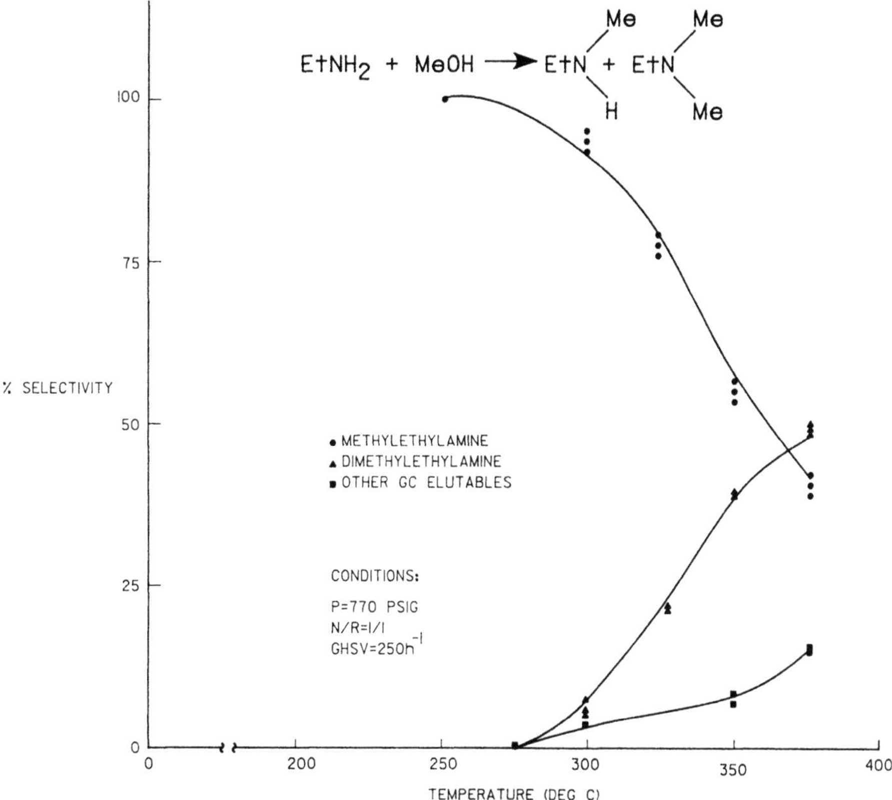

Fig. 2 Influence of temperature on selectivity for ethylamine and methanol; reaction conditions: P = 785 psia; N/R = 1:1; GHSV = 250 hr^{-1}. ● Methylethylamine, ▲ dimethylethylamine, ■ other GC elutables.

formed at low pressure (conversion) and is then converted to dimethylethylamine at higher pressures (conversions).

The influence of amine/alcohol molar feed ratio (N/R) on selectivity was evaluated between 12:1 and 1:12 ethylamine/methanol. Methylethylamine can be prepared selectively at high N/R while dimethylethylamine is the predominant product at low N/R (Fig. 5). Thus, by choice of an appropriate feed ratio, the degree of alkylation can be controlled to give the desired product. At low N/R,

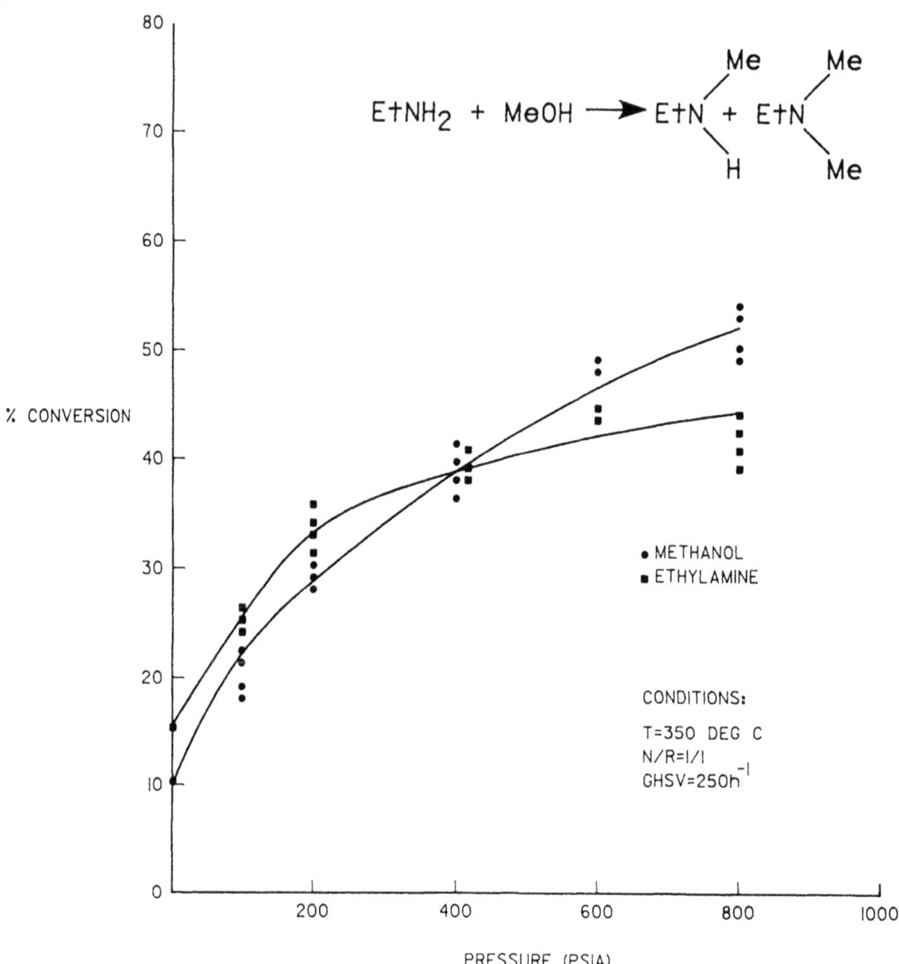

Fig. 3 Influence of pressure on conversion for ethylamine and methanol; reaction conditions: T = 350°C; N/R = 1:1; GHSV = 250 hr^{-1}. ● Methanol, ■ ethylamine.

the proportion of other GC elutables increased, owing to formation of methyl ether, dimethylamine, and trimethylamine. The presence of the latter two compounds indicates that some amine equilibration occurred under alcohol-rich conditions.

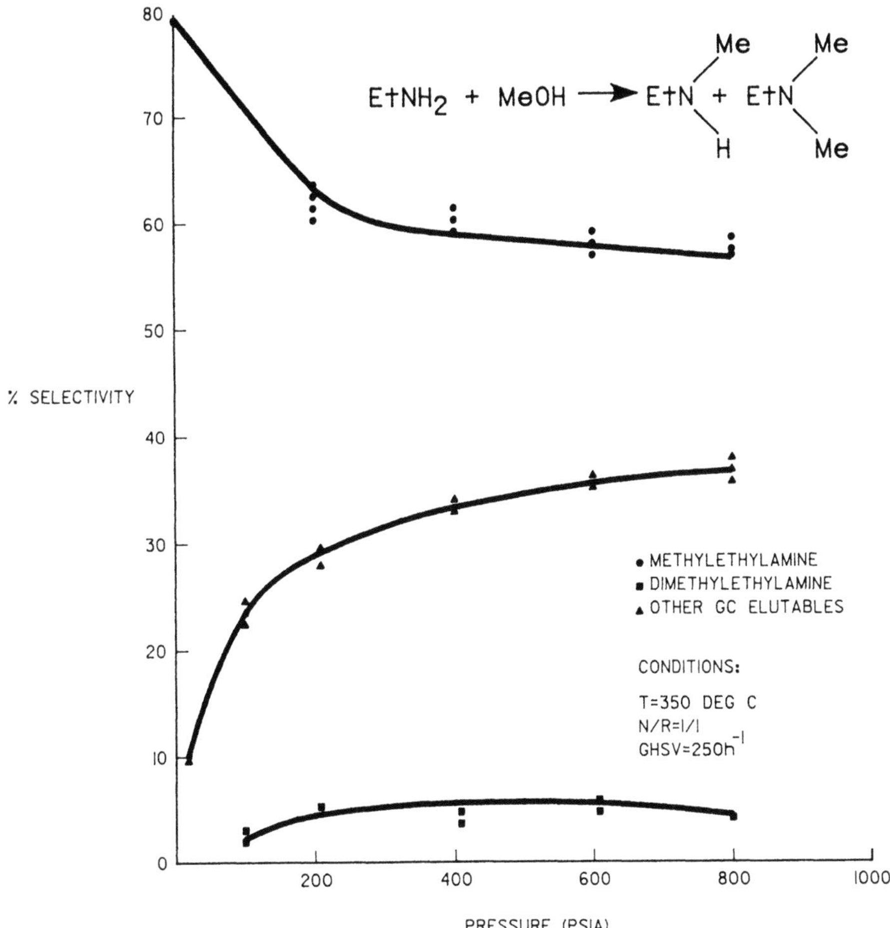

Fig. 4 Influence of pressure on selectivity for ethylamine and methanol; reaction conditions: T = 350°C; N/R = 1:1; GHSV = 250 hr^{-1}. ● Methylethylamine, ▲ dimethylethylamine, ■ other GC elutables.

The effect of space velocity was studied between 60 and 2050 hr^{-1}. Conversion of both methanol and ethylamine decreased smoothly with increasing space velocity (Fig. 6). At low GHSV, dimethylethylamine predominates, whereas methylethylamine is formed with excellent selectivity at high GHSV (Fig. 7). These observations

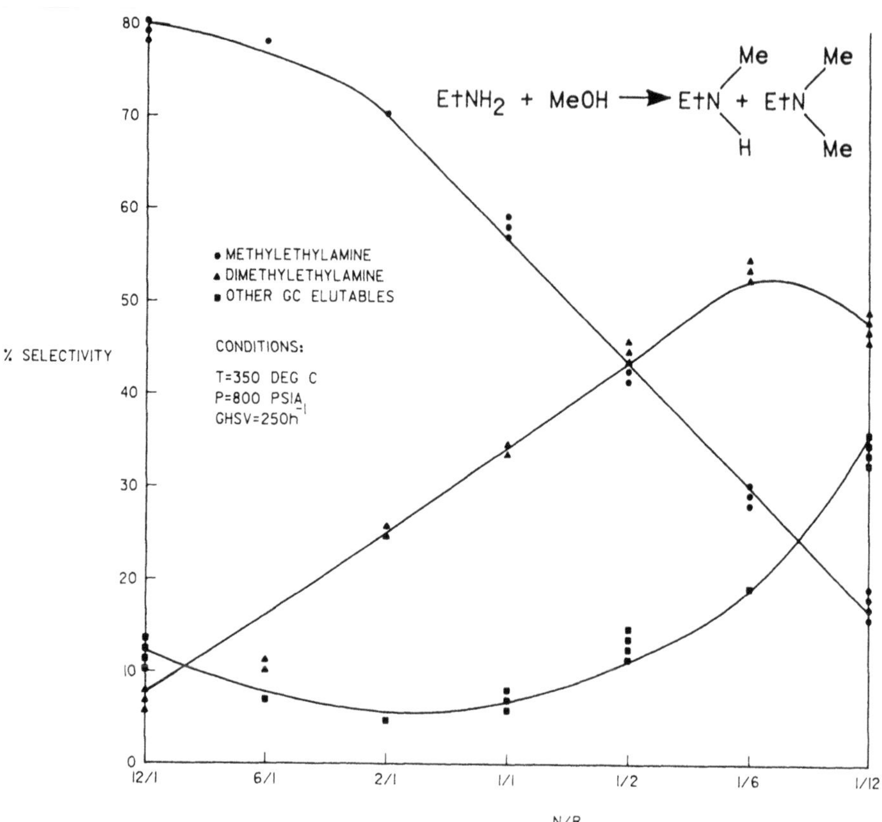

Fig. 5 Influence of N/R on selectivity for ethylamine and methanol; reaction conditions: T = 350°C; P = 800 psia; GHSV = 250 hr^{-1}.
● Methylethylamine, ▲ dimethylethylamine, ■ other GC elutables.

support a stepwise reaction path in which conditions that favor low alcohol conversion (high GHSV) lead to the monomethylated product, while conditions that favor high alcohol conversion (low GHSV) afford further alkylation to the dimethylated product.

The above model reaction demonstrates the potential of metal acid phosphates as catalysts for synthesis of unsymmetric amines. Alkylations of amines with alcohols by metal acid phosphates have the following characteristics:

1. Equilibration of products does not occur to a significant extent.

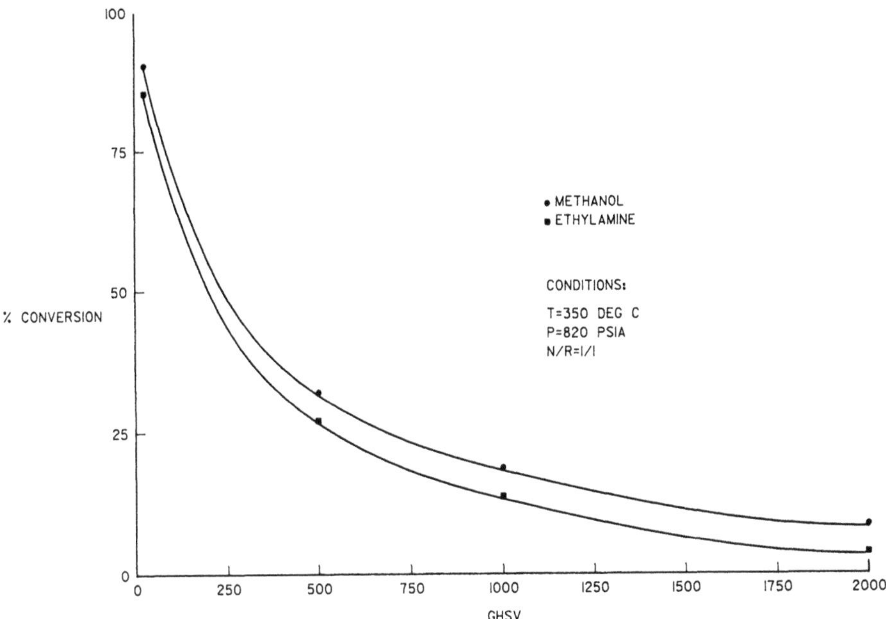

Fig. 6. Influence of space velocity on conversion for ethylamine and methanol; reaction conditions: T = 350°C, P = 820 psia; N/R = 1:1. ● Methanol, ■ ethylamine.

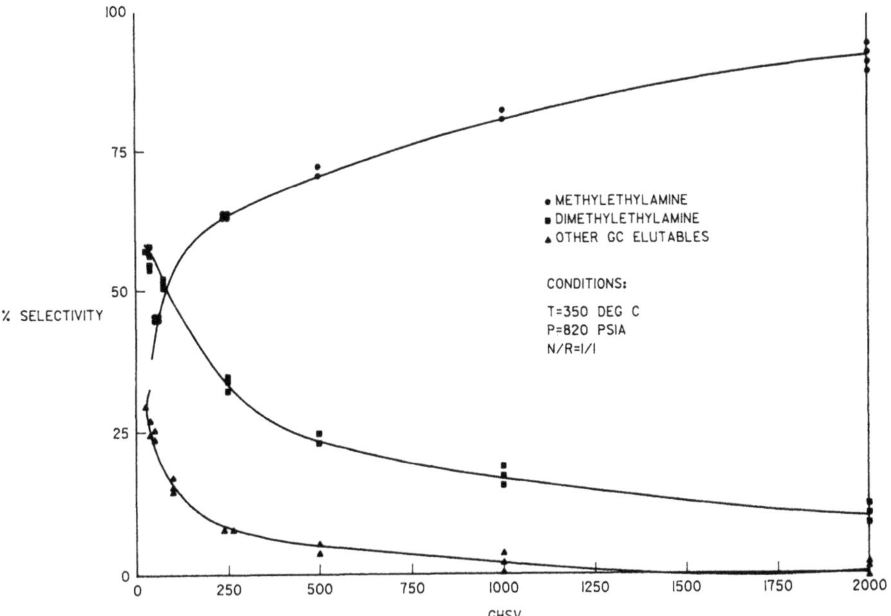

Fig. 7 Influence of space velocity on selectivity for ethylamine and methanol; reaction conditions: T = 350°C; P = 820 psia; N/R = 1:1. ● Methylethylamine, ■ dimethylethylamine, ▲ other GC elutables.

2. Amine alkylation follows a stepwise path.
3. Product selectivity can be controlled with process variables, particularly feed ratio and space velocity.
4. Conversion increases as temperature, space velocity, or pressure increases.

Thus, unsymmetric amines can be prepared by proper selection of process variables, since amine alkylation is a stepwise process and no equilibration occurs.

B. Polyfunctional Amines

To assess the general utility of metal acid phosphates as catalysts for the preparation of mixed amines, synthesis of several polyfunctional mixed amines was undertaken with strontium hydrogen phosphate. Selection of the target amines was based on starting material availability and utility of the product tertiary amines as urethane catalysts. Consequently, simple lower alkylamines and polyfunctional alcohols derived from ethylene oxide were used as starting materials. Good selectivities to both heterocyclic and acyclic unsymmetric amines were obtained under typical operating conditions (Table 1). Reaction of *N*-methyldiethanolamine with dimethylamine provides *N*,*N*,*N'*-trimethyl-*N'*-(2-hydroxyethyl)ethylenediamine (TMHEDA) with complete exclusion of the dialkylation product. However, formation of 1,4-dimethylpiperazine was significant. The feasibility of production of *N*,*N*-dimethylethanolamine and *N*,*N*,*N'*,*N'*-tetramethylethylenediamine from ethylene glycol and dimethylamine was checked under representative conditions (Table 1). Conversion of ethylene glycol was 72%; the combined selectivity to both amine products was 84%. Thus, phosphate-catalyzed alkylation provides a potential route to these amines. With proper selection of reaction conditions, either should be obtained in good selectivity.

In the synthesis of TMHEDA from *N*-methyldiethanolamine and dimethylamine, 1,4-dimethylpiperazine was observed as a major byproduct (18% selectivity). This cyclization exemplifies reaction of a tertiary amine with an alcohol. The reaction path of the cyclization may involve alcohol activation followed by nucleophilic substitution

Table 1 Synthesis of Polyfunctional Amines with $SrHPO_4$

Alcohol	Amine	Product	N/R	LHSV (hr^{-1})[a]	T (°C)	P (psig)	Molar yields	Selectivity
$O(CH_2CH_2)_2NCH_2CH_2OH$	Me_2NH	$O(CH_2CH_2)_2NCH_2CH_2NMe_2$	5	0.12	325	600	70	80
$HOCH_2CH_2NMe_2$	$O(CH_2CH_2)_2NH$	$O(CH_2CH_2)_2NCH_2CH_2NMe_2$	4.1	0.4	325	800	65	80
$(HOCH_2CH_2)_2NMe$	Me_2NH	$Me_2NCH_2CH_2N(Me)CH_2CH_2OH$	10	0.15	260	780	45	68
		$MeN(CH_2CH_2)_2NMe$					13	18
$HOCH_2CH_2OH$	Me_2NH	$HOCH_2CH_2NMe_2$	5	0.15	275	800	38	53
		$Me_2NCH_2CH_2NMe_2$					22	31

[a]Based on amine and alcohol.

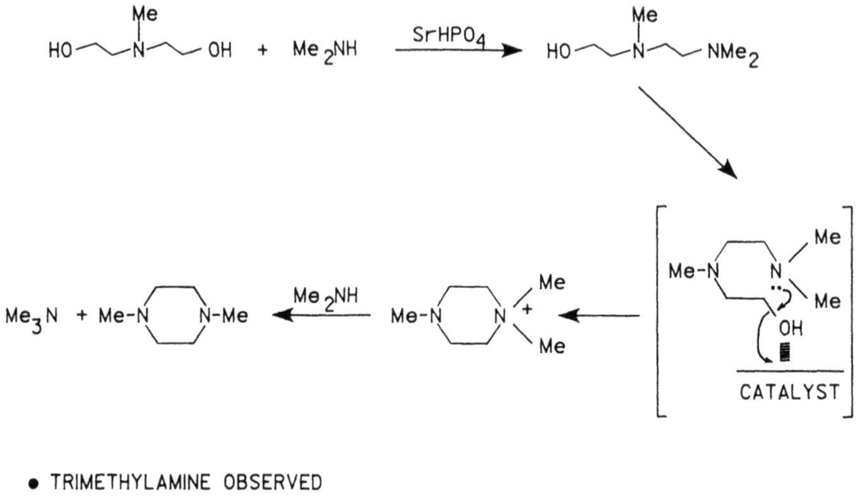

• TRIMETHYLAMINE OBSERVED

Scheme 2 Tertiary amine cyclization over $SrHPO_4$.

by the amine to form a quaternary ammonium intermediate (Scheme 2). Methylation of an amine, alcohol, water, or the catalyst would provide the cyclic amine product. When cyclization occurred, trimethylamine was found as a coproduct.

To confirm that the aminoalcohol was the precursor of 1,4-dimethylpiperazine, TMHEDA was fed over strontium hydrogen phosphate at temperatures between 250 and 350°C (Fig. 8). 1,4-Dimethylpiperazine was formed in up to 61% molar yield, with 62% selectivity at 350°C. The facility of this reaction is shown by the 80% conversion observed at 275°C, a temperature at which alkylations of secondary amines with alcohols usually occur in less than 50% conversion. To demonstrate that the reaction of tertiary amines with alcohols is more facile and selective for those tertiary amines which can form cyclic products, *N,N*-dimethylethanolamine was passed over strontium hydrogen phosphate. At 350°C, 1,4-dimethylpiperazine was obtained in 17% molar yield (32% selectivity). Thus, cyclization to form a six-membered ring is a strong driving force for the reaction of tertiary amines with alcohols, as compared to the intermolecular reaction.

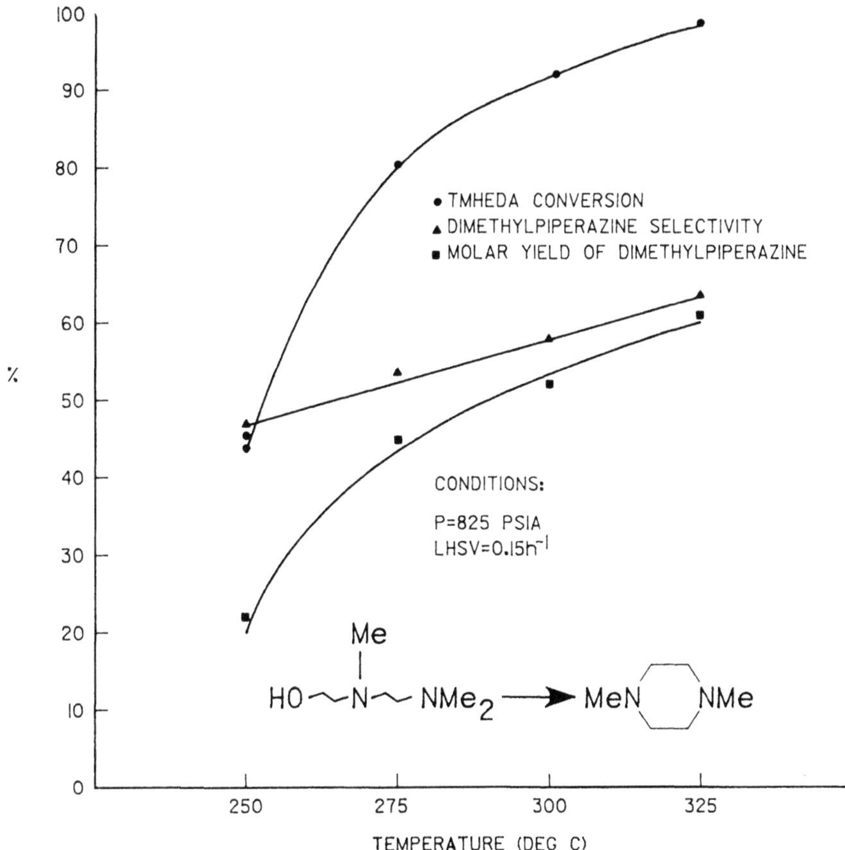

Fig. 8 Cyclization of *N*,*N*,*N*'-trimethyl-*N*'-(2-hydroxyethyl)ethylenediamine (TMHEDA) to 1,4-diethylpiperazine; reaction conditions: T = 825 psia; LHSV = 0.15 hr^{-1}. ● TMHEDA conversion, ▲ 1,4-dimethylpiperazine selectivity, ■ 1,4-dimethylpiperazine molar yield.

C. Mechanism

Acid phosphate-catalyzed alkylation of amines may occur via any of several mechanisms. Although experiments to elucidate the mechanism have not been done, the results of our synthetic studies can be used to suggest a possible pathway. The three most probable mechanistic alternatives are:

1. Phosphate ester. This route entails formation of a phosphate ester on the catalyst surface and nucleophilic attack by an amine. Subsequent desorption would provide the higher amine and regenerate the catalytic site. Phosphate is a good leaving group, and the ability of phosphate esters to function as alkylating agents is well known in solution chemistry.[12]
2. Acid catalysis. Metal acid phosphates may act as selective acid catalysts, in which the acid sites are active for alkylation but not product equilibration.
3. Phosphoramide. Metal acid phosphates may function by formation of a phosphoramide on the catalyst surface. Subsequent attack of an alcohol and hydrolysis of the amide would form the product and regenerate the catalytic site. Reaction of ammonia with phosphates generates surface phosphoramides.[13,14] Further, reaction of a phosphoramide with an alcohol to form an amine has been observed in the solution phase reaction of hexamethylphosphoramide with *p*-chlorobenzyl alcohol.[15]

Our results favor the phosphate ester mechanism (Scheme 3). Despite their high catalytic activity, acid phosphates such as strontium hydrogen phosphate are very weakly acidic (pH ~ 6.0, as measured with Hammett indicators). Therefore, catalyst activity is probably not the result of catalyst acidity. However, the facility with which TMHEDA cyclizes strongly supports alcohol activation as a component of the mechanism. This result also precludes a phosphoramide intermediate since a tertiary amine cannot form a phosphoramide. Consequently, metal acid phosphate-mediated amine syntheses probably

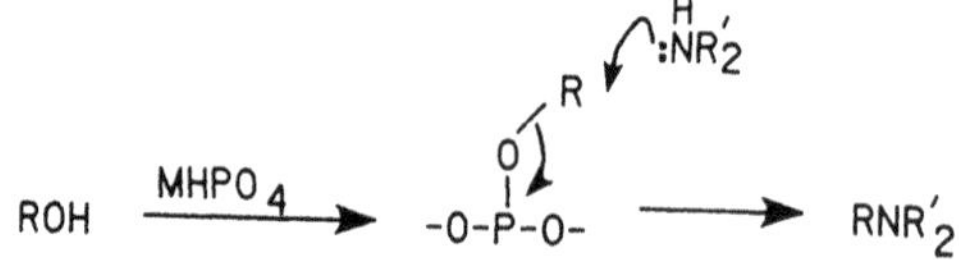

Scheme 3 Phosphate ester intermediate in amine synthesis.

proceed via formation of a phosphate ester with subsequent nucleophilic substitution of the activated alcohol by an amine.

D. Polyethylene Amines

Polyethylene amines are traditionally made by reaction of ammonia with ethylene dichloride.[16,17] Neutralization with aqueous caustic provides ethylenediamine, diethylenetriamine, piperazine, and higher linear, branched, and cyclic homologs. Of these products, diethylenetriamine and higher noncyclics are preferred for a multiplicity of uses. The handling and environmentally acceptable disposal of coproduct sodium chloride (up to 1.9 lb/lb of polyethylene amines) constitute an increasingly severe limitation of this technology.

A liquid phase phosphate-catalyzed route from ethanolamine and ethylenediamine to polyethylene amines was discovered by researchers at Texaco and claimed in a series of patents beginning in 1977. Simple phosphates, such as boron phosphate, are exemplified in the earlier disclosures.[18,19] Recently, liquid phase production of polyethylene amines with supported phosphate catalysts (e.g., on titania or zirconia) was demonstrated.[20-23]

Strontium hydrogen phosphate, lanthanide acid phosphates, and supported phosphoric acid (on silica gel) are effective catalysts for the production of polyethylene amines.[24-26] However, as a result of catalyst solubility in liquid polyethylene amines, continuous operation of this process in a packed-bed reactor requires vapor phase reaction. With vapor phase operation, catalyst integrity is maintained, and phosphate leaching into the product is insignificant.[27,28]

In the phosphate-catalyzed route to polyethylene amines, ethanolamine alkylates ammonia or a lower amine to form the next higher homolog (Scheme 4). Owing to the relatively low nucleophilicity of ammonia, formation of ethylenediamine by reaction of ammonia with ethanolamine is significant only at high pressure (>1000 psig). Reaction of ethanolamine with ethylenediamine forms diethylenetriamine. Chain extension of diethylenetriamine occurs at both the primary and secondary amines to form linear and branched

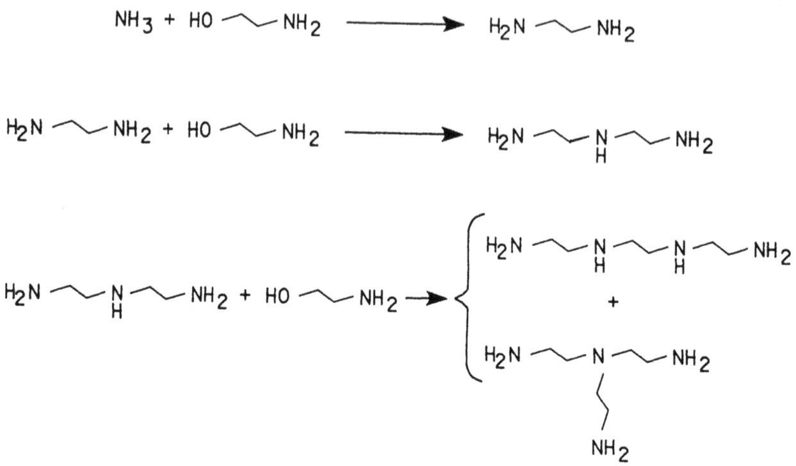

$$NH_3 + HO\text{—}NH_2 \longrightarrow H_2N\text{—}NH_2$$

$$H_2N\text{—}NH_2 + HO\text{—}NH_2 \longrightarrow H_2N\text{—}NH\text{—}NH_2$$

$$H_2N\text{—}NH\text{—}NH_2 + HO\text{—}NH_2 \longrightarrow \begin{cases} H_2N\text{—}NH\text{—}NH\text{—}NH_2 \\ + \\ H_2N\text{—}N(\text{—}NH_2)\text{—}NH_2 \end{cases}$$

- FORMATION OF EDA DEPENDS ON PRESSURE
- DEAMINATIVE CYCLIZATION NOT OBSERVED <275 DEG C

Scheme 4 Reaction pathways in the formation of polyethylene amines from ethanolamine.

tetramines. Further reaction leads to linear and branched pentamines and higher polyethylene amines. In agreement with our results on the synthesis of functionalized amines with strontium hydrogen phosphate, deaminative cyclization, e.g., of diethylenetriamine to piperazine, is not observed at temperatures below 275°C.

In addition to possessing alcohol functionality, ethanolamine also is an amine. Thus, it undergoes self-alkylation to form aminoethylethanolamine. Cyclization of aminoethylethanolamine proceeds readily to form piperazine, the precursor of higher cyclic polyethylene amines via further reaction with ethanolamine. The extent to which this pathway is followed is controlled by the ethylenediamine/ethanolamine feed ratio. In practice, use of molar ethylenediamine/ethanolamine feed ratios of at least 2:1 and reaction temperatures below 270°C provides acceptable product slates.

To determine the effect of selected process variables (temperature, feed ratio, and space velocity) on conversion and selectivity,

Table 2 Effect of Feed Ratio on Polyethylene Amine Preparation

	Molar feed ratio		
Component	E/M = 2:1	E/M = 5.5:1	E/M = 9:1
Piperazine	3.62	2.19	1.60
Aminoethylethanolamine	3.62	1.32	0.48
Diethylenetriamine	50.94	64.92	72.18
Aminoethylpiperazine	5.27	2.44	1.53
Noncyclic-teta[a]	17.16	15.17	12.88
Cyclic-teta[b]	3.93	1.77	1.13
Noncyclic-tepa[c]	6.96	4.34	2.09
Cyclic-tepa[d]	2.48	1.02	0.50
Highers[e]	1.80	—	—
Conversion[f]	42.0	49.0	58.3
Total noncyclics	75.1	84.4	87.2

[a]Total noncyclic isomers of triethylenetetramine.
[b]Total cyclic isomers of triethylenetetramine.
[c]Total noncyclic isomers of tetraethylenepentamine.
[d]Total cyclic isomers of tetraethylenepentamine.
[e]Pentameric and higher oligomeric polyethylene amines.
[f]Based on ethanolamine.

a Box-Behnken statistical design was carried out. Representative conversions of ethanolamine and product distributions for ethylenediamine/ethanolamine feed ratios of 2:1, 5.5:1, and 9:1 are shown in Table 2. As the ethylenediamine/ethanolamine feed ratio is raised, selectivity to diethylenetriamine and total noncyclic products increases.

In conclusion, we have found that metal acid phosphates, and especially strontium hydrogen phosphate, selectively catalyze alkylation of amines. Simple and polyfunctional mixed amines are formed with high selectivity, since product equilibration does not occur. Alkylation occurs via a stepwise process which can be controlled to give the desired unsymmetric amine by proper selection of process variables. Metal acid phosphates are effective catalysts for commercial production of 1,4-diazabicyclo[2.2.2]octane (DABCO), mixed alkylamines, functionalized tertiary amines, and polyethyleneamines.

III. EXPERIMENTAL

A. Materials

Dimethylamine and ethylamine were obtained from Matheson Gases and distilled before use. Morpholine, piperazine, ethylenediamine, ethanolamine, and the alcohol feeds were obtained from Aldrich Chemical and were used without further purification. Diethylene glycol dimethyl ether (gold label grade) was also purchased from Aldrich and used as received. Lanthanum nitrate hexahydrate and ammonium dihydrogen phosphate were obtained from Alfa Products and were used as received. Phosphoric acid on silica (0.3 g/cm^3) was obtained from Davison Chemical. Strontium hydrogen phosphate[10] and lanthanum hydrogen phosphate[29] were prepared according to published procedures.

B. Catalyst Evaluation

The performance of strontium hydrogen phosphate,[11] lanthanum hydrogen phosphate,[29] and phosphoric acid on silica[29] was evaluated in continuous fixed-bed reactors according to the referenced procedures.

Product analyses were carried out with a Varian model 4600 gas chromatograph equipped with a model 8000 autosampler and a thermal conductivity detector; and a Varian model 3700 gas chromatograph equipped with a capillary injection port for use with fused silica columns, model 8000 autosampler, and a flame ionization detector. Separations were effected with a copper 20% UNCONLB 1800x (2% KOH) on Chrom P packed column (20 ft x 1/8 in.) (methanol/ethylamine reaction); a copper Carbowax 20 M on Gas Chrom Q packed column (6 ft x 1/8 in.) (polyfunctional amines); and a fused-silica DB-5 wall-coated open-tubular column (30 m x 0.32 mm; 1.0 mm film thickness) (polyethyleneamines). Quantitation was based on the use of diethylene glycol dimethyl ether as an internal standard. Autosamplers were controlled and the integrations performed by a Varian VISTA 402 chromatography data system. Identities of reaction products were confirmed by gas chromatography/mass spectrometry.

ACKNOWLEDGMENTS

We thank Drs. Dale D. Dixon and Jeff W. Labadie for permission to cite their results on the preparation of mixed alkylamines with strontium hydrogen phosphate. In addition, we are indebted to our coworkers Mark Connor, Dr. Cawas Cooper, Mike Green, Dr. Kathy Hayes, George Korpics, Ken Lumer, Joan Premecz, John Steinmacher, and Mike Turcotte for their many technical contributions.

REFERENCES

1. R. B. Wagner and H. D. Zook, *Synthetic Organic Chemistry*, John Wiley and Sons, New York, 1953, pp. 653-727.
2. C. A. Buehler and D. E. Pearson, *Survey of Organic Synthesis*, Vol. 1, Wiley-Interscience, New York, 1970, pp. 411-494; Vol. 2, New York, 1977, pp. 391-447.
3. J. Volf and J. Pasek, *Studies in Surface Science and Catalysis*, Vol. 27 (L. Cerveny, ed.), Elsevier, New York, 1986, pp. 105-144.
4. A. E. Schweizer, R. L. Fowlkes, J. H. McMakin, and T. E. Whyte, *Encyclopedia of Chemistry and Technology*, Vol. 2 (H. F. Mark, D. F. Othmer, C. G. Overberger, and G. T. Seaborg, eds.), John Wiley and Sons, New York, 1978, pp. 272-283.
5. I. S. Bechara, B. Milligan, and M. H. Ziv, U.S. Patent 3,843,648 (1974).
6. M. E. Brennan, U.S. Patent 4,103,087 (1978).
7. W. H. Brader and R. L. Rowton, U.S. Patent 3,297,701 (1967).
8. W. H. Brader, U.S. Patent 3,342,820 (1967).
9. W. H. Brader, U.S. Patent 3,386,800 (1968).
10. J. E. Wells, U.S. Patent 4,405,784 (1983).
11. J. W. Labadie and D. D. Dixon, *J. Mol. Catal.*, *42*, 367 (1987).
12. A. J. Kirby and W. P. Jenks, *J. Am. Chem. Soc.*, *87*, 3209 (1965).
13. J. B. Moffat and J. F. Neeleman, *J. Catal.*, *39*, 419 (1975).
14. L. E. Kitaev, A. A. Kubasov, and K. V. Topchieva, *Kinet. Katal.*, *17*, 780 (1976).
15. R. S. Monson and D. N. Priest, *J. Chem. Soc. Chem. Commun.*, 1018 (1971).
16. M. Lichtenwalter and T. H. Cour, U.S. Patent 3,484,488 (1969).
17. C. M. Barnes and H. F. Rose, *Ind. Eng. Chem., Prod. Res. Dev.*, *20*, 399 (1981).

18. M. E. Brennan and E. L. Yeakey, U.S. Patent 4,036,881 (1977).

19. M. E. Brennan, P. H. Moss, and E. L. Yeakey, U.S. Patent 4,044,053 (1977).

20. S. H. Vanderpool, U.S. Patent 4,588,842 (1986), and references therein.

21. L. W. Watts and S. H. Vanderpool, U.S. Patent 4,609,761 (1986).

22. T. L. Renken, U.S. Patent 4,612,397 (1986).

23. S. H. Vanderpool, U.S. Patent 4,698,427 (1987).

24. T. A. Johnson and M. E. Ford, U.S. Patent 4,463,193 (1984).

25. M. E. Ford and T. A. Johnson, U.S. Patent 4,503,253 (1985).

26. M. E. Ford and T. A. Johnson, U.S. Patent 4,605,770 (1986).

27. M. E. Ford and T. A. Johnson, U.S. Patent 4,617,418 (1986).

28. M. G. Turcotte, C. A. Cooper, M. E. Ford, and T. A. Johnson, U.S. Patent 4,720,588 (1988).

29. M. E. Ford, T. A. Johnson, J. E. Premecz, and C. A. Cooper, *J. Mol. Catal., 44,* 207 (1988).

15

Amination with Zeolites

M. DEEBA

Engelhard Corporation, Edison, New Jersey

MICHAEL E. FORD and THOMAS A. JOHNSON

Air Products and Chemicals, Inc., Allentown, Pennsylvania

ABSTRACT

Zeolites have attracted considerable attention as catalysts as a result of their high activity and unusual selectivity for acid-catalyzed reactions. However, these properties have been applied almost exclusively to hydrocarbon processing. In this chapter, we summarize initial results in the exploration of a new and potentially broad area of acid catalysis with zeolites, i.e., synthesis of amines. Zeolite-catalyzed amination of ethylene, propylene, and isobutylene to the corresponding primary amines is described. Despite the high reaction temperatures employed, selectivity to total amine products, and specifically to the primary amines, is high. Olefin amination occurs via Markownikoff addition. Reaction is believed to involve a carbocationic intermediate which is formed by the interaction of the olefin with a surface proton or ammonium ion. Catalyst activity is directly proportional to the total number of strongly acidic sites as measured by ammonia chemisorption. The highest activities were obtained with small- to medium-pore acidic zeolites, such as H-clinoptilolite, H-erionite, and H-offretite. The necessity of strongly acidic sites for catalytic amination is demonstrated by the negligible activity of alkali-exchanged zeolites and amorphous silica alumina. Thermodynamic constraints on the reaction are discussed. In the presence of a wide range of acidic zeolites, methanol is aminated to form mixtures of mono-, di-, and trimethylamine. All zeolites examined show higher rates of methanol conversion than amorphous silica-alumina as a result of their higher acidity. High selectivities to mono- and dimethylamine are obtained with the small-pore zeolites H-chabazite and H-erionite. With the latter catalysts, shape selectivity induced by the zeolite structure completely suppresses formation of trimethylamine under typical operating conditions.

I. INTRODUCTION

Conventional routes for the large-scale production of lower alkylamines involve reaction of an alcohol with ammonia in the presence of either acidic or supported metal catalysts.[1] The composition of the product mixture is thermodynamically controlled.[2,3] With the obvious exception of methanol, lower alcohols are typically obtained industrially by acid-catalyzed hydration of the corresponding olefin.[4] In contrast, amination of an olefin to the corresponding amine would provide a more direct route to these commercially useful products. Direct conversion of ethylene to a mixture of ethylamines was initially achieved with alkali metal amide catalysts.[5-8] Satisfactory yields (up to 70%) are obtained with ethylene. However, this process provides low yields of propylamines (less than 19%) owing to the lower susceptibility of propylene to nucleophilic attack.[5-8] With higher olefins, this mode of amination is rendered impractical by complications that include isomerization and/or polymerization of the starting olefin, and disproportionation of the product amines.[5] Despite extended study by many workers,[9] reaction of olefins with ammonia over transition metals or transition metal oxides provides mixtures of the corresponding alkane and the next higher nitrile (e.g., ethane and propionitrile from ethylene and ammonia). Nitrile formation may occur via metal-catalyzed olefin cyanation.[9] Recently, ammonia has been added to simple olefins with ammonium halide[10] and chromium-exchanged borosilicate catalysts.[11] In the former instance, mixtures of the corresponding primary, secondary, and tertiary amines are obtained, probably via the intermediacy of the corresponding alkyl halides. In the latter process, isobutylene is converted to t-butylamine. However, owing to the low reactivity of borosilicates, forcing conditions (300°C/300 atm) are required. Zeolites have attracted considerable attention as catalysts for more than two decades as a result of their high activity and unusual selectivity for acid-catalyzed reactions.[12] However, these properties have been applied almost exclusively to hydrocarbon processing. In this chapter, amination of olefins (ethylene, propylene, isobutylene) and lower alcohols (methanol, ethanol) to the corresponding amines with acidic zeolite catalysts is described. The influence of catalyst parameters (acidity, pore size) and reaction temperature on the conversion and selectivity of amine production has been studied. Thermodynamic limitations of acid-catalyzed olefin amination have been explored.

II. RESULTS AND DISCUSSION—AMINATION OF OLEFINS

A. Effect of Catalyst Pore Structure

Comparison of the effective pore size of representative zeolites with the kinetic diameters of ammonia and the olefin substrates (Fig. 1) indicates that shape-selective amination should be feasible. Thus, isobutylene should undergo amination only with large-pore zeolites, such as H-Y. In contrast, ethylene should be aminated by a wide range of catalysts.

B. Effect of Reaction Temperature

Conversion of ethylene, propylene, and isobutylene to the corresponding amines as a function of temperature over the representative catalysts, H-Y, H-mordenite, and H-erionite, is compared in Figs. 2, 3, and 4, respectively. A minimum temperature of 320°C was required to observe significant ethylene conversion. Mono- and diethylamines were the main products (greater than 98 wt % selectivity) up to 380°C.

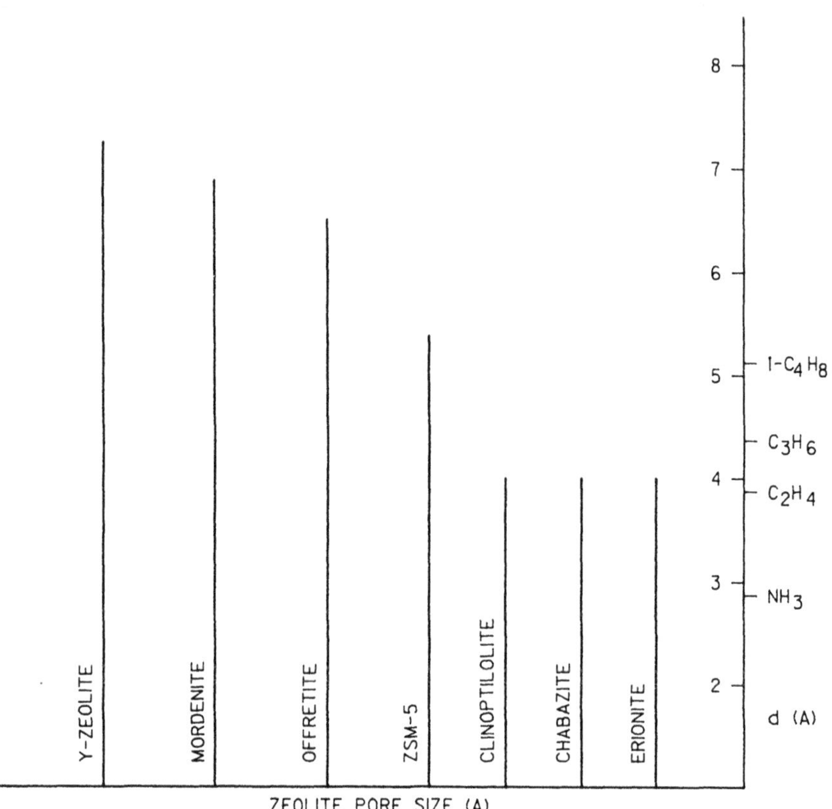

Fig. 1 Steric constraints on zeolite reactivity.

Somewhat better selectivities to monoethylamine were obtained with the small-pore catalysts (H-erionite and H-clinoptilolite) than with the large-pore H-Y (see Table 1). At temperatures above 380°C, formation of nitriles and higher olefins becomes significant. Amination of propylene was observed at 300°C. Monoisopropylamine was the major product (greater than 97 wt % selectivity to total amine; see Table 1). Isobutylene was selectively converted to t-butylamine (greater than 98% yield) between 200 and 300°C. Although ethylene was effectively aminated by all catalysts, small-pore zeolites such as H-erionite or H-clinoptilolite provided low

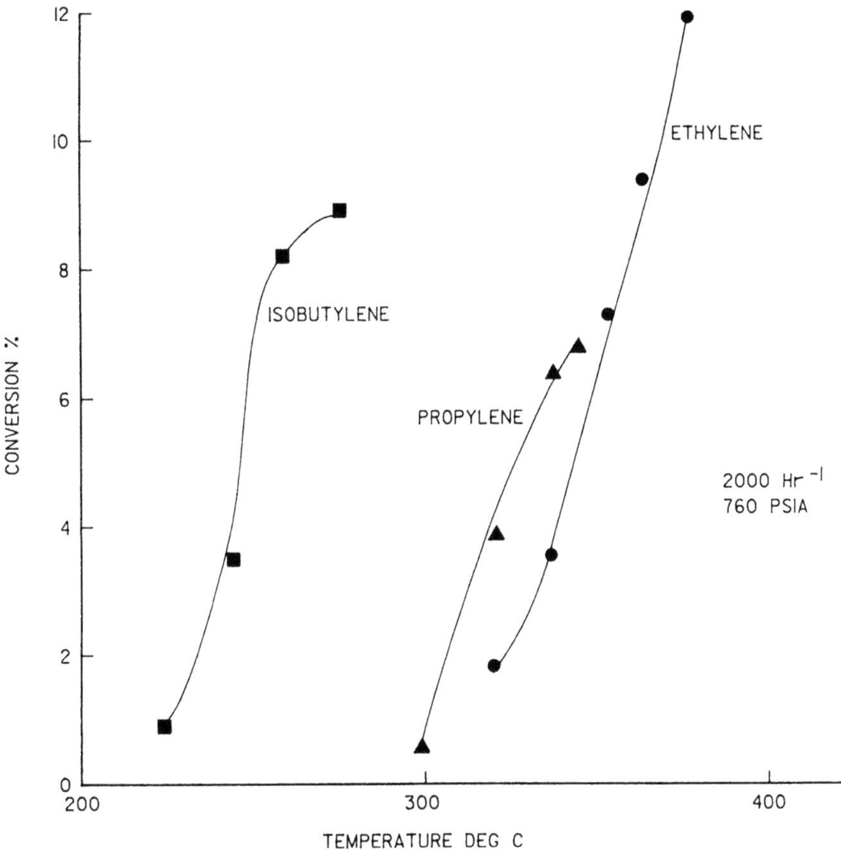

Fig. 2 Conversion of ethylene (●), propylene (▲), and isobutylene (■) over H-Y zeolite at 760 psia with 4:1 ammonia/olefin feed ratio at 2000/hr (GHSV at STP).

conversions of propylene and no detectable conversion of isobutylene (Fig. 4).

C. Correlation of Acidity to Catalyst Activity

Activity of the zeolites investigated was directly proportional to the number of strong acid sites, as determined by ammonia TPD (Fig. 5). The amount of ammonia chemisorbed at 200°C was considered an indicator of the strongly acidic sites. Possibly as a result of

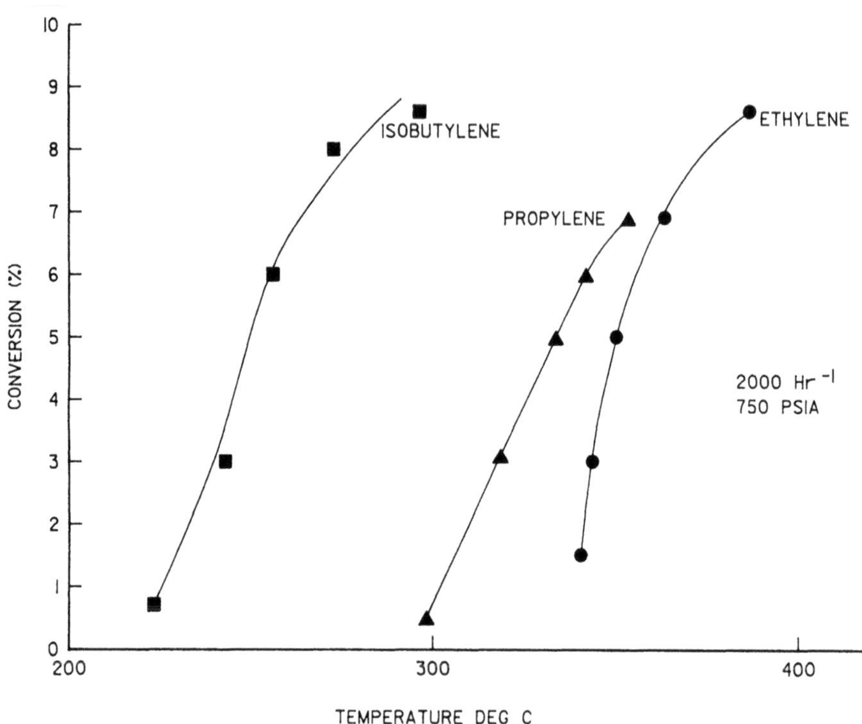

Fig. 3 Conversion of ethylene (●), propylene (▲), and isobutylene (■) over H-mordenite at 760 psia with 4:1 ammonia/olefin feed ratio at 2000/hr (GHSV at STP).

partial plugging of its pore structure by adventitious amorphous alumina, H-mordenite showed lower activity than anticipated. Owing to the larger critical dimensions of propylene, isobutylene, and their reaction products, the dependence of catalyst activity for the production of higher amines was difficult to deduce (cf. Fig. 4).

D. Mechanism of Olefin Amination

Activity of zeolites as catalysts for olefin amination results from the highly acidic nature of proton-exchanged zeolites. At room temperature, ethylene and propylene are reversibly adsorbed by acidic zeolites.[13] Changes in the infrared frequencies of acidic surface

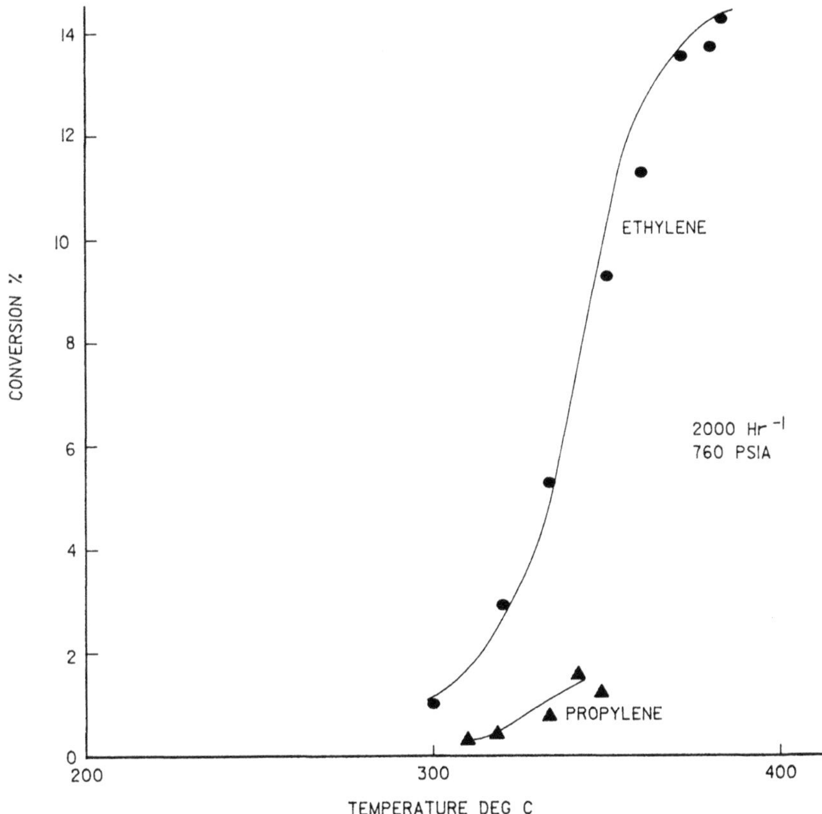

Fig. 4 Conversion of ethylene (•) and propylene (▲) over H-erionite at 760 psia with 4:1 ammonia/olefin feed ratio at 2000/hr (GHSV at STP).

hydroxy groups during this process are indicative of hydrogen bonding and stronger adsorption of propylene (the more basic olefin). Formation of a π complex between the surface hydroxy group and the olefin has been implicated as the mechanism of olefin chemisorption.[13,14] At elevated temperatures, reaction of the π complex with ammonia, adsorbed on the catalyst surface or from the gas phase, forms the adsorbed amine. Subsequent product desorption would regenerate the catalytic site. Intermediacy of a cationic species is further

Table 1 Product Selectivity in Olefin Amination

Amine	Selectivity		Catalyst
	Primary	Secondary	
Ethylamine	87	13	H-Y
	≥93	≥7	H-Offretite
			H-Clinoptilolite
i-Propylamine	≥94	≥6	H-Y
			H-Mordenite
t-Butylamine	≥98	Not observed	H-Y

supported by the relative ease of amination: isobutylene > propylene > ethylene (see Figs. 2 and 3). The necessity of strongly acidic sites, and thus of a protonated intermediate, is demonstrated by the negligible activity of the weakly acidic amorphous silica-alumina (Fig. 5).

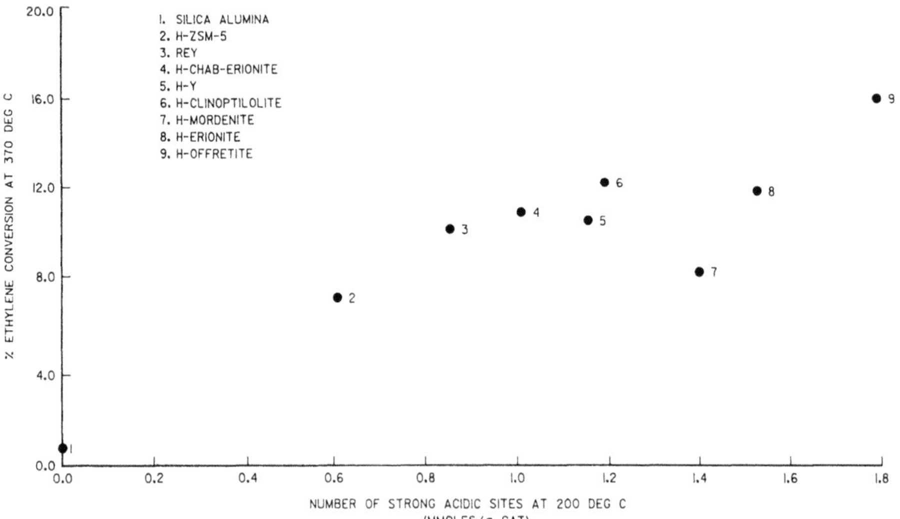

Fig. 5 Correlation of activity for ethylene amination (conversion at 370°C with 4:1 ammonia/ethylene feed; 760 psia; 1000/hr, GHSV at STP) with catalyst acidity (mmol ammonia chemisorbed at 200°C/g of catalyst).

Further, nonacidic sodium-exchanged offretite and Y zeolites were inactive for olefin amination.

E. Thermodynamic Limitation

Conversion of olefins to the corresponding amines is limited by thermodynamic equilibrium between reactants and products. Calculated[15] equilibrium conversions indicate that amination is favored by low temperature, high pressure, and a high ammonia/olefin ratio. However, high reaction temperatures are required to activate simple olefins. The temperature of activation depends on the structure of the olefin. Thus, ethylene is aminated only at high temperatures (over 320°C) by the strongest acid sites. Experimental data for amination of ethylene over H-Y zeolite with a 4:1 molar ratio of ammonia to ethylene at 750 psia show that thermodynamic equilibrium is attained at 390°C (Fig. 6). Propylene, which is more basic and forms a more stable cationic intermediate, is activated by somewhat weaker acid sites (those reactive

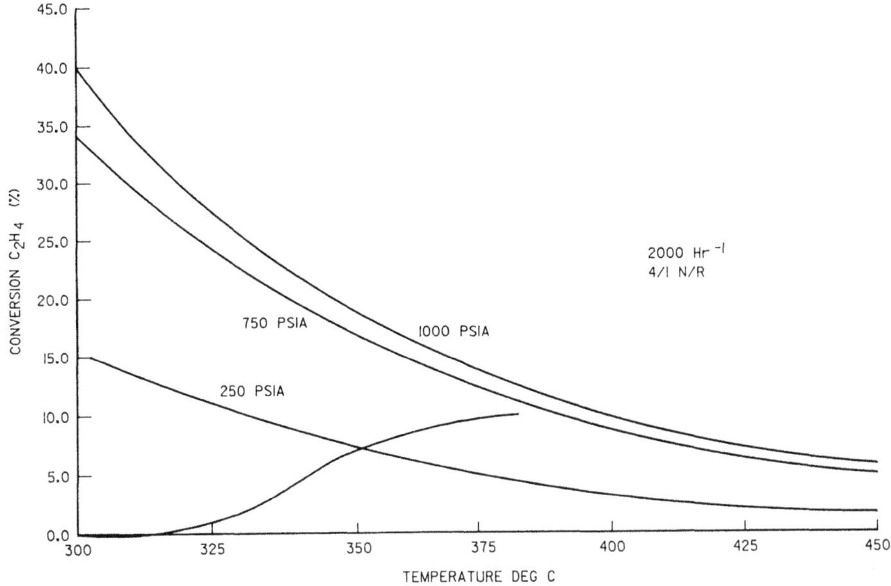

Fig. 6 Comparison of calculated thermodynamic equilibrium conversion with observed ethylene conversion over H-Y zeolite as a function of temperature at 4:1 ammonia/ethylene, 750 psia, and 2000/hr GHSV at STP.

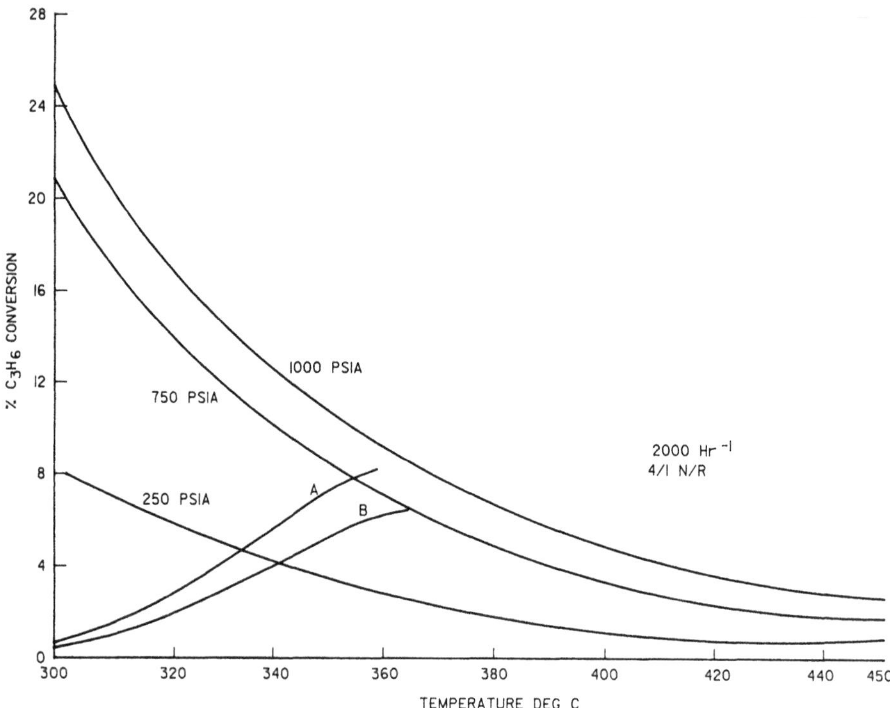

Fig. 7 Comparison of calculated thermodynamic equilibrium conversion with observed propylene conversion over H-Y zeolite as a function of temperature at 4:1 ammonia/propylene. (A) 1000 psia or (B) 750 psia. 2000/hr GHSV at STP.

at or above 300°C). As reaction temperature is increased, propylene conversion increases and approaches thermodynamic equilibrium conversions at 370°C, as shown for reaction over H-Y at 750 and 1000 psia (Fig. 7). Isobutylene, which readily forms a cationic intermediate, is activated at temperatures as low as 220°C. Conversion to *t*-butylamine passes through a maximum at 260-280°C and decreases to the equilibrium limit as the reaction temperature is raised to 300°C (Fig. 8).

F. Catalyst Deactivation and Regeneration

The rate and nature of catalyst deactivation during the ethylene amination depended on the zeolite pore structure. Linear deactiva-

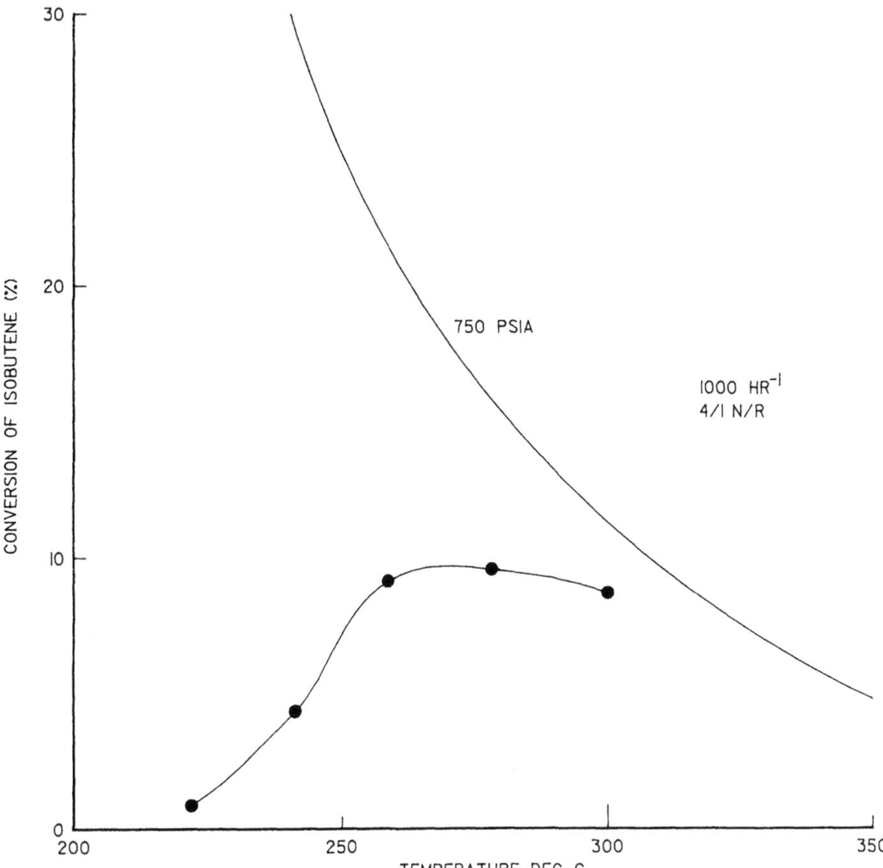

Fig. 8 Comparison of calculated thermodynamic equilibrium conversion with observed isobutylene conversion over rare earth-exchanged Y zeolite (SK-500) as a function of temperature at 4:1 ammonia/isobutylene, 750 psia, and 1000/hr GHSV at STP.

tion with a continuous, concurrent decline in selectivity to total ethylamines was observed with H-Y (Fig. 9). In contrast, after an initial loss of activity, conversion of ethylene over H-erionite remained constant for 90 days (Fig. 10). During this period, selectivity to total ethylamines improved slightly. X-ray diffraction studies of both used catalysts showed no loss of crystallinity. In

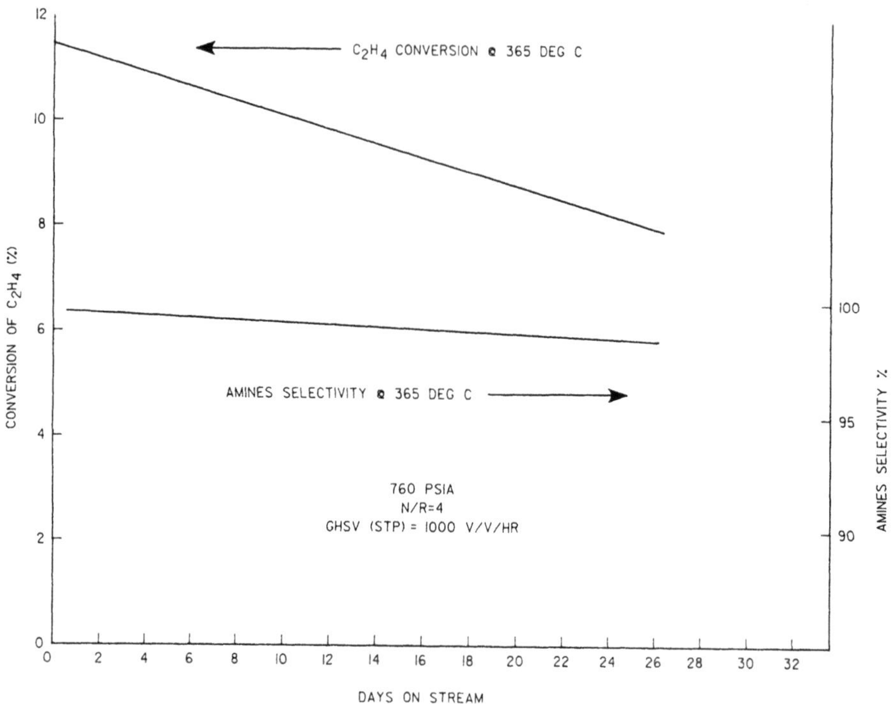

Fig. 9 Dependence of conversion and selectivity of ethylene amination with H-Y catalyst (4:1 ammonia/ethylene, 760 psia, and 1000/hr GHSV at STP).

addition, typical catalyst poisons (such as sodium) were not detected. The cause of deactivation for all zeolites was the acid-catalyzed formation of hydrogen-deficient organic residues within the zeolite structure. Owing to its small-pore structure, which prevents the formation of high molecular weight organic residues, H-erionite showed no ongoing deactivation after 15 days on-stream. In contrast, the large-pore structure of H-Y permitted continued formation of organic byproducts.

Catalyst regeneration was readily accomplished by standard combustion techniques.[16] For example, heating used (90 days operation)

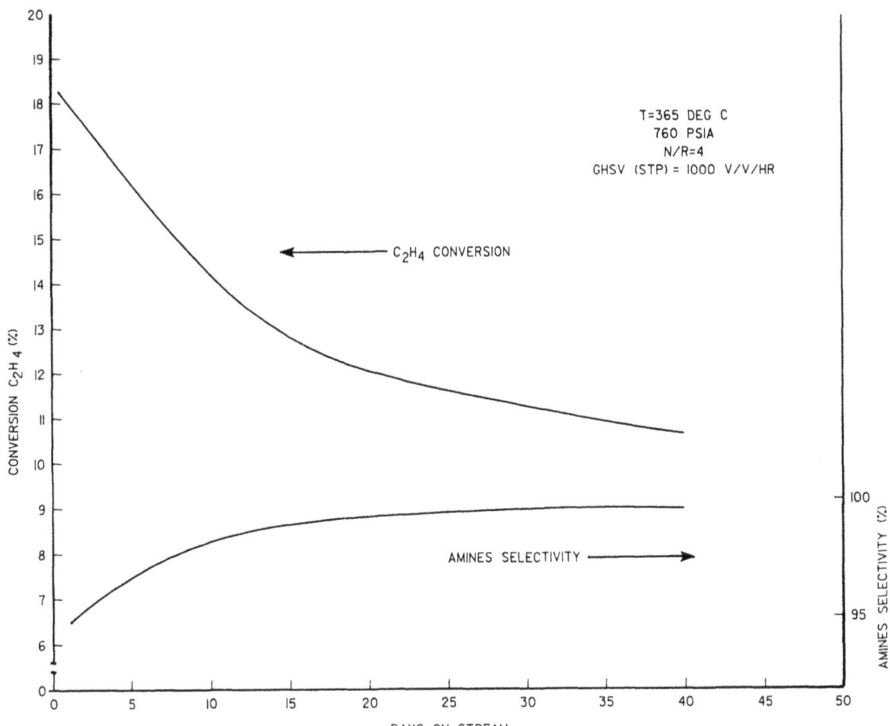

Fig. 10 Dependence of conversion and selectivity of ethylene amination with H-erionite catalyst (4:1 ammonia/ethylene, 760 psia, and 1000/hr GHSV at STP).

H-erionite in air at 500°C restored its acidic and catalytic properties for ethylene amination (Table 2). This treatment removed 3 wt % of carbon from the used catalyst.

Table 2 Regeneration of H-Erionite

	Catalyst		
	Fresh	Used (90 days)	Heat-treated
Acid sites (mmol ammonia/g)	1.3	1.0	1.3
Conversion (ethylene/375°C)	15.0	9.0	13.5

III. RESULTS AND DISCUSSION—AMINATION OF ALCOHOLS

A. Introduction

In olefin amination, zeolites were found to be strongly acidic catalysts for selective production of the corresponding primary amine. In conventional practice with amorphous acidic or metallic catalysts, alcohol amination is slow whereas thermodynamic equilibration of the products is rapid. Recently, however, DuPont researchers reported selective production of dimethylamine via amination of methanol with small-pore zeolites.[2,3] H-Rho, H-chabazite, and HZK-5 were particularly effective. At 325°C, these zeolites provided higher reactivity and selectivity than the amorphous catalysts in current commercial use.

B. Effect of Catalyst Acidity and Pore Structure

To explore the scope of zeolite acidity and shape selectivity for alcohol amination, selected zeolites were evaluated as catalysts for the production of methyl- and ethylamines. Relative to silica-alumina, higher rates of methanol amination were found with representative large-pore (H-Y), medium-pore (H-mordenite), and small-pore (H-erionite) zeolites (Table 3). Product distributions were controlled by zeolite pore structure (Table 4). Thus monomethylamine was the major product obtained with H-erionite; no trimethylamine was formed. With H-mordenite, monomethylamine still predominated. However, the larger intracrystalline pore structure of H-mordenite

Table 3 Relative Rates of Methanol Amination[a]

	Temp. (°F)		
Catalyst	550	600	650
Silica-alumina	1	1	1
H-Erionite	15	9.5	4.6
H-Mordenite	14	12	4.7
H-Y	19	15	7.1

[a]All reactions carried out with 2:1 ammonia/methanol, 280 psia, 9300/hr GHSV at STP.

Table 4 Product Distributions in Methanol Amination[a]

Catalyst	Temp. (°F) 550	600	650
Silica-alumina	(59, 25, 15)	(35, 21, 43)	(31, 27, 42)
H-Erionite	(87, 13, 0)	(81, 19, 0)	(74, 26, 0)
H-Mordenite	(82, 18, 0)	(65, 22, 13)	(57, 25, 18)
H-Y	(58, 20, 22)	(43, 38, 19)	(37, 42, 21)

[a]Triads of numbers indicate selectivities to mono-, di-, and trimethylamines, respectively.

permitted formation of significant amounts of dimethylamine and (at 600 and 650°F) of trimethylamine. Owing to its relatively large-pore structure, little product selectivity was observed with H-Y.

In contrast to the results obtained from methanol amination, only H-Y showed a higher rate of ethanol amination than silica-alumina (Table 5). Lower relative rates of ethanol conversion were found with H-erionite and H-mordenite. Consequently, intracrystalline diffusivity (which is controlled by zeolite structure) is more important than acidity in determining the rate of ethanol amination. While the rates of reaction with H-erionite and H-mordenite were low, high selectivities to monoethylamine were observed (Table 6). Significantly less product selectivity was demonstrated by HY. These

Table 5 Relative Rates of Ethanol Amination[a]

Catalyst	Temp. (°F) 550	650
Silica-alumina	1	1
H-Erionite	0	0.8
H-Mordenite	1	0.6
H-Y	11	2.5

[a]All reactions carried out with 2:1 ammonia/ethanol, 280 psia, 9300/hr GHSV at STP.

Table 6 Product Distributions in Ethanol Amination[a]

Catalyst	Temp. (°F) 550	650
Silica-alumina	(63, 32, 5)	(53, 38, 9)
H-Erionite	—	(82, 18, 0)
H-Mordenite	(100, 0, 0)	(89, 11, 0)
H-Y	(87, 13, 0)	(70, 28, 2)

[a]Triads of numbers indicate selectivities to mono-, di-, and triethylamine, respectively.

results further illustrate and emphasize the importance of zeolite structure in determining the rate and selectivity of alcohol amination.

IV. CONCLUSION

In summary, zeolites are effective catalysts for production of lower aliphatic amines from either the corresponding olefin or alcohol. Direct amination of olefins has been demonstrated with zeolite catalysts. This process is characterized by strong acid catalysis to generate a carbocationic intermediate on the zeolite surface. This intermediate reacts further to form the product amine. With small- to medium-pore zeolites, such as erionite, steric constraints on amination of methyl-substituted olefins are observed. Moreover, owing to the high temperature required for olefin activation, conversion is limited by thermodynamic equilibrium between reactants and products.

Lower alcohols readily undergo zeolite-catalyzed amination. While the reaction is acid-catalyzed, steric constraints imposed by the zeolite pore structure control both rates of reaction and product selectivities. With respect to conventional amorphous catalysts, higher rates are obtained in the zeolite-catalyzed production of methylamines. In addition, the composition of the product mixtures can be varied as desired and is not subject to equilibrium control.

Table 7 Zeolites Evaluated for Amination Activity

Zeolite	Channel system (Å)[a]
Silica-alumina	Amorphous
Rare earth Y	7.4
H-chabazite-erionite	3.6 x 3.7
	3.6 x 5.2
H-Y[b]	7.4
H-clinoptilolite	4.0 x 5.5
	4.4 x 7.2
	4.1 x 4.7
H-mordenite	6.7 x 7.0
	2.9 x 5.7
H-erionite	3.6 x 5.2
H-offretite	6.4
	3.6 x 5.2

[a]W. M. Meier and D. H. Olson, *Atlas of Zeolite Structure Type,* Structure Commission of the International Zeolite Association, Zurich, 1978.
[b]Ion Exchange and Metal Loading Procedures, Union Carbide Catalyst Bulletin F-09.

V. EXPERIMENTAL

A. Materials

The zeolites studied are summarized in Table 7. All were commercially available except H-Y, which was prepared from the sodium form.[17] Ammonia and all olefins were obtained from Air Products (Specialty Gases, Hometown, PA) in high purity (greater than 99.9%). Certified ACS grade absolute methanol was supplied by VWR Scientific (Philadelphia, PA); absolute ethanol was obtained from Pharmco Products (Dayton, NJ). All commercial materials were used as received.

B. Equipment

Catalytic activity for olefin amination was evaluated in a Chemical Data Systems isothermal tubular reactor. The reactor consists of a

316 stainless steel tube (22.9 cm long x 0.64 cm ID) mounted inside a close-fitting metal block which is instrumented for temperature control. A thermowell extending axially next to the reactor was used to measure reactor temperature. With the exception of ethylene, feeds were metered to the reactor as liquids using Isco model 314 high-pressure syringe pumps. Ethylene was introduced as a gas and its flow was controlled with a Brooks model 5841 flow controller. Prior to introduction to the reactor, the feeds were vaporized and mixed in a countercurrent mixer maintained at 150°C. Flow rates of ammonia and the appropriate olefin were adjusted to obtain the desired molar ratio of reactants and total flow rate (GHSV at STP).

Catalytic reactivity for alcohol amination was evaluated in a Berty recycle reactor. Feeds were metered to the reactor as liquids using Isco model 314 high-pressure syringe pumps. Prior to introduction to the reactor, the feeds were vaporized and mixed in a countercurrent mixer maintained at 150°C. Flow rates of ammonia and the appropriate alcohol were adjusted to obtain the desired molar ratio of reactants and total flow rate (GHSV at STP).

Product analyses were carried out by on-line gas liquid chromatography with a Varian model 6000 gas chromatograph equipped with a 6 ft x 1/4 in. 20% Carbowax 20 M on Chromosorb T column and a Vista 402 chromatography data system. Identities of major products were confirmed by GC/MS.

C. Catalyst Evaluation

Catalyst samples (12- to 18-mesh particles) were heated (90°C) in the reactor under nitrogen (10 sccm at 1 atm) for 16-18 hr. The temperature was then raised to the desired level over 3 hr. Nitrogen was shut off. Ammonia was introduced and the desired pressure set with the back pressure regulator. Olefin or alcohol was then introduced to obtain the desired feed ratio.

D. Acidity Measurements

After activation at 400°C under nitrogen in a DuPont model 951 thermogravimetric analyzer, ammonia was adsorbed onto the catalyst at 20°C. TPD indicated the number of strong acid sites, given as the millimoles of ammonia chemisorbed at 200°C per gram of catalyst.

REFERENCES

1. A. E. Schweizer, R. L. Fowlkes, J. H. McMakin, and T. E. Whyte, *Encyclopedia of Chemistry and Technology,* Vol. 2 (H. F. Mark, D. F. Othmer, C. G. Overberger, and G. T. Seaborg, eds.), John Wiley and Sons, New York, 1978, pp. 272-283.

2. F. J. Weigert, *J. Catal., 103,* 20 (1987).

3. M. Keane, G. C. Sonnichsen, L. Abrams, D. R. Corbin, T. E. Gier, and R. D. Shannon, *Appl. Catal., 32,* 361 (1987); L. Abrams, M. Keane, and G. C. Sonnichsen, *J. Catal., 115,* 410 (1989) and references therein.

4. P. D. Sherman and P. R. Kavasmaneck,*Encyclopedia of Chemistry and Technology,* Vol. 9 (H. F. Mark, D. F. Othmer, G. C. Overberger, and G. T. Seaborg, eds.), John Wiley and Sons, New York, 1980, pp. 344-350.

5. B. W. Howk, E. L. Little, S. L. Scott, and G. M. Whitman, *J. Am. Chem. Soc., 76,* 1899.

6. R. D. Clossen, G. M. Napolitano, G. G. Ecke, and A. J. Kolka, *J. Am. Chem. Soc., 79,* 646 (1957).

7. H. Lehmkuhl and D. Reinehr, *J. Organometal. Chem., 55,* 215 (1973).

8. G. P. Pez, U.S. Patent 4,302,603 (1982).

9. D. D. Dixon and W. F. Burgoyne, *Appl. Catal., 20,* 79 (1986), and references therein.

10. D. M. Gardner, P. J. McElligott, and R. T. Clark, European Patent Appl. 200,923 (1986).

11. W. Hoelderich, V. Taglieber, H. H. Pohl, R. Kummer, and K. G. Baur, German Patent 3,634,247 (1987).

12. M. V. Twigg, *Catalysis and Chemical Processes* (R. Pearce and W. R. Patterson, eds.), John Wiley and Sons, New York, 1981, pp. 17-20.

13. B. V. Liengme and H. K. Hall, *Trans. Faraday Soc., 62,* 3229 (1966).

14. N. W. Cant and H. K. Hall, *J. Catal., 25,* 161 (1972).

15. D. R. Stull and H. Prophet, JANAF Thermochemical Tables, U.S. Department of Commerce, National Bureau of Standards, 2nd ed., 1971.

16. C. N. Satterfield, *Heterogeneous Catalysis in Practice,* McGraw-Hill, New York, 1980, pp. 272-274.

17. Ion Exchange and Metal Loading Procedures, Union Carbide Catalyst Bulletin F-09.

16

Ruthenium "Melt"-Catalyzed Oxonation of Terminal and Internal Olefins to Linear Aldehydes/Alcohols

JOHN F. KNIFTON

Texaco Chemical Company, Austin, Texas

ABSTRACT

Terminal and internal olefins may be hydroformylated to give oxo products through the use of ruthenium-catalyzed melt systems. The addition of N and P ligands such as 2,2'-bipyridine, 2,2'-bipyrimidine, and 1,2-bis(diphenylphosphino)ethane allows terminal olefins to be regioselectively oxonated yielding 99% linear oxoalcohols and aldehydes. Through judicious choice of both ligand and low melting point quaternary phosphonium salt, internal olefin substrates may also give predominantly linear oxoalcohols.

Spectral and kinetic data are discussed in terms of a proposed reaction scheme involving ruthenium cluster anions.

I. INTRODUCTION

As part of a program to study the application of melt catalysis to synthesis gas processing using group 8 metals,[1] we describe in this chapter our more recent research into the hydroformylation of internal and terminal olefin substrates to give predominantly linear oxoalcohol/aldehyde products. The principal thrust of this study has been to demonstrate that highly linear oxoalcohols and aldehydes with linear-to-branched (l/b) ratios of 100 may be readily prepared from terminal olefin feedstocks (Eq. 1) and that internal olefin feedstocks may also yield mainly linear oxo derivatives[2] using N- and P-ligand-modified ruthenium catalysts dispersed in low melting point quaternary phosphonium salts.[3,4]

Other objectives have been to:

1. Optimize regioselectivity to linear aldehydes/alcohols.
2. Maximize the yield of desired oxo products.
3. Gain some insight into the mode of ruthenium melt-catalyzed oxonation through reaction parameter studies, etc.

$$RCH{=}CH_2 + CO/H_2 \longrightarrow \begin{array}{c} RCH_2CH_2CHO \\ \\ \begin{array}{l} RCHCHO \\ \;| \\ \;CH_3 \end{array} \end{array} \xrightarrow{H_2} \begin{array}{c} RCH_2CH_2CH_2OH \\ \\ \begin{array}{l} RCHCH_2OH \\ \;| \\ \;CH_3 \end{array} \end{array} \qquad (1)$$

An important practical advantage of this new class of ruthenium hydroformylation catalyst is that, in contrast to analogous homogeneous systems, the ruthenium carbonyl species are stable in the polar quaternary phosphonium salt matrices, even under low-pressure/high-temperature conditions. Consequently, both the desired alcohol and aldehyde product components can be readily isolated from the crude product mix by fractional distillation and the residual melt catalyst recycled without loss of activity—even for higher molecular weight olefin substrates.

II. RESULTS

A. Synthesis

Typical hydroformylation activity is illustrated in Tables 1 and 2 for a series of model C_3-C_8 terminal α-olefin and C_8 internal, linear backbone, olefin mixtures using as catalyst precursor various ruthenium sources dispersed in quaternary phosphonium salts such as tetrabutylphosphonium bromide. Preferably, the ruthenium is employed

Table 1 1-Octene and Propylene Hydroformylation[a,b]

Ex.	Catalyst precursor	Reaction media	α-Olefin substrate	Liquid product composition (%)[c]						Nonanol linearity (%)
						Nonanal		Nonanol		
				Octenes	Octane	Linear	Branched[d]	Linear	Branched[e]	
1	RuO_2-2,2'-bipyridine	Bu_4PBr	1-Octene	9.0	9.1	0.2	0.2	68.7	10.9	86
2	RuO_2-1,10-phenanthroline	"	"	4.2	11.4	0.3	1.7	55.7	18.9	75
3	RuO_2-2,3'-bipyridine	"	"	<1	16.4	0.3	1.5	47.9	22.1	68
4	RuO_2-2,4'-bipyridine	"	"	<1	22.4	0.1	0.1	38.5	28.6	57
5	RuO_2-2,2'-dipyridylamine	"	"	9.7	8.4	10.6	8.7	42.5	7.9	84
6	RuO_2-2,2',2"-tripyridine	"	"	6.0	12.0	10.3	6.4	40.2	14.0	74
7	RuO_2-2,2'-bipyrimidine	"	"	16.6	5.8	0.6	0.3	52.2	9.4	85
8	$Ru_3(CO)_{12}$-2,2'-bipyrimidine	"	"	27.2	5.1	0.1		57.2	3.6	94[f]
9	$Ru_3(CO)_{12}$•2,2'-bipyridine	"	"	9.6	11.3	4.2	4.5	44.9	7.5	86
10	RuO_2-$Ph_2PCH_2PPh_2$	"	"	36.9	16.6	4.5	2.4	29.9	4.6	87
11	RuO_2-$PhP(CH_2CH_2PPh_2)_2$	"	"	91.0	2.8	1.0	0.3	1.9	0.4	
				43.3	17	4.7	0.7	37.5	2.2	94[g]
12	RuO_2-2,2'-bipyrimidine	"	Propylene			6.4[h]	<0.1[h]	89.5[h]	<0.1[h]	>99

[a]Typical run charge: Ru, 6.0 mmol; Ru/N or Ru/P, 1:2; Bu_4PBr, 10.0 g; 1-octene, 200 mmol.
[b]Run conditions: 180°C; 1200 psi (CO/H_2, 1:2); 4 hr.
[c]Liquid products analyzed by GLC, identified by GLC/MS, GLC/FTIR
[d]2-Methyloctanal, 2-ethylheptanal, 2-propylhexanal.
[e]2-Methyloctanol, 2-ethylheptanol.
[f]Run at 160°C.
[g]Two-phase liquid product.
[h]Butanal and butanol products.
Source: Ref. 12.

Table 2 Mixed Internal Octenes Hydroformylation[a,b]

Ex.	Catalyst composition	Media	Octene conv. (%)	Product composition: C_9 Alcohol		C_9 Aldehyde		Total C_9 alcohol + aldehyde yield (mol %)
				Productivity[c] (mol %)	Linearity[d] (%)	Productivity[c] (mol %)	Linearity[d] (%)	
13	$Ru_3(CO)_{12}$	Bu_4PBr	>99	65	42	0.8	50	66
14	$Ru(acac)_3$	"	>99	65	40	1.0	50	66
15	RuO_2-BIPY	"	52	35	69	5.4	68	79
16	RuO_2-DIPHOS	"	29	15	55	2.0	55	59
17	RuO_2-$Co_2(CO)_8$-Bu_3P	"	53	9.1	46	25	46	64
18	RuO_2	Bu_4POAc	56	25	69	4.3	68	52

[a] Typical run charge: Ru, 6.0 mmol; Bu_4PBr, 20.0 g; octenes, 200 mmol; conditions: 180°C, CO/H_2 (1:2).
[b] Internal octenes: 1-octene, 2.5%, *cis* and *trans*-2-octene, 37%, *cis*- and *trans*-3-octene, 38%, 4-octenes, 23%.
[c] Productivity, basis octene charged.
[d] Linearity calculated basis: linear alcohol or aldehyde/total alcohol or aldehyde.
Source: Ref. 10.

in conjunction with certain N- and P-ligand structures (ex. 1-12); optionally a cobalt carbonyl cocatalyst may be added (ex. 17). Hydroformylation is facile under moderate syngas pressures (turnover frequencies of >100 mol/g-atom Ru/hr) and the major products are generally linear alkanols, although aldehydes may dominate under certain conditions (e.g., see ex. 17).

The highest level of α-olefin regioselectivity has been achieved in this program using ruthenium in combination with chelating N ligands such as, for example, the triruthenium dodecacarbonyl-2,2'-bipyrimidine couple dispersed in tetrabutylphosphonium bromide. Under standard screening conditions, propylene is oxonated with this catalyst couple to give butanol in >99% linearity (1/b ratio 100; see Table 1, ex. 12).

In the case of internal olefin substrates, such a mixture of 2-, 3- and 4-octenes, the same class of catalyst precursor also provides linear, primary, alcohols, i.e., 1-nonanol (see ex. 15 and 16, Eq. 2). Other readily available C_{11}- and C_{13}-C_{14} internal olefin fractions may likewise be converted to the corresponding oxoalcohols.[2]

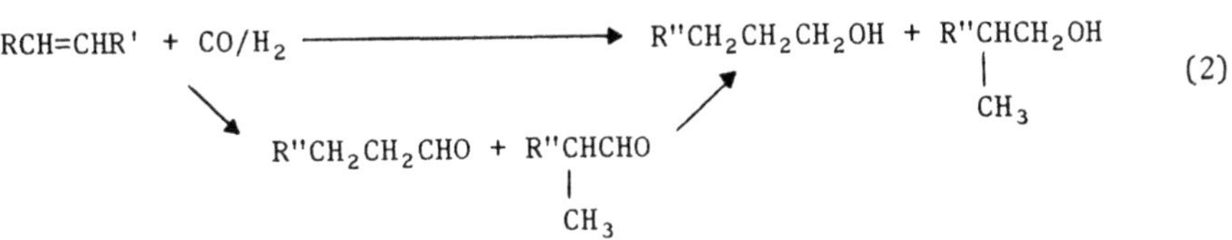

B. Catalyst Structure

With regard to the importance of catalyst structure, we find that mainly linear alcohol/aldehyde syntheses are achieved with various combinations of:[2-4]

1. Ruthenium precursors: e.g., RuO_2, $Ru_3(CO)_{12}$, $Ru(acac)_3$
2. Different N- and P-ligand promoters: e.g., 2,2'-bipyridine (BIPY), 1,10-phenanthroline, 2,2',2"-tripyridine, 2,2'-bipyrimidine, 2,2'-dipyridylamine, 2,4,6-tri(2-pyridyl)-*s*-triazine, 1,2-bis(diphenylphosphino)ethane (DIPHOS), and bis(2-diphenylphosphinoethyl)phenylphosphine.
3. Dispersed in various quaternary phosphonium salts: e.g., Bu_4PBr, Bu_4PI, Bu_4POAc.

Generally, to ensure high alcohol linearity in ruthenium-catalyzed α-olefin oxonation, the N promoter (L-L) should be a bidentate, N-heterocyclic compound capable of forming chelate complexes with ruthenium. This class of amine is best illustrated by 2,2'-bipyridine (I, ex. 1 and 9), 2,2'-dipyridylamine (II, ex. 5), and 2,2'-bipyrimidine (III, ex. 7 and 8).

The importance of N-ligand chelation is exemplified by comparing ex. 1, 3, and 4; here the influence of bipyridine structure on alcohol linearity shows that only the 2,2'-bipyridine (I) provides the desired >80% linearity:

ex. 1	2,2'-bipyridine (I)	nonanol	linearity	86%
ex. 3	2,3'-bipyridine (IV)	"	"	68%
ex. 4	2,4'-bipyridine (V)	"	"	57%

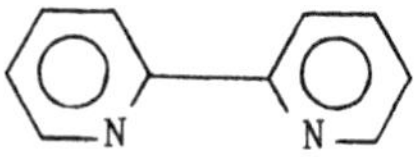

(I) 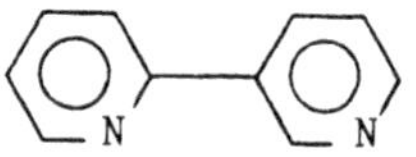 (IV) 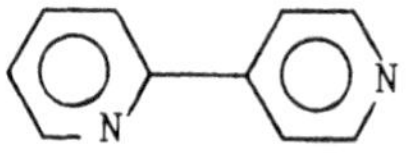 (V)

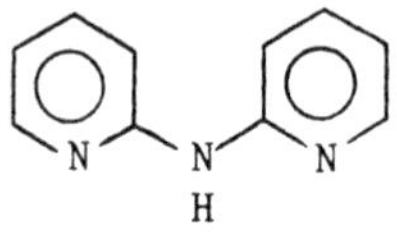

(II) 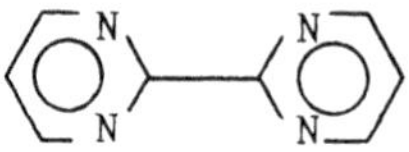 (III)

Much of the ruthenium present in the RuO_2-BIPY/Bu_4PBr catalyst-product solutions is there as $[HRu_3(CO)_{11}]^-$ (ν_{CO}, 1959, 1990, and 2017 cm^{-1}), although other metal-carbonyl bands at 1990, 2030, and 2073 cm^{-1} (see Fig. 1A) may be indicative of the presence of known ruthenium carbonyl bipyridyl clusters.[5,6] In view of such spectral data, the preferred ruthenium-ligand stoichiometry (Ru/BIPY ratios of 1:1.5) likely reflects the equilibria of Eqs. 3 and 4, between the known ruthenium-carbonyl clusters, $[HRu_3(CO)_{11}]^-$, $Ru_3(CO)_{10}(BIPY)$,[5,6] and their derivatives.

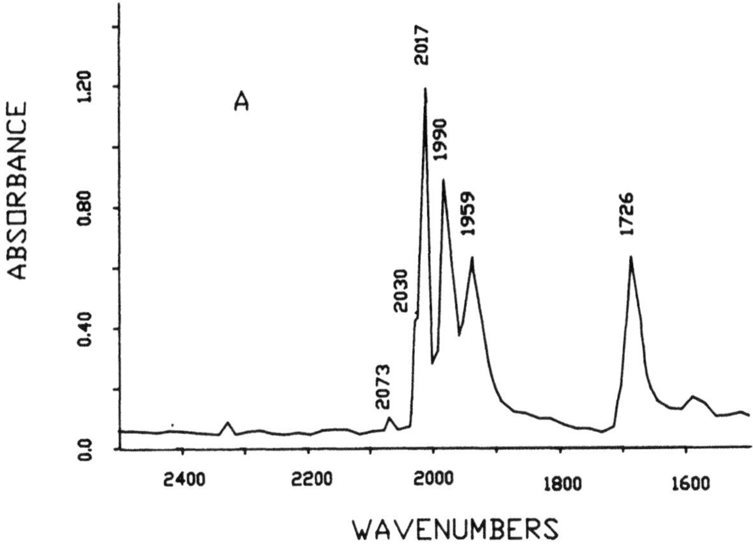

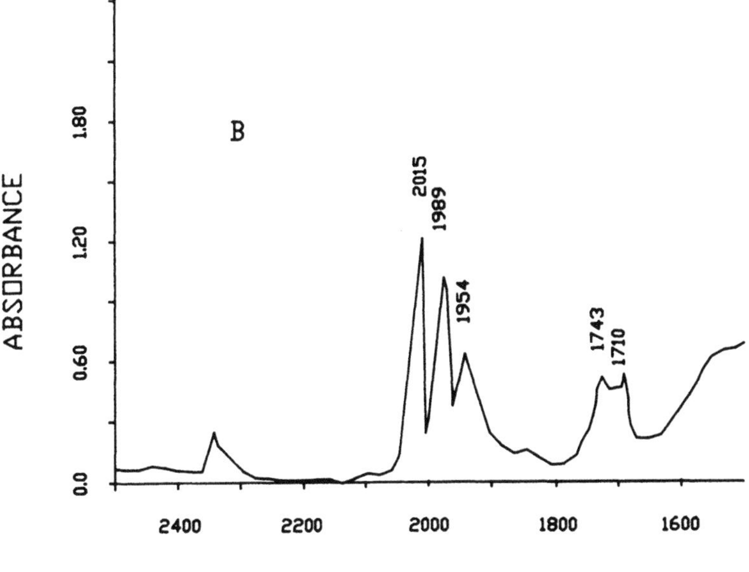

Fig. 1 Oxonation of internal octenes, spectra of typical product solutions: (A) Catalyst precursor RuO_2-BIPY/Bu_4PBr; (B) catalyst precursor RuO_2/BuOAc. (Reproduced from Ref. 10.)

$$[HRu_3(CO)_{11}]^- + BIPY \rightleftharpoons [HRu_3(CO)_9(BIPY)]^- + 2CO \qquad (3)$$

$$Ru_3(CO)_{12} + BIPY \rightleftharpoons Ru_3(CO)_{10}(BIPY) + 2CO \qquad (4)$$

The importance of ligand chelation is illustrated also in another experimental series using RuO_2 in combination with a series of arylphosphines of general structure: $Ph_2P(CH_2)_nPPh_2$ (see Table 3). Productivity is again clearly maximized for the DIPHOS ligand, known to form five-membered ring complexes with ruthenium-carbonyls.[7,8] Similarly, nonanol linearity is acceptable for all Ru-$Ph_2P(CH_2)_nPPh_2$ combinations where n = 1, 2, or 3.

The Ru-DIPHOS reactant solutions are dominated by the $Ru_3(CO)_{10}$(DIPHOS) cluster complex (ir carbonyl stretches at 1970, 1983, 2004(s), 2035, 2066 cm^{-1}, ^{13}C nmr, sharp triplet at 209.9 ppm at r.t.), in line with analogous data reported by Cotton et al.[9] for decacarbonyl[bis(diphenylphosphino)methane]triruthenium. Spectral data indicate that the equilibrium of Eq. 5 is reached rapidly.

Ruthenium-DIPHOS solutions containing excess chelate (e.g., Ru/DIPHOS, 1:5) are catalytically inactive and oxonation invariably leads to the precipitation of most of the ruthenium as mononuclear $Ru(CO)_3$(DIPHOS)-type[7] complexes (Eq. 6).

$$Ru_3(CO)_{12} + DIPHOS \rightleftharpoons Ru_3(CO)_{10}(DIPHOS) + 2CO \qquad (5)$$

$$Ru_3(CO)_{10}(DIPHOS) + 2DIPHOS \rightleftharpoons 3Ru(CO)_3(DIPHOS) + CO \qquad (6)$$

Table 3 1-Octene Hydroformylation Catalyzed by Ruthenium-Bis(Diphenylphosphino)alkane Complexes

Ex.	19	20	21	22	23
$Ph_2P(CH_2)_nPPh_2$,[a,b] n =	1	2	3	4	5
Total nonanol + nonanal, mmol	72.5	108.9	92.3	59.7	62.6
Nonanol linearity, %	86.7	84.5	84.1	66.3	63.8

[a]Run charge: Ru, 6.0 mmol; $Ph_2P(CH_2)_nPPh_2$, 6.0 mmol; Bu_4PBr, 10.0 g; $C_8^=$, 200 mmol.
[b]Operating conditions: 1200 psi (CO/H_2, 1:2); 180°C; 4 hr.
Source: Ref. 12.

While the quaternary phosphonium salts provide the unique reaction media for these oxo syntheses, their structure can also significantly impact both the catalytic activity and selectivity.[10] Starting with internal C_8 olefins, the highest nonanal/nonanol linearity is achieved with the ruthenium(IV) oxide-tetrabutylphosphonium acetate couple. In this case the linearity of the nonanol fraction is 68% (see Table 2; ex. 18). The spectra of the solutions also feature the ν_{CO} bands at 1954, 1989, and 2015 cm^{-1}, characteristic of the ruthenium carbonyl cluster anion, $[HRu_3(CO)_{11}]^-$. The presence of a pair of additional bands at 1710 and 1743 cm^{-1} in the bridging carbonyl region (Fig. 1B) may indicate, however, the importance of both solvent-separated and contact ion-paired species[11] in this media.

III. DISCUSSION

The ruthenium ligand-stabilized melt catalysis illustrated in Tables 1-3 can clearly catalyze both basic oxonation reactions—hydroformylation of α-olefin substrates to predominantly linear oxoaldehydes, and subsequent reduction of these aldehydes to the corresponding alcohol. Our data[10,12] are consistent with a stepwise two-step mode to alcohol formation (Scheme 1), rather than invoking a common intermediate partitioning between alcohol and aldehyde products. The choice of product—be it alcohol or aldehyde (Eq. 1)—is controlled primarily by the operating temperature (Fig. 2). In this regard, the Ru melt catalysis parallels cobalt hydroformylation processes. The apparent higher activation energy for the ruthenium melt-catalyzed aldehyde reduction ensures that at operating temperatures of <140°C, the primary product is the oxoaldehyde, whereas at temperatures of ≥180°C, oxoalcohols are generally favored (see Fig. 2).

The second commercially critical parameter by which this oxo processing must be judged—the linearity of the final aldehyde/alcohol products—has been raised in this study to >99% (l/b >100) by careful control of operating conditions. The primary influencing factors here are the nature of the N- and P-chelating ligand structures (Tables 1 and 3) and the ability of these ligands to form

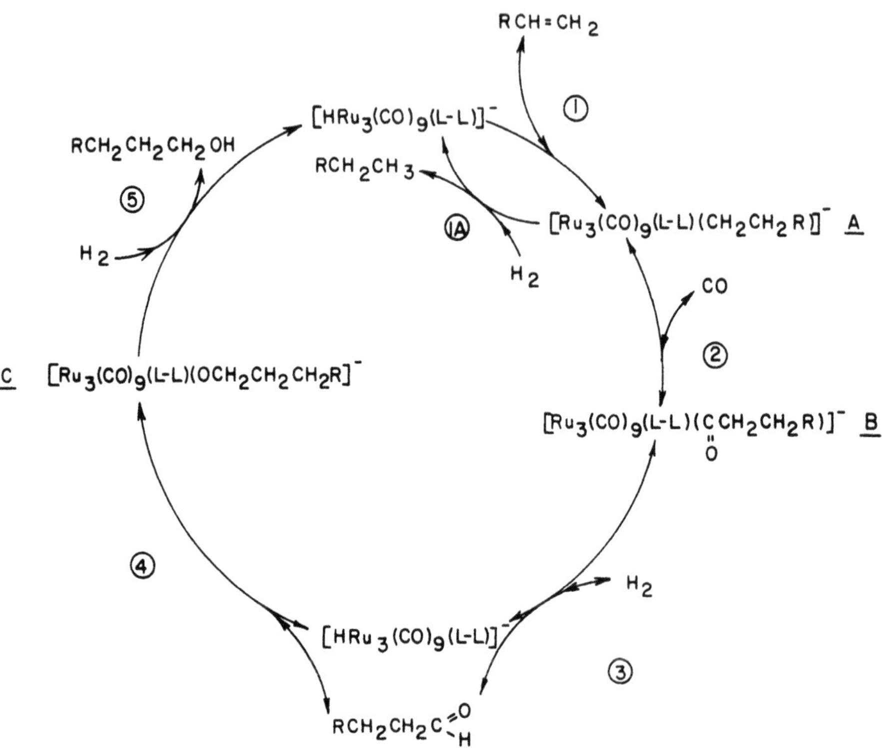

Scheme 1 Proposed mechanism for oxoalcohol and aldehyde formation via ligand-modified ruthenium melt catalysis. L-L is bidentate N or P ligand such as BIPY or DIPHOS. (Reproduced from Ref. 12.)

stable complexes with ruthenium under oxo conditions (e.g., Eq. 3-6). Temperature (Fig. 2) and initial ruthenium-ligand mole ratio[12] also play a role. Below 140°C, for example, both the nonanol and nonanal products show linearities of >97% (l/b ≥30), and at 100°C, the nonanal linearity is 99% (l/b ~100; see Fig. 2).

Total alcohol yield is strongly influenced by syngas composition and pressure (e.g., see Fig. 3). The aldehyde hydrogenation ability of the Ru catalyst decreases with increases in CO partial pressure at 180°C (Fig. 3A). In contrast, an increase in H_2 partial pressure increases competing olefin reduction (Fig. 3B). Because the aldehyde

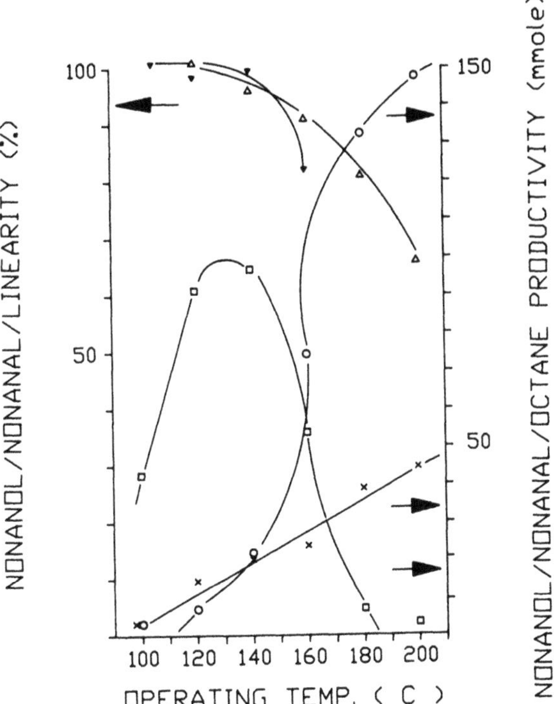

Fig. 2 Effect of operating temperature on oxo activity. Legend: ⊙, total nonanol; ⊡, total nonanal; ▲, nonanol linearity; ▼, nonanal linearity; x, total octane. Conditions as per Table 1. (Reproduced from Ref. 12.)

reduction rate (steps 4 and 5, Scheme 1) is the more sensitive to CO and H_2 partial pressure changes (see Figs. 3A and 3B, data trends ⊙, ⊡ versus x), increases in total CO and H_2 gas pressure at this temperature have the net effect of raising aldehyde yield at the expense of alcohol.[12] Oxonation to aldehyde is also influenced by CO partial pressure at lower temperatures (e.g., 120°C). In fact, the trends evident in Fig. 3, etc. suggest that (1) a number of the critical steps in this oxonation (Scheme 1) have similar rates and (2) there are complex equilibria in these catalyst solutions involving addition/displacement of CO and group 15 ligands bonded to ruthenium (e.g., Eqs. 3-6).

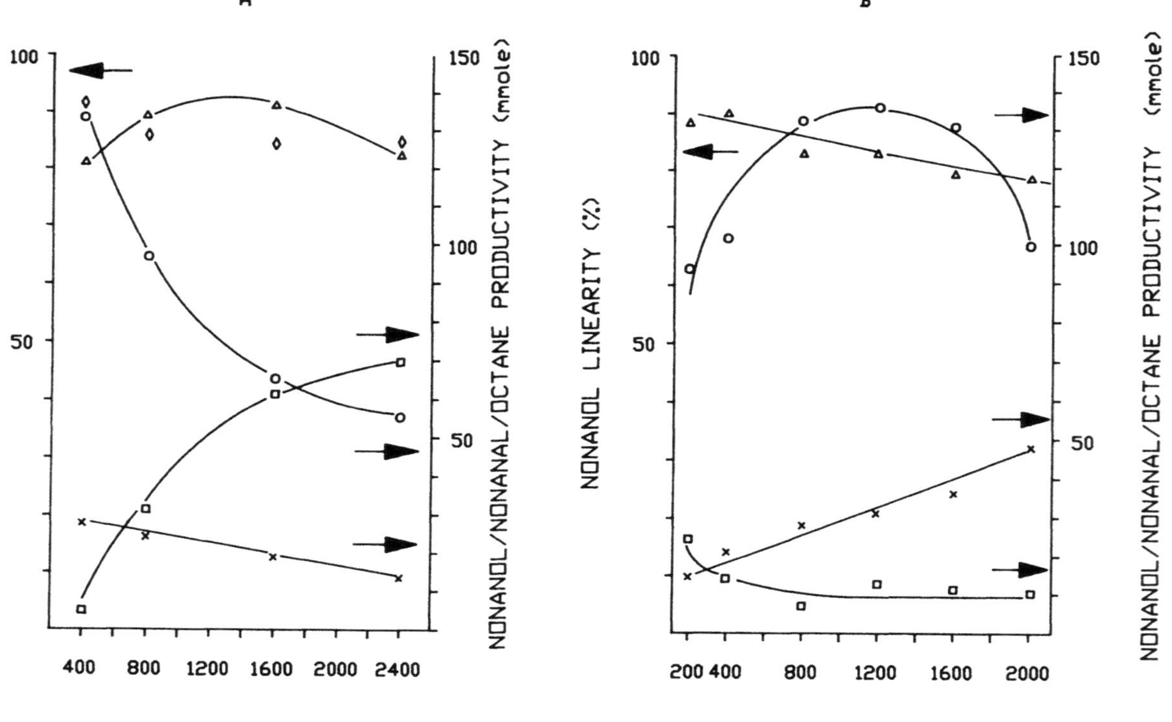

Fig. 3 Effect of CO and H_2 partial pressures on oxo activity. Legend as per Fig. 2, ◈, total nonanol + nonanal. (A) Operating conditions: 800 psi H_2; 180°C; 4 hr; (B) Operating conditions: 400 psi CO, 180°C, 4 hr. (Reproduced from Ref. 12.)

Because H_2 and CO partial pressures only modestly influence the total oxonation activity (nonanal + nonanol, e.g., ◆ data points, Fig. 3) at 180°C, neither oxidative addition of H_2 nor addition of CO is likely rate determining under these particular conditions. The rate determining step for aldehyde formation is most probably association of ligated Ru complex with the 1-octene reactant, or hydride transfer of coordinated alkene to form an alkyl-ruthenium cluster[13] (step 1, Scheme 1). This is consistent also both with the aldehyde productivity at lower temperatures (120°C) being relatively insensitive to H_2 and syngas pressures, and with the apparent first-order rate dependence on olefin charge.[12]

In conclusion, we demonstrated that highly linear alkanols and alkanals can be readily generated through regioselective α-olefin oxonation with ligand-modified ruthenium melt catalysis, while mixed internal olefin substrates can be made to yield predominantly linear products. Although all the interrelationships linking catalyst structure and the various operating parameters are not yet fully understood (particularly in a quantitative sense), the data presented in this chapter do allow the design of high-yield syntheses of either the desired oxoaldehyde or alkanol using essentially the same classes of ruthenium melt catalysts.

REFERENCES

1. J. F. Knifton, *Platinum Metals Rev., 29,* 63 (1985).
2. J. F. Knifton, J. J. Lin, R. A. Grigsby, and W. H. Brader, U.S. Patent 4,451,679 (1984).
3. J. F. Knifton and R. A. Grigsby, U.S. Patent 4,469,895 (1984).
4. J. F. Knifton, U.S. Patent 4,451,680 (1984).
5. M. I. Bruce, M. G. Humphrey, M. R. Snow, E. R. T. Tiekink, and R. C. Wallas, *J. Organomet. Chem., 314,* 311 (1986).
6. T. Venalainen, J. Pursiainen, and T. A. Pakkanen, *J. Chem. Soc. Chem. Commun.,* 1348 (1985).
7. R. A. Sanchez-Delgado, J. S. Bradley, and G. Wilkinson, *J. Chem. Soc. Dalton,* 399 (1976).

8. K. Kurter, D. Ribola, R. A. Jones, D. J. Cole-Hamilton, and G. Wilkinson, *J. Chem. Soc. Dalton*, 55 (1980).

9. F. A. Cotton and B. E. Hanson, *Inorg. Chem., 16*, 3369 (1977).

10. J. F. Knifton, *J. Mol. Catal., 43*, 65 (1987).

11. T. Sakakura, T. Kobayashi, and M. Tanaka, *C_1 Mol. Chem., 1*, 219 (1985).

12. J. F. Knifton, *J. Mol. Catal., 47*, 99 (1988).

13. D. Mani, H. T. Schacht, A. Powell, and H. Vahrenkamp, *Organometallics, 6*, 1360 (1987).

17

Applications of Pyridine-Containing Polymers in Organic Chemistry

GERALD L. GOE, CHARLES R. MARSTON, ERIC F. V. SCRIVEN, and EDWARD E. SOWERS

Reilly Industries, Inc., Indianapolis, Indiana

ABSTRACT

Studies of the application of crosslinked polyvinylpyridine as reagents, catalysts, and as supports for metallic catalysts and reagents are described. The synthesis of polymeric acylation catalysts of the 4-dimethylaminopyridine type and their effectiveness in esterifications and rearrangement are discussed. Process advantages of these catalysts are outlined.

Copolymers of vinylpyridine and styrene crosslinked with divinylbenzene have been used as aids in organic synthesis.[1] Such copolymers, which are insoluble in all solvents, have been used as acid scavengers and when in the conjugate acid form as a proton source. Fréchet and his group pioneered the use of crosslinked polyvinylpyridines as supports for reagents in organic synthesis. They found that poly-4-vinylpyridinium chlorochromate may be used for the oxidation of benzyl alcohol to benzaldehyde.[2] Poly-4-vinylpyridine hydrobromide perbromide is effective for the bromination of alkenes[3] and the formation of α-bromoketones.[4] The use of insoluble polyvinylpyridines in the above and other applications offers the advantage of easy separation of the product from the polymer after completion of reaction. Furthermore, it is usually possible to recycle the polyvinylpyridine with or without regeneration, depending on the circumstances. These polymers may be considered as pyridine in polymeric form, and they are as resistant to oxidation and electrophilic substitution as pyridine itself.

We have developed two polymers, which are available in commercial quantities as Reillex T.M. 402 and 425 (poly-4-vinylpyridine either 2% or 25% crosslinked with divinylbenzene).

We found Reillex T.M. 402 to be an excellent catalyst for the Knoevenagel reaction.[5]

$$RCH_2CHO + CH_2(COOH)_2 \xrightarrow[\text{pip}]{\text{PVP}} RCH{=}CHCOOH \quad CO_2 + H_2O$$

Catalytic amounts of Reillex T.M. 402 hydrochloride (powder form) have been employed for ketal and enamine formation.[5]

$$CH_2OH{-}CH_2OH + CH_3COCH_3 \xrightarrow{\text{PVP·H}^+\,Cl^-} (CH_2O)_2C(CH_3)_2$$

Reillex T.M. 425 hydrochloride (in bead form) acts as a mild acid catalyst for the tetrahydropyranylation of primary, secondary, and tertiary alcohols, and phenols.[6] As in the above example, not only do the polymer-supported catalysts give as good a yield as their soluble equivalents, but they are readily removed from products merely by filtration. Furthermore, this catalyst has been recycled five times without loss in activity in the tetrahydropyranylation of 4-methoxyphenol.

$$\text{dihydropyran} + PhCH_2OH \xrightarrow[86^\circ C,\ 1\cdot 5h]{\text{PVP·H}^+} PhCH_2O\text{-THP} \quad 97\%$$

Table 1 Typical Properties of Reillex[T.M.] Polymers

Property	Reillex[T.M.] 402 off-white granular powder	Reillex[T.M.] 425 off-white beads
Particle size	About 60 mesh	18-50 mesh
Bulk density, g/cm^3	0.45	0.29
lb/ft^3	28	18
Skeletal density, g/cm^3	1.15	1.14
Particle density, g/cm^3	—	0.6
Surface area, m^2/g	About 0.5	About 90
Moisture retained upon filtration, % by weight	36-39	50-60
Approximate pK_a	3-4	3-4
Hydrogen ion capacity in water, meq/g	8.8	5.5
% Swelling from free base to hydrochloride form, in water	100%	52%
% Swelling from dry state to solvent-saturated state		
Methanol	70-75	28-32
Acetone	30-35	32-36
Water	33-37	12-16
Isopropanol	13-17	28-32
Toluene	8-12	18-22
Ethyl acetate	3-6	32-36
Hexane	0	12-16
Temperature stability, maximum	225	260

The remarkable physical properties of these Reillex[T.M.] polymers (Table 1), especially their high thermal stability, led us to explore their use as catalyst supports. We found that it is possible to support metals on these resins in three ways. The resulting supported metal catalysts have been classified as type A, B, or C.

Type A. Coordinately Bound

Pyridine is renowned as a ligand in inorganic chemistry. Therefore, it is not surprising that polyvinylpyridine resins readily form coordination complexes with active metals such as copper. Essentially any metal salt which can form a stable complex with tertiary amines

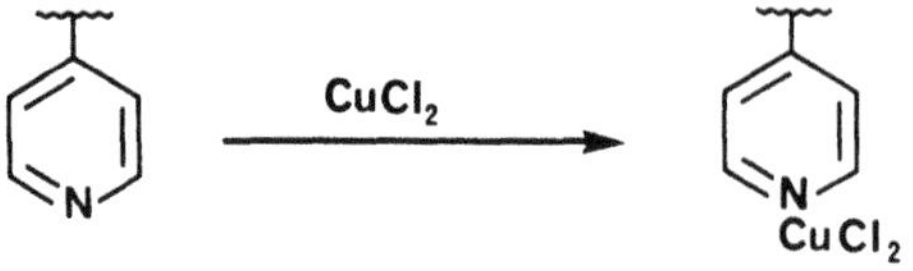

$$\text{Py-resin} \xrightarrow{CuCl_2} \text{Py(N} \cdot CuCl_2\text{)-resin}$$

can be supported by coordination on Reillex.[T.M.] These complexes usually can be prepared by adding an aqueous solution of the inorganic salt to a slurry of the degassed resin.

Type B. Ionically Bound

These are polyvinylpyridinium quaternary salts in which the metal is contained in the anion. Such salts can be formed either by prior or in situ quaternization of the pyridine and subsequent anion exchange.

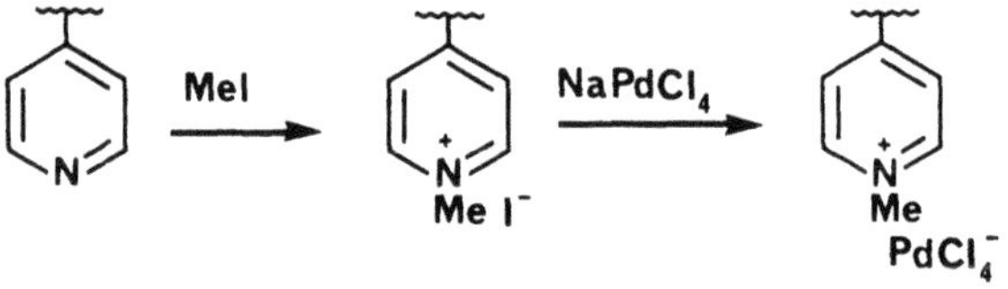

$$\text{Py-resin} \xrightarrow{MeI} \text{Py}^{+}\text{-Me } I^{-} \xrightarrow{NaPdCl_4} \text{Py}^{+}\text{-Me } PdCl_4^{-}$$

Type C. Deposited Metal

Zero valent metal atoms can be dispersed throughout the polymer network by treating type A or B supported catalysts with reducing agents (e.g., hydrazine, formic acid, diborane, or hydrogen).

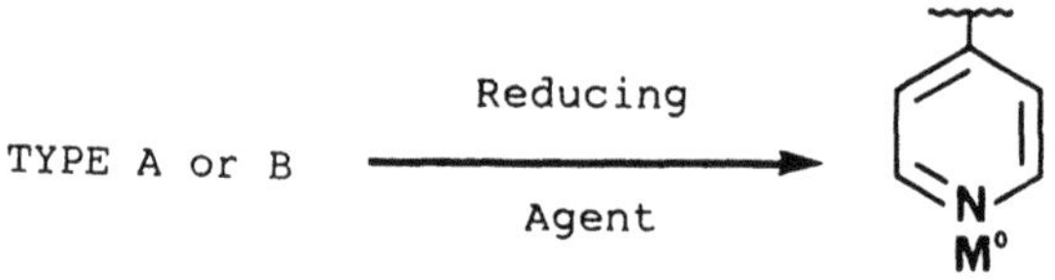

$$\text{TYPE A or B} \xrightarrow[\text{Agent}]{\text{Reducing}} \text{Py(N} \cdot M^{0}\text{)-resin}$$

We have found catalyst loadings of 1-5% (based on the amount of resin) best for the applications described below. Use of such loadings and the choice of less polar reaction media minimize leaching of metal from the resin during reaction. Some applications of the different classes of catalyst will now be considered.

Type A Catalysts

About 35,000 tons of 1,2,3,4-tetrahydronaphthalene-1-one are produced annually in the United States for agricultural and pharmaceutical use. Much of this is produced by the liquid phase oxidation of tetrahydronaphthalene using chromium(III) acetate as a homogeneous catalyst.[7] We carried out this reaction using 2% chromium(III) acetate supported on Reillex$^{T.M.}$ resins.[8]

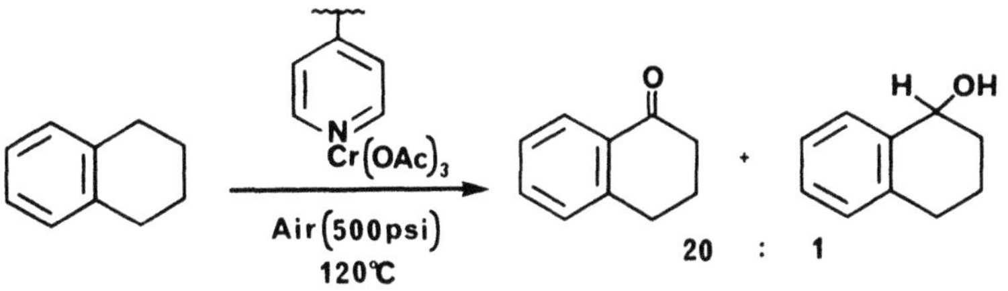

We found a Co(II)-Reillex$^{T.M.}$-supported catalyst to work well for the liquid phase oxidation of cyclohexane to a mixture of cyclohexanol and cyclohexanone. Supporting these LPO catalysts on Reillex$^{T.M.}$ offers the advantage of easy catalyst recovery and recycle. Furthermore, use of Reillex 425 beads would make these catalysts amenable to fixed- or fluid bed processes.

Type B Catalysts

Most acetic acid is now produced by a process discovered by Monsanto in the 1970s. This involves the homogeneous carbonylation of methanol in the presence of a rhodium species and methyl iodide.[9] Addition of Reillex$^{T.M.}$ 425 (19.6 g) to a typical homogeneous reaction mixture of rhodium chloride trihydrate (0.4 g), methyl iodide (28.8 g), and methanol (100 g) in acetic acid (198 g) leads to formation of the heterogeneous catalyst as the reaction vessel is raised to the reaction temperature (180-185°C) and pressurized with carbon monoxide (65-80 bar).[10] Acetic acid was chosen as the solvent for convenience. The catalyst was recovered by filtration from the acetic acid and recycled without further treatment. Conversions of 99% were achieved in most cases, and rates using the heterogeneous catalyst were four times faster than those under comparable homogeneous conditions. Rhodium leaching was not a problem in this work. Anion exchange resins were previously used as supports for acetic acid synthesis catalysts, but only a low temperature and pressure, resulting in impractically low reaction rates.[11]

$$MeOH \xrightarrow[\text{175-200°C, 65-85 bar; CO, MeI, HOAc}]{\text{1-7\% Rh on Reillex}^{TM}\text{ resin}} MeCO_2H$$

The catalyst used in the acetic acid work was also effective for the conversion of methyl formate to acetic acid.[12]

$$HCO_2Me \xrightarrow[\text{175°C, 4h; CO, 65 bar}]{\text{4\% Rh on Reillex}^{TM}\bullet\text{MeI resin}} MeCO_2H$$

Type C Catalysts

Zero valent metals supported on polyvinylpyridine resins have been used in hydrogenations, the hydrolysis of nitriles, and for the oxidative coupling of olefins.[13]

During the 1960s 4-dimethylaminopyridine (DMAP) was found to be a very good catalyst for some of the most difficult acylation reactions.[14] For instance, 1-methylcyclohexanol, a very sterically hindered alcohol, is acetylated quite easily in the presence of a catalytic amount of DMAP, whereas normal acetylation fails.

1-methylcyclohexanol (Me, OH) $\xrightarrow[\text{14h RT}]{Ac_2O}$ 1-methylcyclohexyl acetate (Me, OAc)

pyridine	5%
DMAP/TEA	86%

DMAP is now preeminent among commercially used catalysts for difficult acylations.[15] This has led us to seek a polymeric version of DMAP that would have all the advantages of a heterogeneous catalyst. When we undertook this project several polymeric versions of DMAP had been described. Klotz made the first example by attaching an acid-functionalized dialkylaminopyridine to a polyethyleneimine polymer (Scheme 1).[16] He subsequently made many similar functionalized polyimines[17] and demonstrated their catalytic ability by kinetic

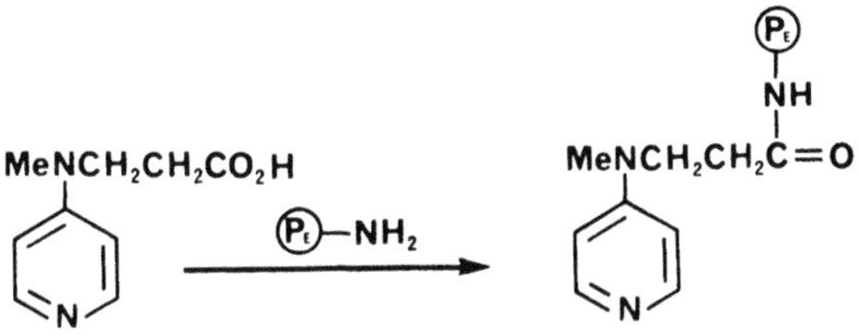

Scheme 1

experiments on the hydrolysis of *p*-nitrophenyl caproate. These polymers suffer from the drawback that the pyridine is attached to the polymer backbone by an amide linkage which is susceptible to scission when, for example, regenerating the resin for reuse using sodium hydroxide.

Verducci and coworkers [18] attached 4-piperidinylpyridine, among others, to a Merrifield resin through an amide bond (Scheme 2). The amide bond in this polymer stands up well on recycle (10 catalyst cycles in the acetylation of 1-methylcyclohexanol, 70°C, 24 hr); however, a high catalyst loading is required.

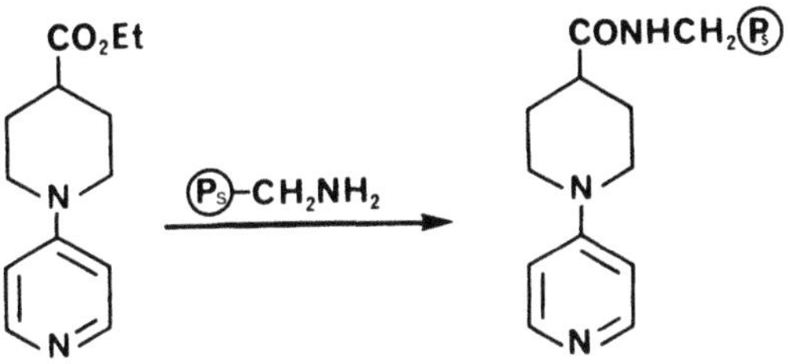

Scheme 2 Ps — 1% DVB XL Styrene

The most popular approaches to DMAP-bound polymers avoid the use of amide linkages. Shinkai and coworkers[19] attached 4-chloropyridine to an aminomethylpolystyrene (Scheme 3) to give BMAP.

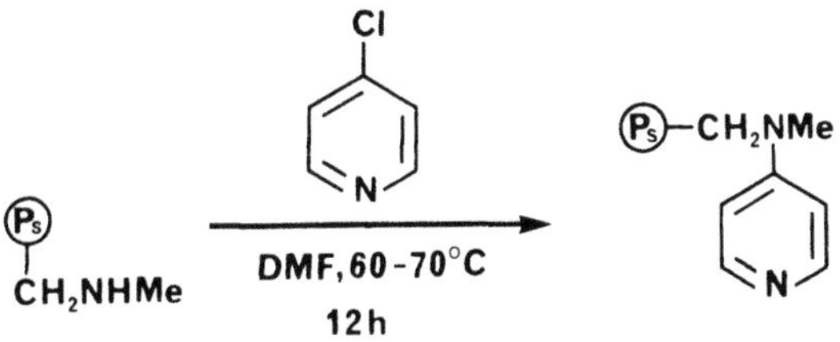

Scheme 3

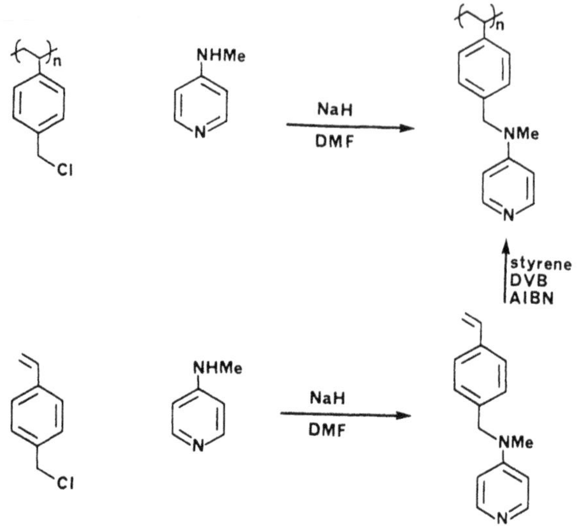

Scheme 4

Tomoi and coworkers[20] compared two other approaches (Scheme 4). They found that the route involving copolymerization of the preformed monomer gave a better catalyst. However, more recently Fréchet[21] found that chloromethylated polystyrenes can be modified "readily and quantitatively." Menger and coworkers[22] were also successful in converting a linear chloromethylpolystyrene resin to BMAP. Their "linear BMAP" is an excellent catalyst for several well-known DMAP-catalyzed processes.

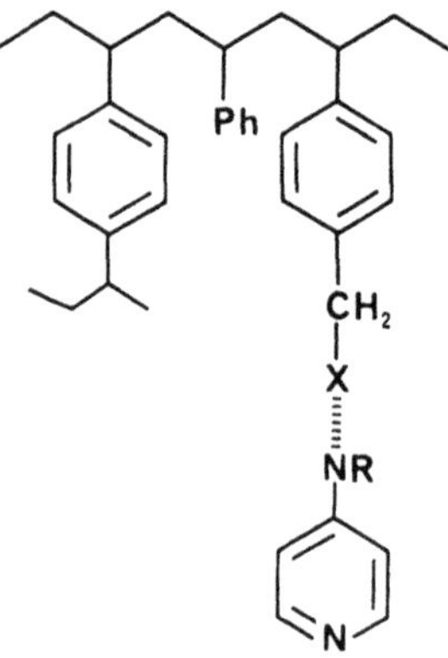

Fig. 1

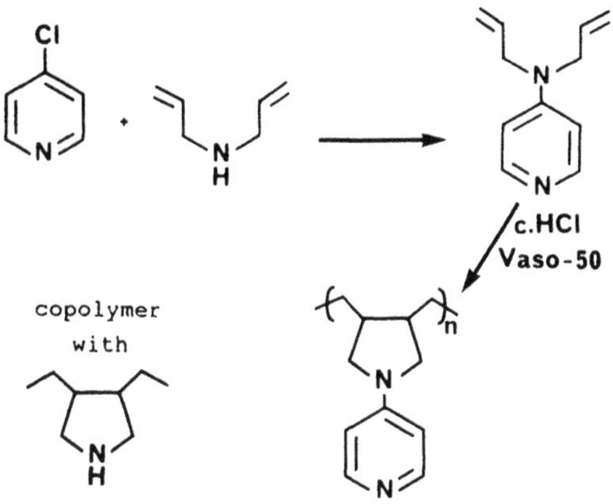

Scheme 5

The groups of Fréchet,[21] Tomoi,[23] Manecke,[24] and Challa[25] studied the effects of variation of the frequency of DMAP to styrene units, variation of the cross-linking, and variation of the length and nature of the spacer arm (Fig. 1). Fréchet made a 34% DVB crosslinked macroreticular resin but this proved to be a much poorer catalyst than 2% crosslinked gel beads for the acylation of 1-methylcyclohexanol.

Mathias and coworkers[26] synthesized polymer-bound analogs of PPY by a different method (Scheme 5). Narang and coworkers[27] prepared a poly-PPY starting with 4-aminopyridine (Scheme 6).

A polymeric form of DMAP has been used in a multipolymer process for the synthesis of organic chemicals.[28] Bird and Karge [29] developed another DMAP polymer which they found effective in catalyzing a Fries-like rearrangement.

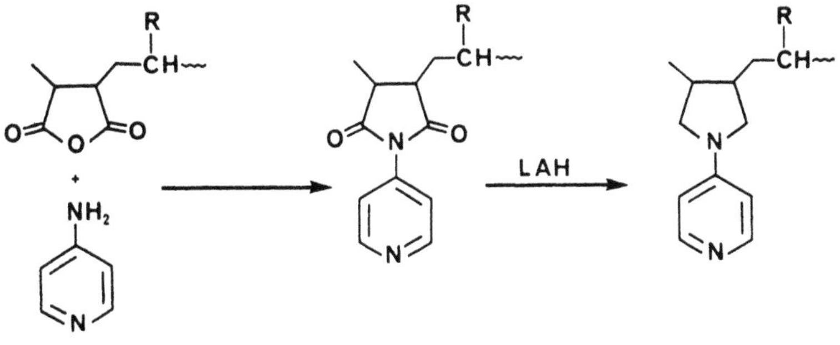

Scheme 6

Table 2 Comparison of Relative Product Yields Using Polymer-Supported DMAP and DMAP

	Relative yield %
Acetylation (6 h) of 1-methylcyclohexanol	86
Acetylation (24 h) of 1-methylcyclohexanol	94
Rearrangement (2 h) of 1	88
Rearrangement (8 h) of 1	100

We have synthesized a number of polymeric DMAP-like catalysts. They have been tested in the acetylation of 1-methylcyclohexanol and the rearrangement of (1) to (2). Yields of products obtained in these reactions using our polymer-supported DMAP compared with an equivalent amount of DMAP are given in Table 2. The results indicate that yields of products are comparable either using DMAP under homogeneous conditions or polymer-supported DMAP under heterogeneous conditions. In the latter case the reaction is a little slower but the heterogeneous catalyst is more easily removed from the desired product. This catalyst can be recycled many times.

PhMe, 100°C

1 → 2

8h 99% (1st cycle)

8h 96% (10th cycle)

Some of the disadvantages encountered with the use of homogeneous catalysts in industrial processes are the necessity for recovery of expensive metal compounds for recycle; the necessity of removing inorganics from residues before incineration; a high inorganic load on effluents; and low space-time yields owing to low catalyst concentrations. We feel that the approaches described above to anchor—"heterogenize"—the homogeneous catalyst onto insoluble supports by means of specific functional group interactions offers a new approach for industrial process technology.

REFERENCES

1. W. T. Ford, *Polymeric Reagents and Catalysts*, A.C.S., D.C., 1986; P. Hodge and D. C. Sherrington (eds.), *Reactions in Organic Synthesis*, John Wiley and Sons, New York, 1980.
2. J. M. J. Fréchet, J. Warnock, and M. J. Farrall, *J. Org. Chem.*, *43*, 2618 (1978).
3. J. M. J. Fréchet, M. J. Farrall, and L. J. Nuyens, *J. Macromol.* *11*, 507 (1977).
4. Reillex[TM] Report 4, Polymeric Reagents, Reilly Tar & Chemical Corporation, Indianapolis, 1986.
5. Pyridine Functionality in Polymer Form, Reilly Tar & Chemical Corporation, Indianapolis, 1980.
6. R. D. Johnston, C. R. Marston, P. E. Krieger, and G. L. Goe, Synthesis, 393 (1988).
7. R. W. Coon, U.S. Patent 4,473,711 (1984).
8. G. L. Goe, T. D. Bailey, and J. R. Beadle, U.S. Patent 4,753,911 (1988).
9. F. E. Paulik et al., U.S. Patent 3,769,329 (1973).
10. G. L. Goe and C. R. Marston, Eur. Pat. Appl. 277,824 (1988).
11. R. S. Drago, E. D. Nyberg, A. El A'mma, and A. Zombeck, *Inorg. Chem.*, *20*, 641 (1981); R. S. Drago and A. El A'mma, U.S. Patent 4,328,125 (1982).
12. H. Hög and G. Bub, Germ. Patent DE 3,333,315 (1985).
13. Reillex[T.M.] Report 3, Catalysts.
14. W. Steglich and G. Höfle, *Angew. Chem. Int. Ed. Engl.*, *8*, 981 (1969); L. M. Litvenenko and A. I. Kirichenko, *Dokl. Akad. Nauk SSSR*, *176*, 97 (1967), *Chem. Abstr.*, *68*, 68325 (1968).
15. A. Höfle, W. Steglich, and H. Vorbrüggen, *Angew. Chem. Int. Ed. Engl.*, *17*, 569 (1978); E. F. V. Scriven, *Chem. Soc. Rev.*, *12*, 129 (1983).
16. M. A. Hierl, E. P. Gamson, and I. M. Klotz, *J. Am. Chem. Soc.*, *101*, 6020 (1979).
17. E. J. Delaney, L. E. Wood, and I. M. Klotz, *J. Am. Chem. Soc.*, *104*, 799 (1982); I. M. Klotz, S. E. Massil, and L. E. Wood, *J. Polymer Sci. Chem. Ed.*, *23*, 575 (1985).

18. F. Guendouz, R. Jacquier, and J. Verducci, *Tetrahedron Lett.*, *25*, 4521 (1984).

19. S. Shinkai, H. Tsuji, Y. Hara, and O. Manabe, *Bull. Chem. Soc. Jpn.*, *54*, 631 (1981).

20. M. Tomoi, Y. Akada, and H. Kakiuchi, *Macromol. Chem. Rapid Commun.*, *3*, 537 (1982).

21. A. Deratani, G. D. Darling, D. Horak, and J. M. J. Fréchet, *Macromolecules*, *20*, 767 (1987).

22. F. M. Menger and D. J. McCann, *J. Org. Chem.*, *50*, 3928 (1985).

23. M. Tomoi, M. Goto, and H. Kakiuchi, *J. Polym. Sci. Polym. Chem.*, *25*, 77 (1985).

24. W. Storck and G. Manecke, *J. Mol. Catal.*, *30*, 145 (1985).

25. C. E. Koning, J. J. W. Eshuis, F. J. Viersen, and G. Challa, *Reactive Polymers*, *4*, 293 (1986).

26. L. J. Mathias, U.S. Patent 4,591,625 (1986).

27. S. C. Narang and R. Ramharack, *J. Polym. Sci. Polym. Lett.*, *23*, 147 (1985).

28. Y. Shai, K. A. Jacobson, and A. Potchornik, *J. Am. Chem. Soc.*, *107*, 4249 (1985); A. Patchornik, Y. Shai, and S. Pass, U.S. Patent 4,552,922; A. Patchornik, *Chemtech*, 58 (1987).

29. G. J. Bird and R. L. Karge, Austral. Patent 52992 (1988).

Part VI

New Catalytic Chemical Processes

18

Selective Oxidative Dehydrogenation of Light Alkanes over Vanadate Catalysts

MAYFAIR C. KUNG, KIMMAI THI NGUYEN, DEEPAK PATEL, AND HAROLD H. KUNG

Northwestern University, Evanston, Illinois

ABSTRACT

It has been observed that a mixed oxide of magnesium and vanadium is a rather selective oxidative dehydrogenation catalyst for light alkanes. Selectivities of 50% and higher could be obtained at an alkane conversion of about 50%. Spectroscopic analyses of the catalysts showed that the active component was magnesium orthovanadate. Other orthovanadates were equally selective. However, magnesium pyrovanadate and magnesium metavanadate were nonselective. This was explained by the different bonding of oxygen ions in these compounds.

The reaction mechanism was consistent with the assumption that the first step of the reaction was the breaking of a C-H bond to form an alkyl radical. At lower reaction temperatures, hexane oxidation yielded a substantial amount of hexanone. At higher temperatures, butane and propane yielded only dehydrogenation products but no oxygenates. It was confirmed that some of the alkyl radicals were desorbed into the gas phase for further reaction in the postcatalytic volume.

I. INTRODUCTION

Selective oxidation of light saturated hydrocarbons is an attractive way to utilize this cheap source of feedstock. This is particularly true when the selective oxidation products are high-valued specialty chemicals. For example, butane may be oxidized to butenes, butadiene, butanediol, crotonaldehyde, furan, maleic anhydride, and other products in addition to carbon oxides. Practically all of these partial oxidation products are very useful. However, an economically viable process must achieve high selectivity in the production of only one of these possible products. In general, partial oxidation products can be classified into two groups: dehydrogenation products in which oxidation is achieved by removal of hydrogen atoms from the molecule, and oxygenates in which oxygen atoms are incorporated into the molecule. In this chapter, the emphasis is on the production of dehydrogenation products.

Dehydrogenation reactions can be carried out in the absence of gaseous oxygen, and there are commercial processes using Cr-Al-O catalysts. In these processes, coking is usually quite severe, and the catalysts need to be regenerated every several minutes. The process is also limited by the thermodynamic equilibrium yield such that high temperatures in excess of 600°C are required for high conversions. At the high temperatures, cracking of the hydrocarbon to smaller hydrocarbons also takes place readily. Thus there is room for improvement in the process of selective dehydrogenation of alkanes.

Catalytic dehydrogenation reactions can also be carried out in the presence of gaseous oxygen. Then other competing reactions are present, including the formation of oxygenates and combustion to carbon oxides. In addition, gas phase oxidation reactions may also be important at sufficiently high temperatures.

Since dehydrogenation does not involve formation of a C-O bond while the formation of oxygenates does, it is possible that by regulating the reactivity of the oxygen ions of a catalyst, selective production of dehydrogenation products or oxygenates can be controlled.

In this chapter, advances in our laboratory toward understanding the effect of the nature of a catalyst on selectivity is presented.

II. SELECTIVE OXIDATIVE DEHYDROGENATION CATALYSTS

A selective oxidative dehydrogenation catalyst must (1) be able to activate alkane, presumably by breaking a C-H bond; (2) interact weakly with alkenes so that the dehydrogenation products do not undergo further reactions; and (3) discriminately break C-H but not C-C bonds.

Vanadium oxide is among the few oxides that are selective in oxidation reactions. However, as shown in Table 1, it is not selective in the oxidation of alkanes at 500°C at conversions higher than a few percent.[1,2] This may indicate that the lattice oxygen of vanadium oxide is too active. Therefore, it seems possible that by

Table 1 Conversion and Selectivity in Oxidative Dehydrogenation of Propane and Butane over V-Mg-O Catalysts[a]

Catalyst	Weight[b] (g)	Alkane conv. (%)	Selectivity (C_3 or C_4 basis)					
			CO	CO_2	C_3H_6	C_4H_8	C_4H_6	Dehyd.[c]
			Oxidation of Butane					
V_2O_5	0.4	5.1	0	54	0	18	0	18
3V-Mg-O[d]	0.3	35.5	24.9	50.7	4.3	8.3	4.3	12.6
8V-Mg-O	0.3	50.9	19.7	42.7	1.8	11.1	20.4	31.5
19V-Mg-O	0.1	34.2	12.4	24.3	4.9	24.8	30.8	55.6
24V-Mg-O	0.1	56.2	17.0	30.1	0.8	12.9	37.2	50.1
54V-Mg-O	0.1	31.3	15.9	21.9	5.1	29.3	24.2	53.5
			Oxidation of Propane					
V_2O_5	0.4	22.0	64.8	17.2	18.0			
19V-Mg-O	0.3	35.8	18.9	36.4	42.4			
24V-Mg-O	0.1	28.9	19.0	37.2	42.4			
40V-Mg-O	0.2	33.4	23.1	33.7	41.9			

[a]Reaction conditions: temperature, 540°C; feed, 4% alkane, 8% O_2, and 88% He; flow rate, 100 ml/min.
[b]The catalysts were diluted with twice the weight of silica.
[c]Total selectivity to form dehydrogenated products.
[d]The number denotes the weight percent of V_2O_5 in the catalyst.

reducing the reactivity of the lattice oxygen in vanadium oxide, a more selective catalyst can be obtained.

One method to accomplish this is to use a mixed oxide of vanadium with another component that is less active in oxidation and which should interact weakly with alkenes. Since alkenes (and dienes) are Lewis bases because of their high π-electron density, a basic oxide may be a desired component.

When a mixed oxide of vanadium and magnesium was tested with a $C_4H_{10}/O_2/He$ mixture of 4:8:88, a rather selective production of butenes and butadiene was obtained, as shown by Table 1. Furthermore, it can be seen that the production of dehydrogenation products increased with increasing amounts of vanadium until the catalyst contained about 20 wt % of vanadium oxide. Beyond this the selectivity of the catalyst remained unchanged with vanadium content, although the activity changed primarily because of changes in the surface area.

When these catalysts were tested with propane instead of butane, selective production of propene was also obtained as shown in Table 1. For both propane and butane, the selectivity for dehydrogenation decreased with increasing conversion. The decrease was more severe for propane than for butane.

To understand the interaction between vanadium and magnesium oxide that led to the enhanced selectivity for oxidative dehydrogenation, the catalysts were characterized with various spectroscopic techniques. Figure 1 shows the Raman spectra of V_2O_5 and a 40% V_2O_5/MgO catalyst. The spectrum for V_2O_5 shows the characteristic peaks such as those at 996, 703, and 530 cm^{-1}. These peaks were missing in the spectrum for V-Mg-O. It is clear that no crystallites of V_2O_5 were present in the selective catalysts. In fact, the spectrum could be interpreted by the formation of magnesium orthovanadate, $Mg_3(VO_4)_2$.

The formation of magnesium orthovanadate in the V-Mg-O catalysts has been confirmed by other techniques including x-ray diffraction, infrared spectroscopy, and Auger electron spectroscopy.[1] Furthermore, it was shown that the high selectivity of the mixed oxide could be

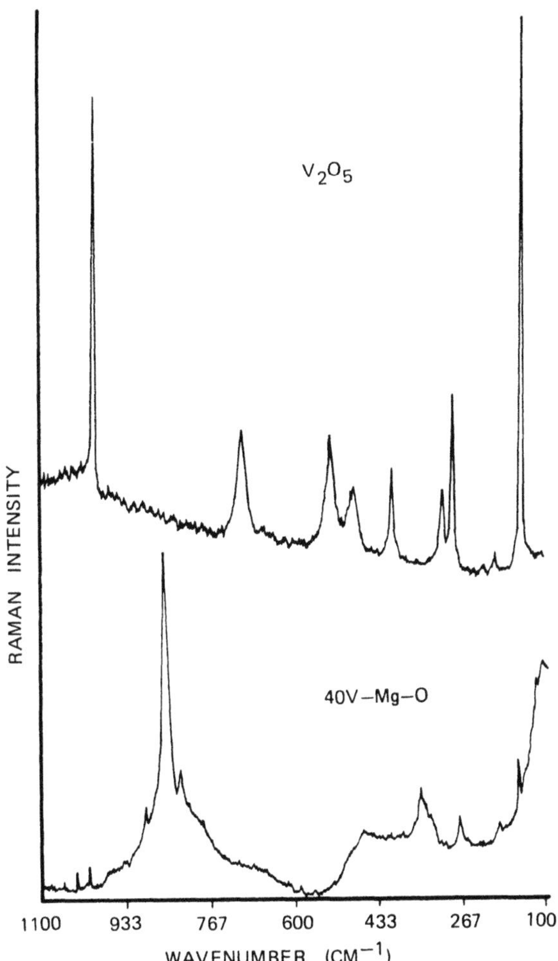

Fig. 1 Raman spectra of V_2O_5 and 40V-Mg-O.

correlated with the appearance of the orthovanadate. Thus it appeared that this compound was the active component.

To substantiate this, stoichiometric magnesium orthovanadate samples were prepared and tested in the reaction. As shown in Table 2, the catalytic behavior of the stoichiometric sample was very similar to those of the mixed oxides. Thus the active component was clearly identified.

Table 2 Selectivity in Oxidative Dehydrogenation of Butane on Vanadate Catalysts (540°C, butane/O_2/He = 4:8:88)

Catalyst	Alkane conv. (%)	Dehydrogenation select. (%)		
		Butenes	Butadiene	Total
MgV_2O_6	13	14.1	0	14.1
$Mg_2V_2O_7$	20	14.6	1.5	16.1
$Mg_3(VO_4)_2$	17	43.3	12.7	56.0
$NdVO_4$	14	49.8	7.7	57.5

III. ROLE OF THE STRUCTURE OF VANADATES

Magnesium and vanadium oxides form three different compounds: magnesium orthovanadate, $Mg_3(VO_4)_2$, magnesium pyrovanadate, $Mg_2V_2O_7$, and magnesium metavanadate, MgV_2O_6. In addition to their different stoichiometries, they also differ in their crystal structures. In particular, their vanadate units are different. If the reactivity of the lattice oxygen is important, these three compounds should have very different catalytic properties because they have different V-O bonds.

Stoichiometric compounds of these three vanadates were prepared and tested in the oxidation of butane. The results are shown in Table 2. Only the orthovanadate was selective in dehydrogenation. The pyrovanadate and metavanadate were both quite nonselective. These data could be interpreted as follows. The pyrovanadate unit is formed by two corner-sharing VO_4 tetrahedrons. The metavanadate is made up of chains of edge-sharing highly distorted VO_6 units. In both structures, there are oxygen ions shared between vanadium ions. Perhaps these oxygen ions could be readily removed. They are highly reactive. The orthovanadate structure has isolated VO_4 units. None of the oxygen ions are shared between vanadium ions. It is possible that they are much less reactive. This results in the high dehydrogenation selectivity of this oxide.

If indeed the vanadate structure is the important factor, it should be able to replace magnesium with other ions that form orthovanadate without much change in the catalytic behavior. This conjecture has been successfully confirmed. As shown in Table 2, the selectivity for dehydrogenation remained essentially the same when magnesium was substituted by neodymium. The compound $NdVO_4$ has the same orthovanadate structure. Since the metal cation-to-vanadium ratio is 1:1 in this compound instead of 3:2 in magnesium orthovanadate, the data also show that the structure of the oxide is much more important than the cation ratio.

A catalyst of high dehydrogenation selectivity was also obtained by substituting Mg with Ba to form $Ba_3(VO_4)_2$, barium orthovanadate. These data further substantiated the above conclusion. However, this catalyst showed deactivation because of the formation of $BaCO_3$ that was stable under the reaction conditions.

IV. REACTION MECHANISM

The formation of alkenes from alkanes involves the breaking of two C-H bonds. It is most likely that the breaking of the first C-H bond to form an alkyl species is the rate-limiting step. Indeed, this has been shown to be the case in the oxidation of butane on vanadyl pyrophosphate, as was shown by kinetic isotope studies.[3] Whether this applies to the magnesium orthovanadate catalysts was studied by comparing the rates of reaction of various alkanes. The results are shown in Table 3.[4] It can be seen that the rates of reaction decreased as 2-methylpropane > butane > propane > ethane. This trend can be understood if the rate reflects primarily the reaction involving the

Table 3 Kinetics of Oxidation of Alkanes on 40V-Mg-O Catalyst

Alkane	Rate at 500°C (mol/min-g cat)	E_{act}[a] (kcal/mol)
Ethane	3.6×10^{-5}	31.5 ± 3
Propane	10.8×10^{-5}	34 ± 3
Butane	19.2×10^{-5}	22 ± 3
2-Methylpropane	24.8×10^{-5}	27 ± 3

[a]Determined over the range 475-540°C.

weakest C-H bond in the molecule. The bond energy of a primary, secondary, and tertiary C-H bond is 410, 395, and 375 kJ/mol, respectively.[5] Thus 2-methylpropane, which has a tertiary carbon, reacts the fastest. Butane, which has two secondary carbon atoms, reacts faster than propane which has only one secondary carbon. Ethane, which has only primary carbon, reacts the slowest.

In addition to the rate, it may be expected that the activation energy also parallels the C-H bond energy. There may exist a rough correlation. 2-Methylpropane and butane showed lower activation energies than propane and ethane (Table 3).[4]

These kinetic data are consistent with the mechanism that the first step of the reaction is the breaking of a C-H bond. This is very different from the gas phase pyrolysis of alkanes where alkane molecules are activated by breaking C-C bonds, but similar to the gas phase oxidation of alkanes.[6,7] In the gas phase oxidation reactions, alkyl radicals and HO_2 species are formed. It seems reasonable that the first step on a magnesium orthovanadate catalyst is the formation of an alkyl radical and a surface hydroxyl group. At sufficiently high temperatures (500°C and higher), the hydroxyl groups desorb rapidly as water. Whether the alkyl radical stays on the catalyst surface to react further or desorb into the gas phase remains to be determined.

At a lower reaction temperature, desorption of surface hydroxyl groups is less rapid. It becomes possible that they react with an adsorbed hydrocarbon species. This happened when hexane was oxidized at 414°C. As shown in Table 4, unlike the oxidation of lower alkanes where no oxygenates were observed, a substantial amount of hexanone was formed. It should be mentioned that at this temperature under our reaction conditions, gas phase oxidation of hexane also yielded oxygenates. However, most of the oxygenates were products of cracking reactions, i.e., they were oxygenates of less than C_6.

The oxidation of hexane most likely proceeds by first breaking a C-H bond to form an alkyl species. To form hexanone, this alkyl species may react with a surface hydroxyl to form an alcohol, which

Table 4 Product Distribution in the Oxidation of Hexane on 40V-Mg-O (414°C, hexane/O_2/He = 2.7:7.5:89.8, total flow = 100 ml/min)

Hexane conv.	14.2%
Product select., %	
CO	12.3
CO_2	26.2
Hexenes + hexadiene	29.3
Benzene	trace
2,3-Hexanone	30.2

is dehydrogenated to form a ketone. Alternatively, it may react with oxygen (or HO_2) to form a peroxy species (or peroxide) which upon breaking the O-O bond forms a ketone.

V. ROLE OF GAS PHASE REACTION

It has been reported in the literature that free radicals were desorbed from the catalyst in the oxidation of propene on bismuth molybdate,[8,9] and the oxidative coupling of methane.[10] These free radicals then reacted in the gas phase. Although the data reported here in Tables 1-4 were obtained under conditions where there were negligible reactions in the absence of the catalyst, it was not obvious whether there were any reactions in the postcatalytic volume. Experiments were conducted to test this possibility.

A special reactor was used in which a catalyst wafer could be placed at one end of the reactor. The volume next to the wafer could be varied by packing the reactor to different levels with quartz chips. It had been tested in separate experiments that if the reactor was packed with quartz chips, there was little conversion of propane in a stream of oxygen and propane. Then experiments were performed to determine the conversion in a reactor that contained a 15-ml empty volume, i.e., a 15 ml section of the reactor was not packed with quartz chips, as well as the conversion in a reactor that contained a catalyst wafer but no empty volume. The results are shown in Table 5.

Table 5 Postcatalytic Gaseous Reaction in the Oxidation of Propane on 19V-Mg-O Catalyst (570°C, propane/O_2/He = 4:8:88, total flow rate = 100 ml/min)

Reactor condition	Propane conv. (%)
1. Catalyst alone	14.0
2. Reactor with empty volume	7.9
3. Empty volume preceding catalyst	22.9
4. Catalyst preceding empty volume	28.7

The next experiments were performed with a catalyst wafer in the reactor and an empty volume adjacent to the wafer. When the feed entered the reactor from the side of the empty volume, the conversion of propane was the sum of the conversion due to the empty volume and the wafer separately. When the feed entered from the side of the catalyst wafer, the conversion of propane was substantially higher than when the feed entered from the opposite direction. These data are shown in Table 5. This can be interpreted as follows. Under these conditions, free radicals were desorbed from the catalyst wafer. If they were desorbed into an empty volume, free radical chain-branching reactions took place in the gas phase, which resulted in substantial enhancement in conversion. If the free radicals were desorbed into a volume packed with quartz chips, the chain-branching reactions were greatly inhibited.

These results clearly demonstrated the dual function of the magnesium orthovanadate catalyst. This catalyst served to activate alkane by abstracting a hydrogen atom. The alkyl radical may be desorbed into the gas phase to initiate a chain reaction; it may also react on the surface. In the latter case, the catalyst also serves to complete the oxidative dehydrogenation reaction. Compared to the gaseous reaction which activates alkane by C-C bond breaking, the catalytic reaction generates alkyl radicals without cracking the carbon chain. It should be a more desirable reaction.

VI. CONCLUSION

It has been found that orthovanadates are good oxidative dehydrogenation catalysts. It is believed that the high selectivity to dehydrogenation versus combustion is due to the presence of isolated VO_4 groups in these compounds and the absence of oxygen ions bridging two vanadium ions. Such bridging oxygen ions are too active for the selective oxidation of alkanes.

On the orthovanadate catalysts, the oxidation of alkanes proceeds by first breaking a C-H bond to form alkyl radicals. Under certain conditions, especially at higher temperatures, some of these radicals are desorbed into the gas phase to participate in gaseous free radical chain reactions. The remaining radicals stay on the surface for further oxidation. At lower temperatures when there are substantial surface hydroxyl groups, alcohols may be formed by the reaction between the surface hydroxyl groups and the alkyl radicals. This eventually leads to the formation of ketones. Alternatively, the ketones are formed from the peroxide species derived from the gaseous reaction between an alkyl radical and oxygen. Thus it seems possible to control the selectivity of the oxidation reaction by controlling the structure of the catalyst, the reaction conditions, and the reactor geometry.

ACKNOWLEDGMENT

Support of this work by the Division of Chemical Sciences, Office of Basic Energy Research, Department of Energy, grant DE-FG02-87ER13725 is gratefully acknowledged.

REFERENCES

1. M. Chaar, D. Patel, M. Kung, and H. Kung, *J. Catal., 105,* 483 (1987).
2. M. Chaar, D. Patel, and H. Kung, *J. Catal., 109,* 463 (1988).
3. M. Pepera, J. Callahan, M. Desmond, E. Milburger, P. Blum, and N. Bremer, *J. Amer. Chem. Soc., 107,* 4883 (1985).

4. D. Patel, M. Kung, and H. Kung, Proc. 9th Intern. Cong. Catal., *4* 1954 (1988).
5. H. Kung, *Ind. Eng. Chem. Prod. Res. Dev., 25,* 171 (1986).
6. S. Layokun, *Ind. Eng. Chem. Process Des. Devel., 18,* 241 (1979).
7. M. Cathonnet, J. Boettner, and H. James, Eighteenth Int. Symp. Combustion, Proceedings, 1981, p. 903.
8. W. Martir and J. Lunsford, *J. Am. Chem. Soc., 103,* 3728 (1981).
9. C. Daniel and G. Keulks, *J. Catal., 24,* 529 (1972).
10. D. Driscoll and J. Lunsford, *J. Phys. Chem., 89,* 4415 (1985).

19

The Selective Hydrogenation of Acetylene in the Vinylchloride Manufacturing Process

KLAUS M. DELLER

Degūssa AG, Hanau, Federal Republic of Germany

ABSTRACT

During the thermal cracking of 1,2-dichlorethane (EDC)—the basic reaction of the vinylchloride manufacturing process—2 mol of HCl can be cleaned in consecutive steps and acetylene is formed in small amounts.

In order to use the HCl efficiently for the production of EDC, the acetylene should be removed nearly completely. The easiest way to do this is by hydrogenation of acetylene. From an economic point of view, the selective formation of ethylene would be most attractive.

Now, a new catalyst for this type of application has been developed, which consists of an acid-resistant silica-based carrier on which palladium is deposited in a thin layer. The support material itself is unporous, so that these uncontrollable parallel and consecutive reactions can hardly occur. In pilot plant runs it was demonstrated that the activity of this catalyst is as high as that of the conventional Al_2O_3-based catalysts and that the selectivity with regard to the formation of ethylene is much higher (65-75%).

The results have been proven over several years by running a large-scale hydrogenation reactor in a vinylchloride plant of Hoechst AG (Gendorf, West Germany).

I. INTRODUCTION

Vinylchloride (VC) is one of the most important monomers for the production of numerous polymers of significant commercial interest. Since the beginning of its production in the 1930s, it has achieved an enormous market which has led to an annual world production of about 18.5 million tons.

Vinylchloride is produced worldwide in modern plants according to combined porcesses which have replaced nearly completely the old process based on addition of HCl to acetylene.[1]

Figure 1 shows the currently used VC production scheme. In Fig. 2 the corresponding reaction equations are summarized. In this process the common EDC production from ethylene and chlorine is combined with the oxichlorination of ethylene. The crude EDCs coming from both reaction steps are purified together in a separate EDC purification.

The thermal dehydrochlorination of the EDC to VC is carried out in a pyrolysis reactor at 500-600°C and 25-30 bar. After the VC separation the HCl of the top of the distillation column is recycled to the oxichlorination, so that all the HCl formed in the EDC pyrolysis can be reused.

This HCl contains, in addition to traces of VC and residual ethylene, mainly acetylene in amounts up to 2000 ppm as impurities. Acetylene is formed in the pyrolysis of EDC in a consecutive reaction by HCl elimination from vinylchloride.

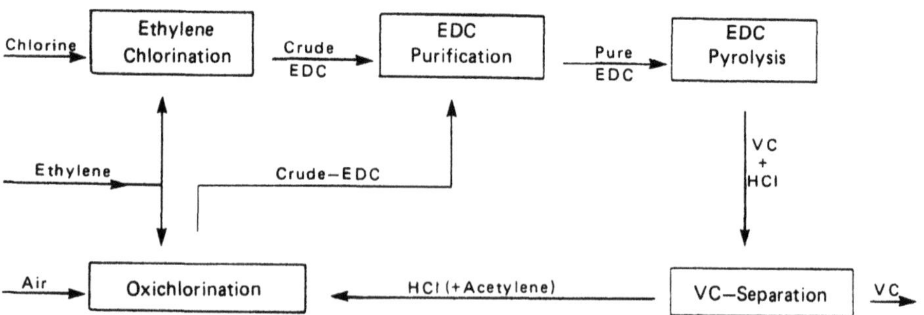

Fig. 1 Scheme of the VC manufacturing process.

1. Ethylene-Chlorination:

$$H_2C=CH_2 + Cl_2 \longrightarrow ClH_2C-CH_2Cl \qquad \Delta H_R^\circ = -180\ kJ/mol$$

2. Oxichlorination:

$$H_2C=CH_2 + 2HCl + 1/2\ O_2 \xrightarrow{CuCl_2\text{-}Cat.} H_2ClC-CH_2Cl + H_2O \qquad \Delta H_R^\circ = -239\ kJ/mol$$

3. EDC-Pyrolysis:

$$ClH_2C-CH_2Cl \xrightarrow{500-600^\circ C} H_2C=CHCl + HCl \qquad \Delta H_R^\circ = +71\ kJ/mol$$

4. Side Reaction:

$$H_2C=CHCl \longrightarrow HC\equiv CH + HCl \qquad \Delta H_R^\circ = +99\ kJ/mol$$

Fig. 2 Reaction scheme of the VC-manufacturing process.

The acetylene causes trouble, particularly in the oxichlorination step as it is partially oxidized to CO_2 and, also, it brings about the formation of higher chlorinated ethanes.

This results in additional costs for distillation and treatment of the residues as well as higher consumption of chlorine and ethylene in the total process.

In order to improve the economy of the VC production process, the installation of an additional hydrogenation step has been applied successfully by hydrogenating acetylene selectively to ethylene. By doing this, not only can the loss of chlorine resulting from the presence of acetylene be minimized, but also the raw material may be regained (Fig. 3) and the content of acetylene can be reduced below 10 ppm.

II. SELECTIVE HYDROGENATION OF ACETYLENE IN THE VINYLCHLORIDE MANUFACTURING PROCESS

A. Some General Aspects of Reaction Mechanism and Thermodynamics of Hydrogenation of Acetylene

The selective hydrogenation of acetylene to ethylene has been described quite frequently in the literature.[2-4] The reaction is carried out with heterogeneous catalysts by adding hydrogen in

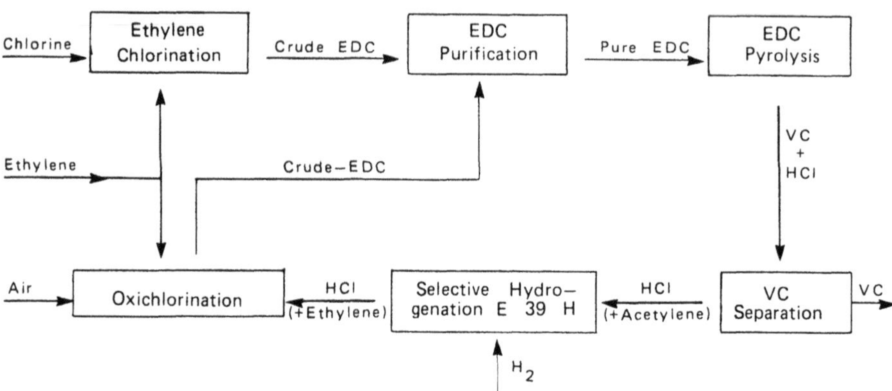

Fig. 3 Scheme of the VC manufacturing process with selective hydrogenation of acetylene.

greater than stoichiometric amounts relative to acetylene. The mechanism can be described as shown in Fig. 4.[5]

H_2 is dissociatively adsorbed on the surface of palladium and is present in the form of atomic bonded surface species. The H atoms are highly reactive and, according to the low-site exchange energy,

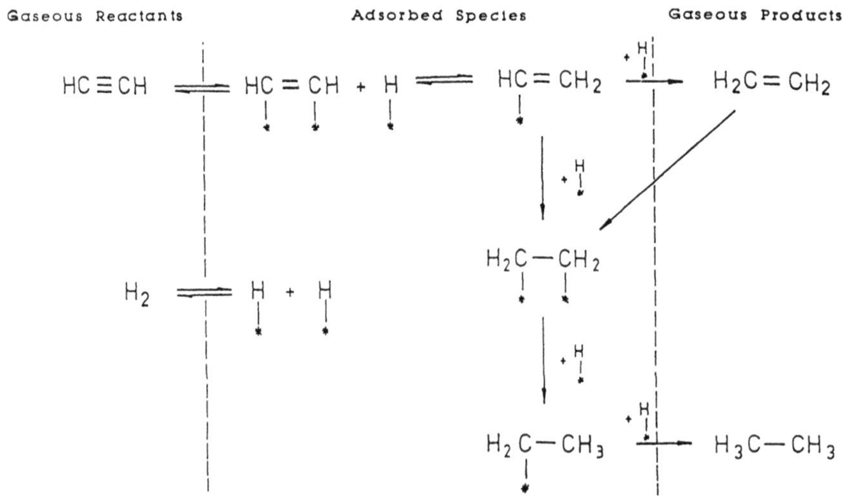

Fig. 4 Mechanism of the acetylene hydrogenation with precious metal catalysts.

have a high mobility on the catalyst surface. Acetylene is adsorbed on the palladium surface at its two C atoms in the form of a double center bond at the same time the triple bond converts to a double bond.

In the first step of the surface reaction the addition of an adsorbed H atom to an acetylene molecule takes place, and as an intermediate compound a partially hydrogenated vinyl radical is formed. This vinyl radical can be converted to the desired end product, ethylene, by addition of another H atom, or by consecutive H addition it can be hydrogenated completely to ethane.

In Fig. 5, besides the desired conversion of acetylene to ethylene and the undesired complete saturation to ethane, some other possible side reactions are listed:

a.) Hydrogenation Reactions

1. $HC{\equiv}CH + H_2 \rightleftharpoons H_2C{=}CH_2$
2. $H_2C{=}CH_2 + H_2 \rightleftharpoons H_3C{-}CH_3$

b.) Cracking Reactions

Reactions of Feed:

3. $HC{\equiv}CH \rightleftharpoons 2C + H_2$
4. $2\,HC{\equiv}CH \rightleftharpoons H_2C{=}CH_2 + 2C$
5. $2\,HC{\equiv}CH \rightleftharpoons CH_4 + 3C$

Reactions of Products:

6. $H_2C{=}CH_2 \rightleftharpoons 2C + 2H_2$
7. $H_2C{=}CH_2 \rightleftharpoons CH_4 + C$

c.) Dimerisation Reactions

8. $2\,HC{\equiv}CH \rightleftharpoons HC{\equiv}C{-}CH{=}CH_2$
9. $2\,H_2C{=}CH_2 \rightleftharpoons H_2C{=}CH{-}CH_2{-}CH_3$

Fig. 5 Catalytic hydrogenation of acetylene: main and some possible side reactions.

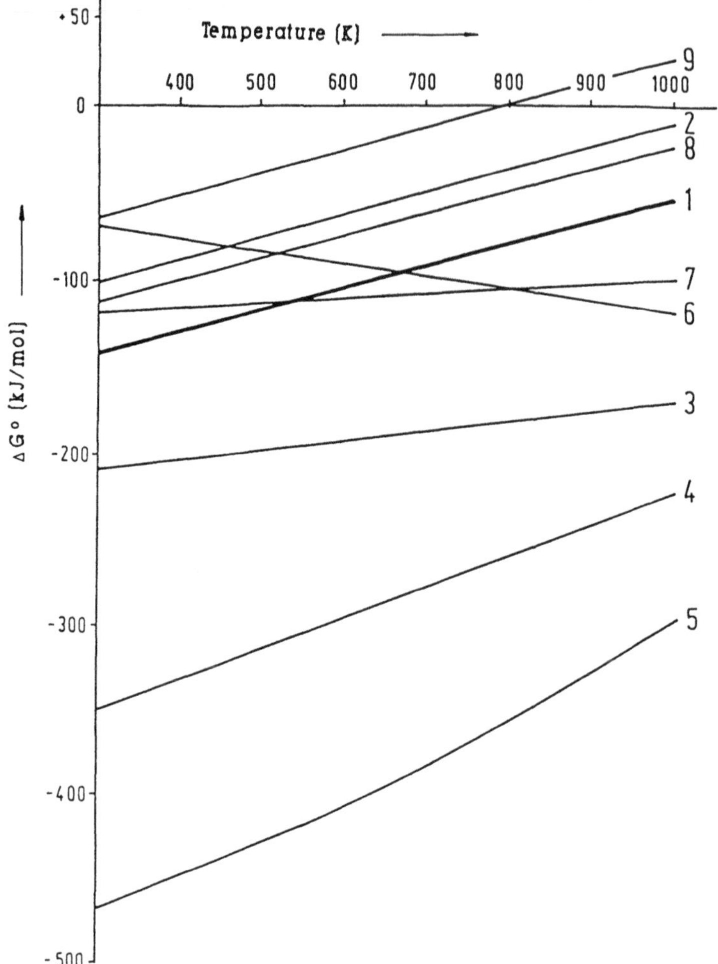

Fig. 6 Catalytic hydrogenation of acetylene Gibbs energy of main and some side reactions as a function of temperature.

Cracking reactions of feed, products and byproducts which can lead to carbon or carbon-rich coke products on the catalyst surface.

Dimerization reaction of acetylene and ethylene, which can easily continue to form higher polymerized products.

Besides these mechanistic considerations one can take thermodynamic aspects into account. If one looks at the dependence of the free reaction enthalpy on temperature, which is shown in Fig. 6, it follows that the selective hydrogenation of acetylene to ethylene (Fig. 5, Eq. 1) is favored over the hydrogenation of ethylene (Eq. 2). In both reactions the thermodynamic probability decreases with increasing temperature, so that a process temperature as low as possible is advantageous. But according to the nearly parallel curves, neither reaction can be influenced dramatically by choice of temperature. This graph shows on the other side that especially the cracking reactions (Eqs. 3-5) are favored at low temperatures.

The consecutive reactions of the products (Eqs. 6 and 7) as well as the dimerization reaction (Eq. 8)—an exception is dimerization reaction 9—are lying thermodynamically within the range of the desired reaction and it is difficult to predict a priori what reaction path will occur. Since the goal is to direct the reaction behavior selectively in the direction of the formation of ethylene, it is important to find an adequate working catalyst system.

Usually for this hydrogenation palladium catalysts on porous Al_2O_3 supports have been used. Indeed, as fresh catalysts they have a sufficiently high activity. However, the activity is quickly decreased by deposition of carbon black coke products and oily substances on the catalyst surface. Using this type of catalyst only a short lifetime of about 1 year has been achieved, and also these catalyst systems produce too much ethane.

B. Choosing the Adequate Catalyst

1. *Selection of the Active Components*

In choosing the most active component in the hydrogenation of acetylene, it emerges from numerous empirical studies that palladium is the most suitable precious metal for both activity and selectivity (Fig. 7).[6]

In comparison to other precious metals, palladium is very special as it can adsorb large amounts of hydrogen. It is well

1. Activity

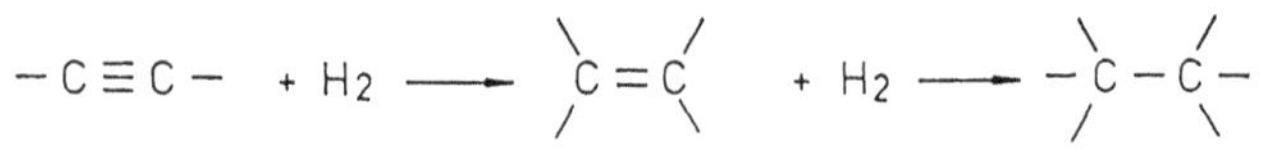

$$-C\equiv C- \; + H_2 \longrightarrow \;\rangle C=C\langle \; + H_2 \longrightarrow -\overset{|}{C}-\overset{|}{C}-$$

Pd > Ni > Pt ≈ Rh > Fe, Cu, Co, Ir > Ru, Os

2. Selectivity

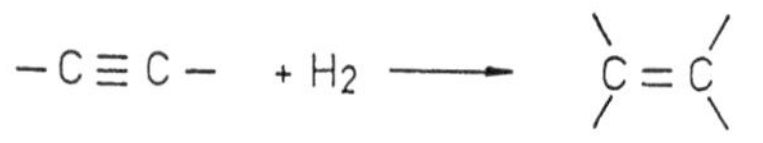

$$-C\equiv C- \; + H_2 \longrightarrow \;\rangle C=C\langle$$

Pd > Rh, Pt > Ru > Ir, Os

Fig. 7 Catalytic hydrogenation of acetylene activity and selectivity of different metals.

known that this is done by formation of a H_2-poor α phase (up to Pd $H_{0.03}$) as well as a H_2-rich β phase (up to Pd $H_{0.68}$).

Whereas the α phase is able to selectively hydrogenate acetylene to ethylene without any secondary reaction to ethane, the β phase unselectively hydrogenates acetylene to ethylene.[7]

The selectivity in the hydrogenation of acetylene is also influenced by the stronger adsorption of acetylene in comparison to ethylene on palladium, a fact by which the further hydrogenation of ethylene is suppressed.

2. Choosing the Right Support

A suitable support material, which will be covered with palladium, must be chosen carefully, especially with regard to the acidity of the material. To avoid polymerization and cracking reactions, the support material should have no acidic surface centers.

From the undesired consecutive hydrogenation reaction of ethylene to ethane it is obvious that a nonporous support material should be used on which the precious metal should be deposited in the form of a thin layer. Figure 8 shows that by using a porous

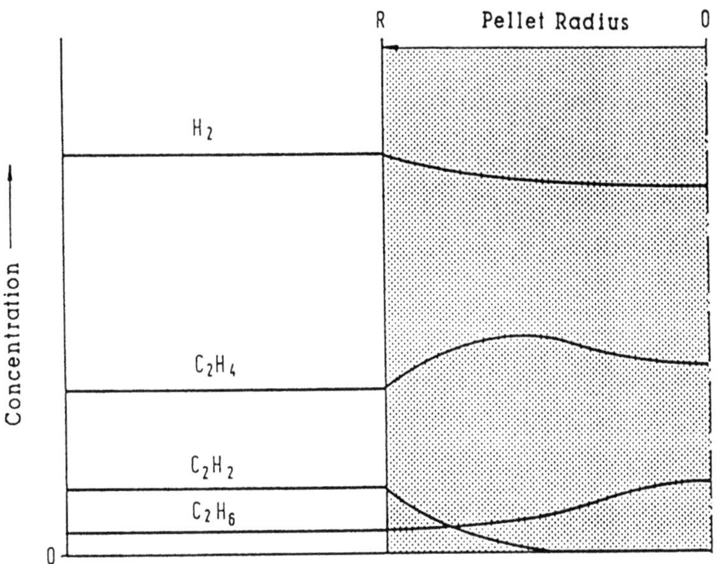

Fig. 8 Radial concentration profile in a porous catalyst pellet.

support the decrease of the acetylene concentration in the inner pores leads preferably to ethane as a consecutive product of ethylene.

New investigations indicate that Al_2O_3 as support material strongly favors the formation of ethane by a spillover effect of H_2 in the presence of carbon depositions which are formed by aging of the catalyst.[8]

During our development work we quickly gave up the use of Al_2O_3 supports, even in calcined forms, for this application since catalyst systems based on Al_2O_3 showed a dramatic decrease both in activity and selectivity within the first days of use.

From all available support materials based on SiO_2 a very unporous acid-resistant SiO_2 granulate with a low BET surface area seemed to be the most promising.

C. Catalyst E39H 0.15% Palladium on SiO_2

Based on the previous considerations, a new catalyst for the selective hydrogenation of acetylene in the production of VC was developed.

Pd- Content	0,15 %
Support	SiO_2-Granules (acid resistant) Ø 3-5 mm SiO_2)99 % Traces of Fe, Ca, Al: (0,5 %
Bulk Density	1,5 - 1,6 kg/dm^3
Void Fraction	35 - 40 %
Specific Suface Area	(1 m^2/g
Crushing Strength	50 N/cm^2
CO-Adsorption	0,04 ml/g
Crystallite Size	0,1 - 0,5 μm

Fig. 9 Typical data of catalyst E39H for selective hydrogenation of acetylene.

By using a special preparation method it is possible to deposit the precious metal palladium as a thin layer highly dispersed on an acid-resistant SiO_2 support with low BET surface area. Figure 9 shows the typical physicochemical data. A picture of the catalyst is shown in Fig. 10. Figure 11 shows the layer of palladium crystals covering the surface of the support nearly completely, so that undesired side reactions caused by the surface chemistry of the support are minimized. The thickness of the layer is on average 8 μm, and the palladium particles have a crystallite size of 0.1-0.5 μm.

Fig. 10 Catalyst E39H 0.15% Pd on SiO_2 granules for selective hydrogenation of acetylene.

Magnification 10.000 : 1

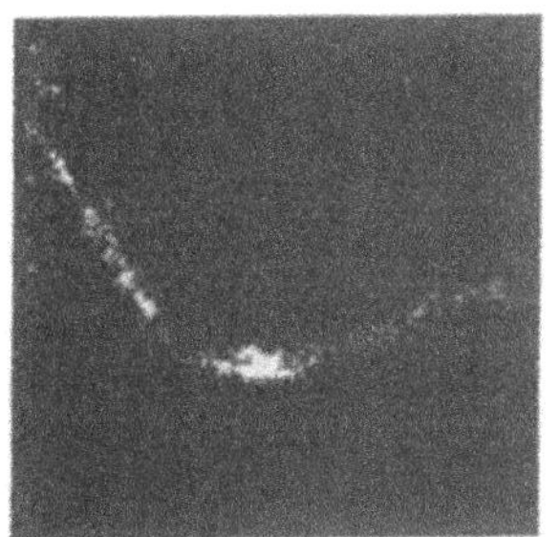

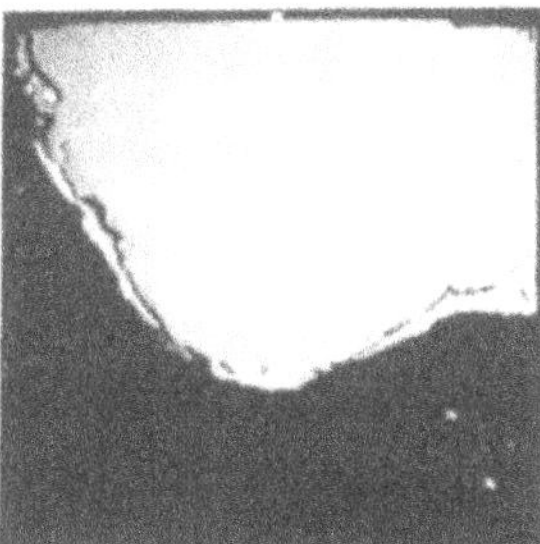

Magnification 225 : 1

Fig. 11 Picture of the precious metal distribution of catalyst E39H.

D. Results from Laboratory and Pilot Plant Experiments

After the lab experiments and screening tests in which we found out that catalyst E39H shows the best results, a pilot scale slipstream reactor was installed at the industrial plant of the VCM unit of Hoechst AG, West Germany. Figure 12 shows a flow scheme of the reactor. The reactor tube itself contained 15 liters of catalyst with a length of 3000 mm and a diameter of 82.9 mm.

The reaction mixture coming from the top of the industrial reactor is heated in a steam-jacketed preheater. The feed flow is

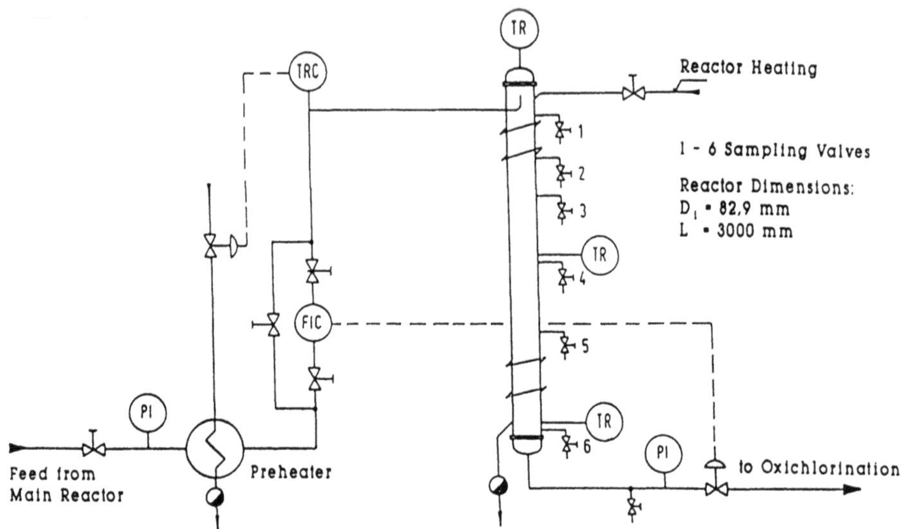

Fig. 12 Pilot plant for the selective hydrogenation of acetylene.

regulated with a valve after the reactor exit. The reactor is equipped with three temperature control points before and after the catalyst bed. In addition, six sampling valves are installed to measure the concentration profile at the entrance and exit and in the catalyst bed at different heights.

Since the concentration of the reactants is low, the heat generated by the reaction in the industrial reactor causes a temperature increase of only about 4-6°C. To come to a similar temperature profile conditions expected commercially, the pilot reactor was equipped with an additional heating system which allowed running the reactor nearly isothermally.

In a test program the reaction conditions as T, P, GHSV, and H_2/C_2H_2 ratio were varied within the limits of the data of the main reactor. Figure 13 shows the reaction conditions under which the catalyst was run. The GHSV was varied up to 2200 hr^{-1}. Acetylene concentrations in the off-gas of about 10 ppm can also be achieved at higher GHSV but with a corresponding temperature increase. The

Gas Hourly Space Velocity (GHSV)	up to 2200 h^{-1}
Temperature	130 - 180° C
Pressure	6 - 8 bar
H_2/C_2H_2 Ratio	3 : 1 to 5 : 1
Composition Feed Gas:	
HCl	) 99.5 %
C_2H_2	1200 - 2000 ppm
C_2H_4	100 - 200 ppm
VC	5 - 20 ppm

Fig. 13 Process conditions for catalyst E39H.

temperature should not exceed 180°C because with increasing temperature cracking and coking reactions can be observed which lead to catalyst deactivation. The ratio of H_2 to acetylene was varied in the range of 3-5 times the stoichiometric amount. If the H_2 quantity is too small, we observe less conversion; too much H_2 favors the formation of ethane by complete hydrogenation.

A typical concentration profile of the reactants acetylene, ethylene, and ethane along the length of the catalyst bed is shown in Fig. 14. The increase of ethane is related on the one hand to the decrease of acetylene and on the other hand to the concentration of ethylene. For the formation of ethane apparently two different mechanisms are responsible:

1. In the linear range of the concentration profile of acetylene, which indicates a reaction rate of zero order,[3,9] the active centers of the catalyst are preferably blocked by the adsorbed acetylene, so that the adsorption of gaseous ethylene cannot occur. However, the reaction product ethylene formed on the catalyst surface can be partially hydrogenated further to ethane before its desorption. This has been described in the reaction mechanism before.

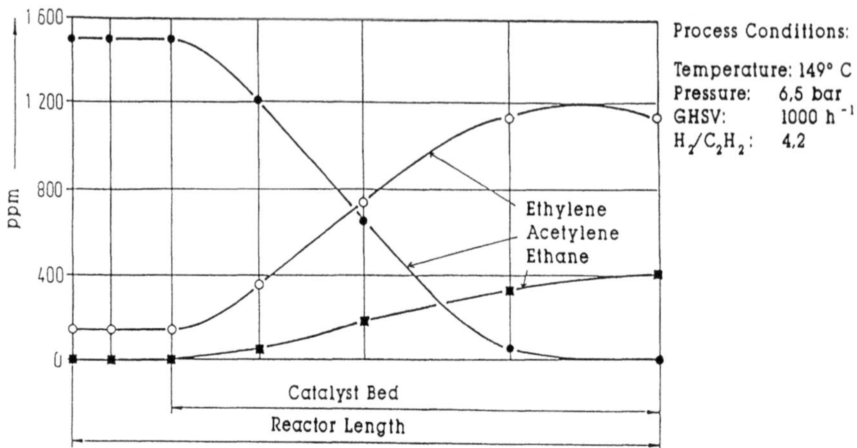

Fig. 14 Axial concentration profiles in the hydrogenation reactor.

2. In the region of the reactor exit there is almost no acetylene present; additional ethylene is present. The active centers are now also available for ethylene adsorption out of the gaseous phase which can lead to more ethylene hydrogenation to ethane. The hydrogenation of ethylene under the described reaction conditions is considerably lower than the hydrogenation of acetylene, which is highly advantageous for the desired high selectivity. At lower temperatures (20-80°C) apparently the opposite effect can be observed. For the commercial operation of the hydrogenation reactor the temperature should be chosen so that at changing throughputs the acetylene is nearly completely converted at the exit of the reactor. This is necessary to achieve the highest possible ethylene yield.

In Fig. 15 the reaction temperature which is necessary to achieve the desired 99% acetylene conversion (inlet concentration 1500 ppm; outlet residual content of about 10 ppm) is plotted in relation to the GHSV. The results are based on four times the stoichiometric excess of H_2. By adjusting the temperature at

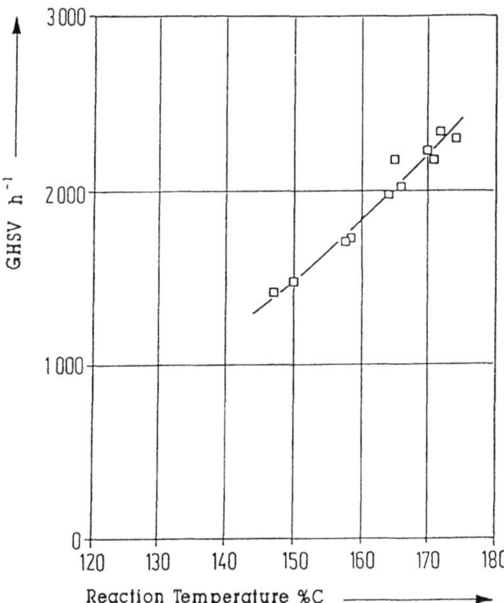

Fig. 15 Optimum GHSV as a function of reaction temperature.

changing throughputs it is possible, as the graph in Fig. 15 shows, to maintain as constant the residual concentration of acetylene in the outlet gas stream.

E. Operating the Catalyst in an Industrial Reactor at Hoechst AG, FRG

Catalyst E39H was tested for the first time in a large-scale hydrogenation reactor of the VC plant of Hoechst, Gendorf, West Germany. Figure 16 shows the catalyst behavior in the first period of the technical trial. It can be seen that in the first months the C_2 balance of the gas phase shows a distinct deficiency. This can be explained by coke deposition on the catalyst surface caused by the previously mentioned side reactions.

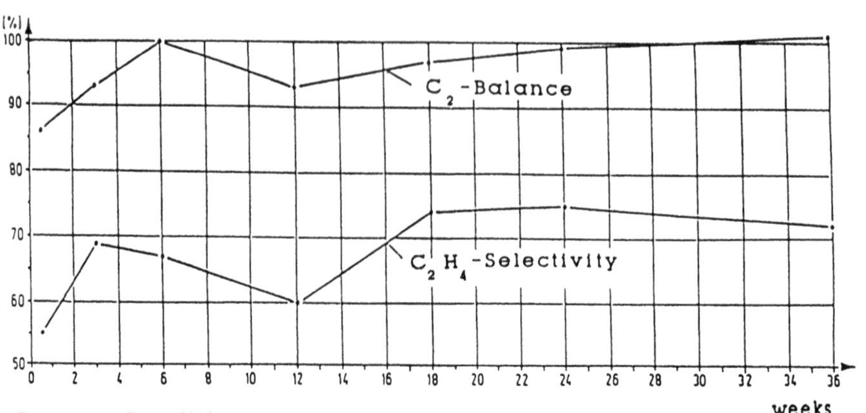

Process Conditions:

Temperature:	150 - 170° C	C_2H_2 input:	1200 - 1600ppm
Pressure:	6.5 bar	C_2H_2 output:	< 5 ppm
GHSV:	1200 - 2200 h^{-1}	H_2/C_2H_2 :	3 - 5

Fig. 16 C_2- balance and ethylene selectivity as a function of operating time.

The selectivity widely varies in the first month in the region of 55-75% and follows the C_2 balance curve.

A catalyst induction period of about 2 years occurs in which coke deposition takes place until constant conditions at the catalyst surface are obtained. The selectivity then reaches its maximum at about 75%.

In Fig. 17 the catalyst behavior in the Hoechst plant is shown during the whole lifetime of the catalyst. The lifetime of the catalyst has been about 9 years, although the throughput varied and to some extent has been quite high. The conversion has always been higher than 99%. The selectivity increases to a maximum of about 76% in the middle of the catalyst life and decreases again at the end of the lifetime.

In addition, the activity starts to decrease beginning with the eighth year of life and the temperature has to be increased. As a temperature of 180°C was necessary to achieve the desired residual acetylene concentration, the catalyst was replaced. The analysis

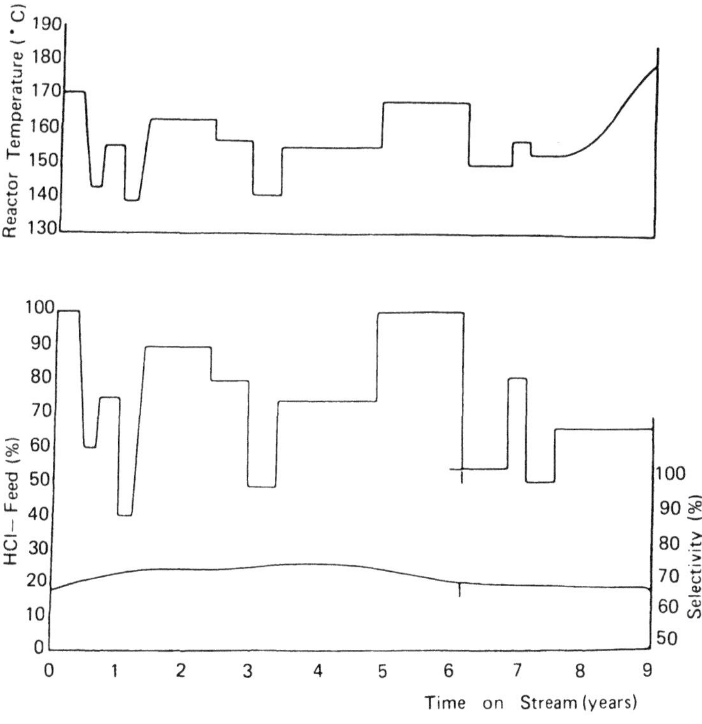

Fig. 17 Performance of catalyst E39H in an industrial plant as a function of time on stream.

of the used catalyst showed that about 3% coke depositions on the catalyst surface led to the final activity loss. The palladium content of the used catalyst had been still high, so that it can be assumed that the active centers were blocked by coke formation. Furthermore, such heavy metals as Fe and Ti from the reactor materials, Cu from the oxichlorination, and Hg from the H_2 feed can lead to a quicker deactivation of the catalyst.

The precious metal of the catalyst can be reclaimed in high yield. If the deactivation of the catalyst mainly is caused by carbon deposition, the catalyst also can be regenerated.

III. CONCLUSIONS

In summary, it can be stated that catalyst E39H 0.15% palladium on a SiO_2 support has shown to be a suitable catalyst for the selective hydrogenation of acetylene in the VCM process. This catalyst has been used worldwide in different VCM processes. In comparison to conventional catalysts based on Al_2O_3, this catalyst shows that

High selectivity at a high activity level is obtained which means high ethylene recycling and less chlorine consumption.

Long catalyst life up to 9 years can be reached. Under the usual process conditions the catalyst life is greater than 6 years.

High precious metal recovery from spent catalyst is observed.

ACKNOWLEDGMENTS

The author thanks Dr. W. Kühn, Mr. Brandstetter, Dr. Oberrauch, Dr. Riedl, Mr. Schwarzmaier, and Mr. Trost of Hoechst AG, Gendorf, West Germany for fruitful cooperation and for running most of the plant and pilot plant experiments. The author also thanks Dr. G. Vollheim, Dr. H. Müller, Dr. Kastenhuber, Mr. H. Mösinger, and Dr. A. Strätz from Degussa AG for collaboration and valuable discussions.

REFERENCES

1. *Ullmann's Encyklopädie der Technischen Chemie,* Bd. 9, Butadien bis Cytostatica, VCH-Verlag, Weinheim/Bergstr (1975).
2. G. C. Bond, *Catalysis by Metals,* Academic Press, London, 1962.
3. P. Mars and M. J. Gorgels, 3rd European Symposium on Chemical Reaction Engineering, Amsterdam, 1964, S. 55.
4. W. T. Mac Grown, C. Kemball, D. Whan, and M. S. Scurell, *Trans. Faraday Soc., 73,* 632 (1977).

5. G. C. Bond, *Heterogeneous Catalysis: Principles and Applications,* Clarendon Press, Oxford, 1974.

6. B. Imelik, G.-A. Martin, and A.-J. Renonprez, *Catalyse par les metaux.* Editions du Centre National de la Recherche Scientifique, Paris, 1984.

7. Gmelin, *Handbook of Inorganic Chemistry*, 8th ed., Pt. Supplement, Vol. A1, Springer-Verlag, Berlin, 1986, p. 203.

8. A. H. Weiss, S. le Viness, V. Nair, L. Guczi, A. Sarkany, and Z. Schay, The Effect of Pd Dispersion in Acetylene Selective Hydrogenation, 8th International Congress on Catalysis, Berlin (West), 1984, Proceedings, Vol. 5, p. 591.

9. E. G. Schlosser, *Heterogene Katalvse*, VCH-Verlag, Weinheim, 1972.

20

A Chemical Model for the Amoco "MC" Oxygenation Process to Produce Terephthalic Acid

WALT PARTENHEIMER

Amoco Chemical Company, Naperville, Illinois

ABSTRACT

One major method for producing terephthalic acid (1,4-dicarboxybenzene) is via the reaction of *p*-xylene and dioxygen in acetic acid. The catalysts used for this reaction are soluble forms of cobalt(II), manganese(II), and bromide salts. The number of elementary reactions that occur between bromide, Co(II), Co(III), Mn(II), Mn(III), acetic acid, and *p*-xylene have been characterized. These are used to rationalize why the Amoco MC (Mid-Century) process is more active and selective than all of the current alternative processes to make terephthalic acid.

Kinetic and thermodynamic arguments can be used to explain why *p*-xylene does not react with dioxygen. This problem is overcome by forming a highly unstable radical of *p*-xylene. The formation of this radical (the initiation step) can occur in acetic acid via the reaction of cobalt(II) acetate with dioxygen. This initiation sequence becomes greatly amplified via a propagation step with the *p*-xylene radicals. Using only cobalt as the catalyst produces very poor yields of terephthalic acid because of the electron-withdrawing nature of the carboxy group. The yield cannot be improved by use of high temperatures due to the decarboxylation of acetic acid by cobalt(III). This has led to alternative methods of terephthalic acid manufacture such as partial oxidation and esterification (Witten process), changing the cobalt catalyst and cooxidation (Eastman process). A large increase in activity and selectivity (at a given temperature) and an elimination of a thermal barrier occurs when bromide is added to cobalt. A series of reactions and their relative rates are proposed to explain why this occurs. The addition of manganese to cobalt and bromide results in increased activity and selectivity because of a rapid electron transfer reaction between cobalt(III) and manganese(II) and because manganese(III) decarboxylates the solvent at a rate much less than that of cobalt(II).

I. INTRODUCTION: DEFINITION, ORIGIN, AND USES OF THE AMOCO MC PROCESS

About 97% of the terephthalic acid (1,4-dicarboxybenzene) produced is used to produce polyethylene terephthalate:

$$\underset{\text{terephthalic acid}}{1,4\text{-}C_6H_4(COOH)_2} + \underset{\text{ethylene glycol}}{HOCH_2CH_2OH} \longrightarrow \underset{\text{polyethylene terephthalate}}{HO(\text{-}CH_2CH_2O_2CC_6H_4CO_2)_nCH_2CH_2OH} \quad (1)$$

This polyester is used for fibers, films, bottle resins, and various engineered plastics.[1]

The cheapest reagent for the preparation of terephthalic acid is *p*-xylene (1,4-dimethylbenzene), which is readily available from oil refineries, and dioxygen, which can be obtained from air. The resulting autooxidation[2] of *p*-xylene leads to a mixture of predominately *p*-toluic acid and terephthalic acid:

$$\underset{p\text{-xylene}}{1,4\text{-}C_6H_4(CH_3)_2} + O_2 \xrightarrow{HOAc} \underset{p\text{-toluic acid}}{CH_3C_6H_4COOH} + \underset{\text{terephthalic acid}}{HOOCC_6H_4COOH} \quad (2)$$

Up to 1958 the best reported yields of terephthalic acid was only approximately 15%[3] using cobalt(II) salts as a catalyst. In 1958, A. Saffer and R. A. Barker of Scientific Design claimed[4] a new, highly active autooxidation catalyst consisting of soluble salts of cobalt(II), manganese(II), and bromide. The yields of terephthalic acid were in excess of 90% as compared to 15% obtained using just cobalt(II) salts. Amoco (then Standard Oil of Indiana) purchased the holding company of this patent, Mid-Century Corporation,[5] and hence the patent rights to this catalyst.

The major use of the Co/Mn/Br catalyst is for the production of terephththalic acid; however, it works well for oxidizing most other methylated benzenes to aromatic acids. Amoco also produces isophthalic acid from *m*-xylene, trimellitic acid from pseudocumene (1,2,4-trimethylbenzene), and a number of other "fine" acids such as trimesic and 2,6-dicarboxynaphthalene.

For the remainder of the chapter, I will attempt to rationalize why we use cobalt, manganese, bromide, and acetic acid in this process. This will be, of course, a retrospective look since the original discovery almost certainly was not obtained in this way. The approach will be looking at a series of elementary reactions and then piecing them together into a cohesive model. There is no guarantee that the model is "right" or complete. All of the roles of the various catalyst components may not have been found. Any person acquainted with autooxidation is very aware of its many synergisms and idiosyncrasies.[6] General references on autooxidation are available for more details.[7]

II. WHY ARE COBALT AND ACETIC ACID USED IN THIS PROCESS?

A. Kinetic and Thermodynamic Barriers Preventing the Reaction of *p*-Xylene with Dioxygen

We found experimentally that toluene will not react with dioxygen even at high temperatures and pressures:

$$C_6H_5CH_3 + O_2 \xrightarrow[\text{HOAc}]{205°\text{C, 27 bar}} \text{no evidence for oxygen uptake} \quad (3)$$

The oxygen atom, a species having a triplet spin state, does not react with nitrogen or carbon monoxide, both singlet species, because of the rule of "conservation of spin."[10] This produces a large activation energy barrier which makes the reactions slow. In a similar way, dioxygen, with its triplet ground spin state, does not react with hydrocarbons since they also exist in a singlet ground state. Consistent with this concept is the fact that singlet oxygen rapidly reacts with hydrocarbons.[8]

One could initiate an autooxidation if we could produce a radical of the *p*-xylene since it would rapidly react with triplet dioxygen at room temperature[12]:

$$R\cdot + O_2 \quad RO_2\cdot \tag{4}$$

Radicals of hydrocarbons, however, are not easily prepared since C-H bonds are 70-100 kcal/mol:

$$C_6H_5CH_3 \longrightarrow C_6H_5CH_2\cdot + H\cdot \qquad \Delta H = 85\ \text{kcal/mol} \tag{5}$$

$$C_6H_5CH_3 + C_6H_5CH_3 \longrightarrow 2C_6H_5CH_2\cdot + H_2 \qquad \Delta H = 66\ \text{kcal/mol} \tag{6}$$

Thus, the high bond dissociation energies of most hydrocarbons prevent their formation under 500°C.[11]

The other possible way of preparing the radical of a hydrocarbon is by reacting it with oxygen. However, it is often not realized that the initial one-electron reduction potential of dioxygen potential is only -0.2 V. Thus dioxygen itself is a very weak oxidizing agent.[8] The heats of reaction of dioxygen with hydrocarbons to form radicals have been estimated by Benson,[13] and he finds them to be strongly endothermic and negligibly slow below 300°C.

$$RCH_3 + O_2 \longrightarrow RCH_2\cdot + HO_2\cdot \qquad \Delta H = +38\ \text{kcal/mol} \tag{7}$$

$$2RCH_3 + O_2 \longrightarrow 2RCH_2\cdot + H_2O_2 \qquad \Delta H = +41\ \text{kcal/mol} \tag{8}$$

B. The Reaction of Transition Metal Acetates with Acetic Acid

A method for modifying the spin state and reactivity of dioxygen is to form a metal dioxygen complex in solution. We did a series of simple experiments to see if transition metals would react with dioxygen in acetic acid by placing them in the solvent and detecting any uptake of dioxygen. Table 1 illustrates that, indeed, there is

Table 1 Rates of Oxygen Uptake, ml O_2/min, for Metal Acetates with and without *p*-Toluic Acid in Acetic Acid

Catalyst	HOAc alone[a]	*p*-Toluic acid[b]
None	0.0	0.0
$Ca(II)(OAc)_2$	0.0	0.0
$Ti(IV)O(ACAC)_2$	0.0	0.0
$V(IV)O(OAc)_2$	0.0	0.0
$Cr(III)(OAc)_3$	0.0	0.0
$Mn(II)(OAc)_2$	0.05	0.1
$Fe(II)(OAc)_2$	0.0	0.0
$Co(II)(OAc)_2$	0.35	1.6
$Ni(II)(OAc)_2$	0.0	0.1
$Cu(II)(OAc)_2$	0.08	0.0
$Zn(II)(OAc)_2$	0.0	0.0

[a]At 113°C, flow rate of air 25 ml/min. Metal concentration 0.120 M except for Ti, V, and Cr which were saturated (solids present in the flask).
[b]At 95°C, 45 ml/min of air, 25.0 g *p*-toluic acid, 75 ml of HOAc, concentration of metals 0.02-0.04 M. Reactions were initiated with azobis(methylpropylnitrile).

oxygen uptake in three cases, with by far the highest uptake occurring with cobalt. (The results for manganese and copper are close to experimental error.) If one attempts to oxidize *p*-toluic acid in acetic acid using these same metal salts, one again finds the highest oxygen uptake occurring with cobalt. This correlation suggests that we might obtain some important clue to the importance of cobalt by understanding the nature of the cobalt-dioxygen interaction in acetic acid.

C. The Reaction of Cobalt(II) Acetate with Acetic Acid

We characterized this reaction in four ways:

1. We find that the reaction of cobalt(II) acetate, acetic acid, and dioxygen forms carbon dioxide, much less methane, and traces of carbon monoxide; see Fig. 1.

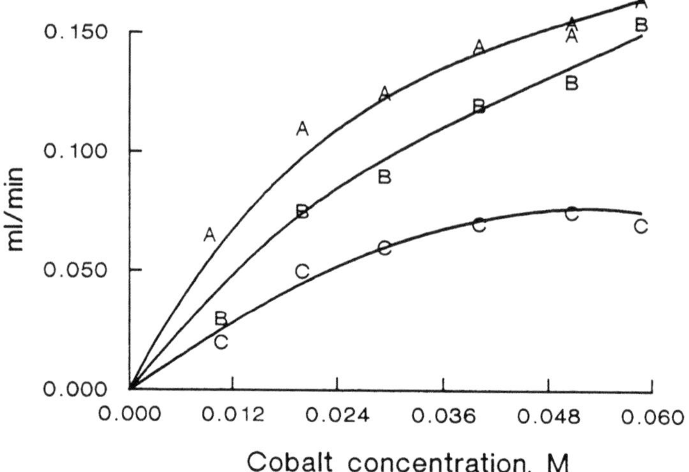

Fig. 1 Vent gases formed when cobalt(II) acetate is placed in acetic acid at 113°C; 30 ml/min air into reactor. A = rate of reaction of dioxygen, B = rate of carbon dioxide formation, C = 10 x rate of methane formation.

2. Measurement of the UV-VIS spectrum of a solution of cobalt(II) acetate in acetic acid, with and without air, is shown in Fig. 2. Using an authentic sample of cobalt(III) acetate as a standard, we performed a multicomponent fit of the spectrum assuming that Co(II) and Co(III) acetate are present. 1.2% of the cobalt present exists as cobalt(III).
3. An argentometric titration of the cobalt(II) acetate solution heated with air gives a value of 0.89% cobalt(III).
4. Decomposition of an authentic sample of cobalt(III) in acetic acid at 100°C and subsequent vent analysis illustrates that the products are carbon dioxide, methane, and a trace of carbon monoxide.

The above is consistent with the following sequence of reactions:

$$Co(II)(OAc)_2 + O_2 \longrightarrow Co(III)(OAc)_3 + ? \quad (9)$$

$$Co(III)(OAc)_3 \longrightarrow Co(II)(OAc)_2 + CH_3COO\cdot \quad (10)$$

$$CH_3COO\cdot \longrightarrow CH_3\cdot + CO_2 \quad (11)$$

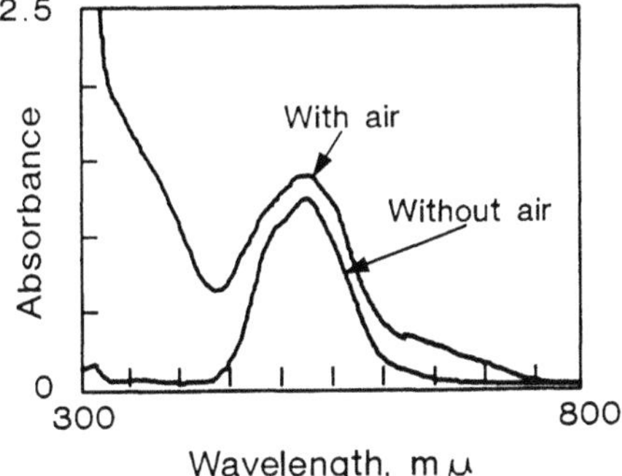

Fig. 2 UV-VIS spectrum of a 0.080 M solution of cobalt(II) acetate heated to 113°C prior to measurement with and without a flow of air (30 ml/min in 100 ml HOAc).

$$CH_3 + HOAc \longrightarrow CH_4 + \cdot CH_2CO_2H + ? \quad (12)$$

$$CH_3\cdot + O_2 \longrightarrow CH_3O_2\cdot \quad (13)$$

$$CH_3O_2\cdot + Co(II)(OAc)_2 + HOAc \longrightarrow Co(III)(OAc)_3 + ? \quad (14)$$

The sequence is a chain reaction involving the cooxidation of cobalt(II) and methyl radicals using $CH_3\cdot$ and $CH_3O_2\cdot$ as chain carriers. It produces the experimentally observed Co(III), carbon dioxide, and methane. We postulate the chain reaction because the initial reaction of Co(II) acetate with dioxygen is probably thermodynamically unfavorable in acetic acid. The chain reaction amplifies the amount of cobalt(III) so that it can be experimentally seen.

D. The Initiation of the Oxidation of *p*-Xylene

The characterization of the reaction of cobalt with acetic acid allows us to make the radical of *p*-xylene in two ways;

$$Co(III)\ (OAc)_3 + H_3CC_6H_4CH_3 \longrightarrow Co(II)\ (OAc)_2 + H^+ + H_3CC_6H_4CH_2\cdot \quad (15)$$

$$CH_3\cdot + H_3CC_6H_4CH_3 \longrightarrow CH_4 + H_3CC_6H_4CH_2\cdot \quad (16)$$

The reaction of cobalt(III) with methylbenzenes has been extensively studied.[14] We commonly use cobalt(III) acetate to initiate autooxidation in our laboratory. In reaction 16, the hydrogen atom abstraction by the methyl radical may not be very important, however. This is because the methyl radical can favorably attack in a thermodynamic sense nearly any C-H bond present including acetic acid and the ring hydrogen atoms.[15]

E. Amplification of the Reaction of *p*-Xylene with Dioxygen

The formation of the radical of *p*-xylene allows us to invoke the classical chain reaction scheme, which has been extensively studied[16]:

Initiation:

$$Co(II)(OAc)_2 + HOAc + O_2 \longrightarrow Co(III)(OAc)_3 + ? \quad (17)$$

$$Co(III)(OAc)_3 + H_3CC_6H_4CH_3 \longrightarrow Co(II)(OAc)_2 + H^+ + H_3CC_6H_4CH_2\cdot \quad (18)$$

Propagation (amplification)

$$H_3CC_6H_4CH_2OO\cdot + O_2 \longrightarrow H_3CC_6H_4CH_2OO\cdot \quad (19)$$

$$H_3CC_6H_4CH_2OO\cdot + H_3CC_6H_4CH_3 \longrightarrow H_3CC_6H_4CH_2OOH + H_3CC_6H_4CH_2\cdot \quad (20)$$

Chain Branching (further amplification)

$$H_3CC_6H_4CH_2OOH + Co(II)(OAc)_2 \longrightarrow OH^- + H_3CC_6H_4CH_2O\cdot + Co(III)(OAc)_3 \quad (21)$$

The chain-branching reaction results in the formation of the alkoxy radical of *p*-xylene which can then form the first observable product in the autooxidation of *p*-xylene—the alcohol:

$$H_3CC_6H_4CH_2O\cdot + H_3CC_6H_4CH_3 \longrightarrow H_3CC_6H_4CH_2OH + H_3CC_6H_4CH_2\cdot \quad (22)$$

The alcohol can now enter into the chain reaction sequence to form aldehydes and aromatic acids, which has been discussed elsewhere.[16]

Some of the roles of the metal catalyst are illustrated in the top of Fig. 3. Peroxides and other potential oxidants, such as hydroxyl radicals, are indicated by (o) in this figure. One role of the cobalt is to decompose the peroxides. At the same time, the cobalt may be oxidized to Co(III). The Co(III) can then initiate (or propagate) the chain by reacting with the methyl aromatic to form the radical. Included in this figure are two important side reactions—the decarboxylation of acetic acid and *p*-toluic acid (or any other aromatic acid present) by cobalt(III). The decarboxylation

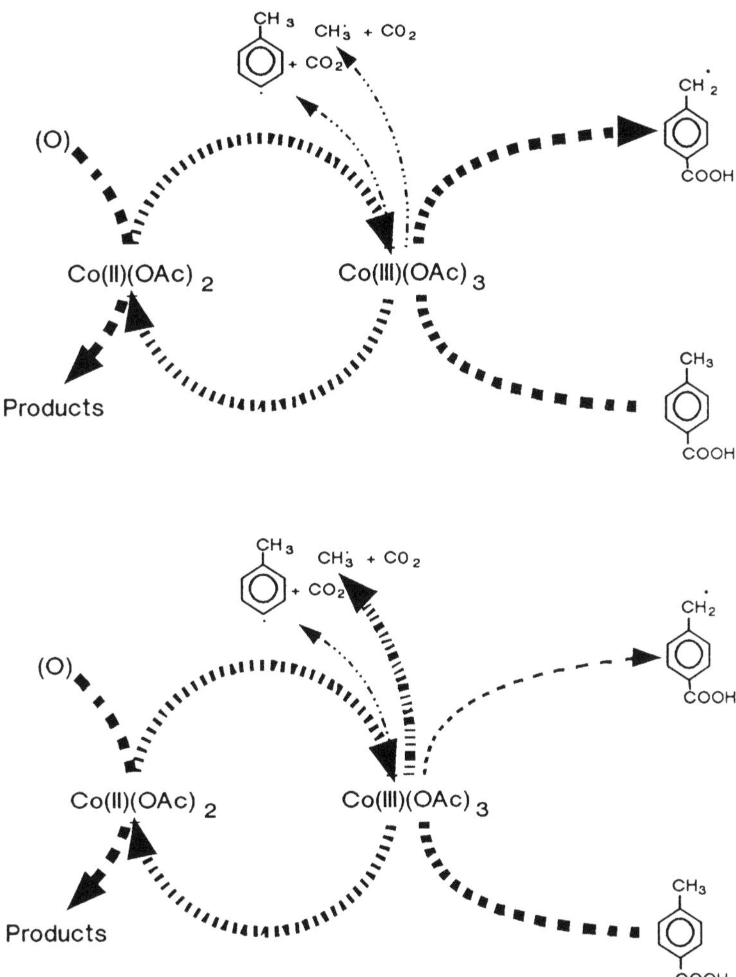

Fig. 3 Catalytic cycle in which the generated peroxides propagate the formation of the radical of *p*-toluic acid. The relative broadness of the lines indicates their relative importance. The bottom half of this figure represents higher temperatures than the top half.

of acetic acid can produce three of the most important observed byproducts of autooxidations: carbon dioxide, methyl acetate (from the methyl radical), and trimellitic acid (via attack of the methyl radical on the aromatic ring).[18] Decarboxylation of aromatic acids can produce benzoic acid, phenols, and biphenyls.[18] Tracer studies in *p*-xylene oxidations have confirmed that the carbon oxides come from both *p*-xylene and the acetic acid.[19]

F. Summary

There are kinetic and thermodynamic barriers which prevent the reaction of dioxygen with *p*-xylene at temperatures lower than 300°C. Among the first row transition metals, cobalt appears unique in its reaction with the solvent, acetic acid. This answers our question of why acetic acid and cobalt are used in this process. This weak reaction produces a small amount of cobalt(III) which initiates the oxidation. The reaction then becomes greatly amplified using a classical chain reaction scheme which has been extensively characterized.[16] The conservation of spin rule has been avoided by having dioxygen initially interact with the cobalt. We also assume that there is a transient cobalt(II)-dioxygen complex formed in acetic acid in very low concentration,[17] although we are not aware of any evidence for its formation.

III. WHY ONE OBTAINS LOW YIELDS OF TEREPHTHALIC ACID USING A COBALT CATALYST IN ACETIC ACID

A. The Electronic Effect When the First Methyl Group Is Oxidized

From the Hammett study of Heiba et al., the rate of reaction of Co(III) in acetic acid of *p*-toluic acid is 26 times slower than that of *p*-xylene.[20] Consistent with this is the fact that the oxidation of *p*-xylene occurs in steps:

$$\underset{p\text{-xylene}}{H_3CC_6H_4CH_3} + \frac{3}{2}O_2 \longrightarrow \underset{p\text{-toluic acid}}{H_3CC_6H_4COOH} + H_2O \tag{23}$$

$$H_3CC_6H_4COOH + \frac{3}{2}O_2 \longrightarrow \underset{\text{terephthalic acid}}{HOOCC_6H_4COOH} + H_2O \tag{24}$$

Since the rate of reactions of autooxidations is directly proportional to the rate of initiation of radicals,[24] the rate of Reaction 23 is much faster than that of Reaction 24.

B. Deactivation Mechanisms in Autooxidations

The slow rate of *p*-toluic acid oxygenation presents obvious economic problems (long residence times) but doesn't explain why the reaction normally stops at low yields of terephthalic acid. There are at least four deactivation mechanisms that can cause the reaction to terminate:

1. Poor mixing or "dead spots" in the reactor causing the radicals of *p*-toluic acid to dimerize[25] rather than to react with dioxygen. This is a termination step in the chain reaction. The loss of radicals interferes in the propagation sequence, Reaction 19.
2. Decarboxylation of carboxylic acids by Co(III). The effect of this can be visualized in Fig. 3. The desirable reaction of Co(III) with *p*-toluic acid to give its radical is in competition with the undesirable decarboxylation reactions of Co(III). The rate of *p*-toluic acid formation will be decreased as the rate of decarboxylation reactions increase.
3. The phenols that form (presumably by decarboxylation and decarbonylation in autooxidations)[18] are extraordinarily strong oxidation inhibitors.[22]
4. The formation of the byproduct trimellitic acid[18] (and other high-boiling acids) can deactivate and precipitate the catalyst.[23]

C. The Temperature Barrier Caused by Decarboxylation of Acetic Acid

An obvious way to overcome the large rate decrease caused by the carboxylic acid in *p*-toluic acid is to increase the temperature and pressure. Interestingly, this does not work because the observed rate of oxidation decreases above 100°C (see Table 2). We have also performed pilot plant Co/2-butanone catalyzed oxidations of *M*-xylene and obtained a 57% yield of isophthalic acid at 104°C which decreased to 12% yield at 208°C. Extrapolation of the reaction rates in Table 2 to 150°C gives essentially zero rates. There is a temperature

Table 2 Rates of Oxidation of Various Hydrocarbons with Cobalt Catalysts in Acetic Acid as a Function of Temperature

Temperature (°C)	Relative rates			
	p-Xylene[a]	Benzaldehyde[b]	Cyclohexane[c]	*p*-Xylene[e]
60	1.4	—	—	—
70	3.9	14.9	—	—
80	4.4	13.2	—	—
90	1.8	15.4	Fastest rate (dark green)	—
95	—	15.1	—	—
100	1.0	14.3	—	0.6
105	—	12.9	—	—
110	—	11.5	—	—
120	—	—	—	2.2
130	—	—	Decreased rate (reddish orange)	—
140	—	—	—	2.3
160	—	—	—	0.8
extrapolated[d]				
150	.02	.05	—	

[a]William F. Brill, *Industrial and Engineering Chemistry, 52*(10), 837 (1960). Promoted with 2-butanone. Relative rates.
[b]Measured in our laboratory. Maximum rate of a batch oxidation using a concentration of 0.94 M in benzaldehyde and 0.020 M in cobalt(II) acetate.
[c]Kyugo Tanaka, *Chemtech, 556* (Sept. 1974). He reports that the color changes to reddish orange at 130°C in contrast to a dark green below 100°C. This is indicative of the steady-state concentration of cobalt(III) decreasing as the temperature increases.
[d]Extrapolated for *p*-xylene by taking the log of the last three points as a function of temperature, correlation coefficient 97%. Extrapolated value for benzaldehyde taking the last three points, correlation coefficient 100%.
[e]*Source*: From Ref. 27.

barrier which limits the rates at which cobalt-catalyzed autooxidations can occur.

The temperature barrier is illustrated on Fig. 3. At the top of the figure we have autooxidations where the decarboxylation of the solvent is not important. This is indicated by the thickness of the arrows; most of the cobalt(III) is being used to make the radical of *p*-toluic acid. The bottom of the figure illustrates the effect of

Table 3 Rates of Reduction of Cobalt(III) in Acetic Acid and Pertinent Oxygenation Data on Benzaldehyde[a]

$$Co(III)(OAc)_3 \xrightarrow[N_2]{HOAc} Co(II)(OAc)_2 + CH_3COO \qquad (1)$$

$$C_6H_5CHO + O_2 \xrightarrow{Co} C_6H_5COOH + H_2O \qquad (2)$$

	Reaction (1)[a]	Reaction (2)[b]			
Temp. (°C)	Half-life (min)	Rate (ml O_2/min)	% Co(III) (ss)	COx (mmol)	CH_4 (mmol)
70	—	14.9	88	3.3	0.063
80	145	—	—	—	—
90	—	15.4	—	6.7	.151
95	—	15.1	68	6.4	.18
100	12	14.3	—	12.3	.47
105	—	12.9	51	6.6	.34
110	—	11.5	—	12.1	1.1
		extrapolated[b]			
150	0.1	.5	8	—	—

[a]Reaction conditions for benzaldehyde given in Table 2.
[b]Rate of reaction of Co(III) with acetic acid using an activation energy of 33.1 kcal/mol. Extrapolation of rate data as described in Table 2. Steady-state value obtained by plotting Co(III)ss versus temperature, correlation coefficient 90%.

going to too high a temperature. Acetic acid now competes successfully for the cobalt(III) moiety with *p*-toluic acid. Most of the cobalt(III) is now being wasted in the decarboxylation of the solvent. Table 3 gives the half-life of Co(III) as a function of temperature in acetic acid. At 80°C, half of the Co(III) is decomposed in 145 min while at 150°C this value is 0.10 min. This model predicts decreasing steady-state concentration of cobalt(III), increased vent carbon oxides, and increased vent methane as the temperature is increased for a given reaction. This is indeed observed for benzaldehyde (see Table 3).

It is often implied that the selectivity of a Co-catalyzed oxygenation is higher than a Co/Mn/Br oxidation since the former are normally run at lower temperatures. This is not true. As will be seen, the vent oxides of a Co-catalyzed oxidation drastically decreases when bromide is added to it. Co/Mn/Br oxidations are run at higher temperatures than Co-catalyzed oxygenations because there is a drastic change in mechanism (see later) which allows them to be. On the other hand, Co-catalyzed oxygenations are run under relatively mild conditions because they have to be to avoid excessive solvent decomposition.

IV. TEREPHTHALIC ACID PROCESSES BASED ON COOXIDATION

A. Cooxidation with *p*-xylene and Esterification: The Witten Process

Often, but not always, one can enhance the rate of autooxidation of an unreactive substrate by oxidizing it with a more reactive one. We have seen that *p*-xylene is much more reactive than *p*-toluic acid hence *p*-xylene could be used as a cooxidant to further oxidize a given amount of *p*-toluic acid to terephthalic acid. We can visualize a process as follows:

Step 1: Oxidize *p*-xylene to a mixture of *p*-toluic acid and terephthalic acid. The yield of terephthalic acid will be around 15%.

Step 2: Take the above mixture and add *p*-xylene to it and oxygenate the mixture again. This is a cooxidation. The amount of terephthalic acid will increase in the mixture. The yield of terephthalic acid based on the original *p*-xylene added will be much higher.

Step 3: Separate the terephthalic acid from the *p*-toluic acid.

Step 4: Take the *p*-toluic acid, separated in step 3, and cooxidize with *p*-xylene, i.e., go back to step 2.

The process described above does not work well since *p*-toluic and terephthalic acids are not easily separated. The separation does become relatively easy if one esterifies the mixture in step 3 since distillation can now be used. This process, called the Witten process, produces dimethylterephthalate which competes significantly with the Amoco MC process.[26]

B. Cooxidation with Aliphatic Aldehydes and Ketones

The Witten process, described above, uses *p*-xylene to cooxidize the *p*-toluic acid to produce terephthalic acid. An alternative route is to add a reagent that oxidizes the cobalt(II) to cobalt(III), which keeps the steady-state concentration of cobalt(III) high and pushes the reaction to completion. This in effect, replaces the cobalt(III) lost due to decarboxylation reactions. Acetaldehyde, paraldehyde, and 2-butanone have been used to cooxidize the cobalt.[27,28]

C. Effect of Catalyst Concentration on the Decarboxylation of Acetic Acid

The cobalt-catalyzed autooxidations are characterized by high concentrations of cobalt, usually around 0.1 M.[27,28] It is well established that cobalt(III) exists in a number of different forms in acetic acid[29] and that these involve monomers and polymers of cobalt(III) and mixed-valence Co(II)/Co(III) complexes. The data in Fig. 4 suggest one reason why such high cobalt concentrations are used. Addition of cobalt(II) acetate to cobalt(III) greatly decreases the rate of reaction of cobalt(III) with the solvent. Essentially, cobalt(II) has modified the catalyst structure of cobalt(III), presumably by polymer formation, and decreased the decomposition of solvent.

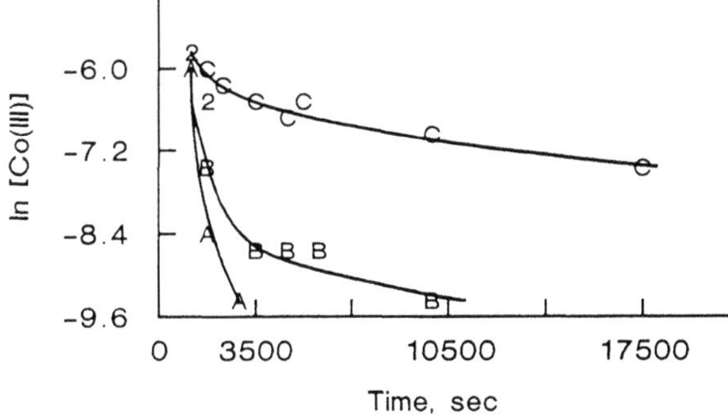

Fig. 4 The effect of Co(II) on the rate of Co(III) acetate decomposition in acetic acid. Initial Co(III) conc. = 0.004 M at 95°C. A = no Co(II) added, B = 0.016 M Co(II) added, C = 0.080 M Co(II) added.

V. EFFECT OF BROMIDE ON COBALT-CATALYZED OXIDATIONS

A. Effect of Bromide Addition to a Cobalt-Catalyzed Oxygenation

We oxygenated *p*-xylene using a cobalt(II) acetate catalyst in a glass flask.[33] After 2.5 hr an equal molar amount of sodium bromide to cobalt is added to the flask. A spectacular change occurs:

	Before bromide addition	After bromide addition
Color of solution	Dark green	Blue
% of Co as Co(III)[30]	11	0.6 (approx.)
Rate of oxygenation, ml O_2/min	1.3	5.0
Vent carbon dioxide, %	0.36	0.061

The color change and reduction in the steady-state Co(III) concentration suggests a significant change in either degree or kind in the mechanism. The rate of reaction increases by a factor of 3.8 while the vent carbon dioxide decreases. The latter suggests that the selectivity of the oxygenation has also increased. This is because many of the byproducts formed in autooxidations are caused at least initially by decarboxylation and decarbonylation reactions (see Sec. III.B).

B. Relative Rates of Co(III) with Acetate, *p*-Xylene, and Bromide

Figure 5 illustrates that the rate of reduction of cobalt(III) is essentially instantaneous with HBr at 80°C, while acetate and acetate with *p*-xylene react at least 200 times slower.

C. A Rationalization of the Observations

Co(III) acetate in acidic water, having a redox potential of 1.9 V, is one of the strongest oxidants known in chemistry. Part of the efficacy of cobalt as an autooxidation catalyst is its strong oxidizing power since it must produce a highly unstable methylbenzene radical in order to initiate the oxidation. However, in acetic acid, cobalt is strongly bound with acetate ligands and, presumably, acetic acid ligands as well. This intimate contact of the ligands with such a "hot" metal leads to a significant competition of the cobalt(III)

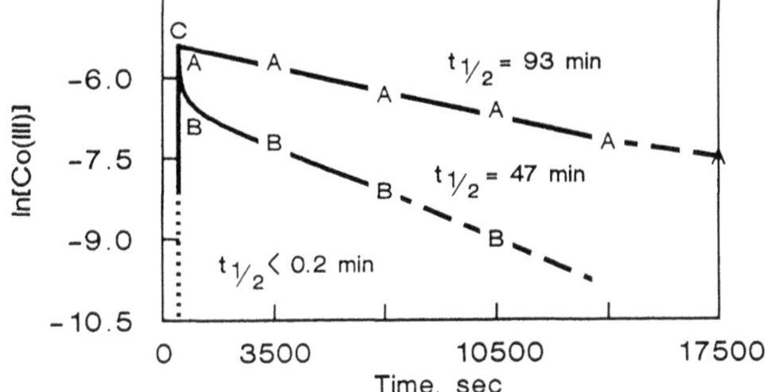

Fig. 5 The rates of Co(III) reduction in (A) acetic acid, (B) with *p*-xylene present, (C) with hydrobromic acid. At 80°C in a nitrogen atmosphere. Initial concentrations: Co(III) = 0.0040 M, *p*-xylene = 0.81 M, HBr = 0.0056 M.

with the solvent as well as with methylbenzene. The activation energies are apparently such that the cobalt(III)-acetate oxidation becomes more favorable than the cobalt(III)-methylbenzene oxidation at temperatures higher than about 130°C (the temperature barrier). What we need is a stable ligand that will complex with cobalt(III) and rapidly become oxidized so that the hot cobalt(III) lives for only a short period of time in acetic acid.

Measured stability constants in acetic acid of cobalt(II) with bromide are large[9] so that most of the bromide is coordinated to the cobalt:

$$Co(II)(OAc)_2 + HBr \rightleftharpoons Co(II)(OAc)(Br) + HOAc \qquad K = 800 \tag{25}$$

We postulate that the oxidation of the cobalt bromide complex will produce a transitory Co(III) complex which rapidly undergoes a intra-molecular electron transfer to a Co(II)bromine atom complex.[40] Either this dissociates or reacts directly with the methylbenzene to produce the radical:

$$HOAc + Co(II)(OAc)(Br) + (O) \longrightarrow Co(III)(OAc)_2Br \tag{26}$$

$$Co(III)(OAc)_2Br + HOOC_6H_4CCH_3 \longrightarrow Co(II)(OAc)_2 + HBr + HOOCC_6H_4CH_2\bullet \tag{27}$$

The important point is that this is a very rapid process as compared to cobalt(III) oxidation of either acetate or *p*-xylene (Fig. 5). The observations then can be rationalized:

1. The color changes from green to blue because the spectrum in the visible region changes from domination by Co(III) complexes to that by tetrahedral Co(II) complexes.[32]
2. The steady-state Co(III) concentration decreases because of the rapid intramolecular electron transfer to bromide.
3. Fewer carbon oxides are seen in the vent since the steady-state cobalt(III) concentration is decreased.
4. The catalyst is more active since fewer byproducts are formed via decarboxylation, which caused deactivation to occur (Sec. III.B). It is highly probable that the bromide radical (or its complex with cobalt or manganese) becomes a chain carrier and also increases the rate.[34]
5. The decreased steady-state concentration of cobalt(III) means that the temperature barrier can be raised. Indeed, cobalt/bromide oxidations typically operate in commercial conditions at 190-210°C while cobalt/aldehyde oxidations operate at 100-140°C.[3]
6. The much greater activity and higher temperature barrier means that one can use much lower catalyst concentrations in cobalt/bromide oxidations than in cobalt/acetaldehyde oxidations.

VI. EFFECT OF MANGANESE ON COBALT/BROMIDE-CATALYZED OXIDATIONS

A. Observations

Why is manganese added to Co/Br-catalyzed autooxidations in the oxidation of polyalkylbenzenes? When some of the cobalt was replaced by an equal molar amount of manganese(II) acetate, the following effects were reported (if one compares Secs. VI.A and V.A, one sees that the effect of bromide addition to cobalt is very similar to substitution of some of the cobalt by manganese):

1. There is a synergistic increase in the rate of oxidation. This was first reported by Ravens[36] using *p*-xylene as the reagent. Data from our own laboratory is given on Table 4. Using *p*-toluic acid as the reagent, a 4.3-fold rate increase

Table 4 Effect of Manganese Substitution for Cobalt on the Rate of Oxidation of *p*-Toluic Acid[a]

(Co), M	(Mn), M	(Br), M	Rate, ml O_2/min
0.013	0.0	0.013	0.66 ± 0.22
0.0	0.013	0.013	0.58 ± 0.10
0.0065	0.0065	0.013	2.83 ± 0.25

[a]At 95°C, *p*-toluic acid concentration 2.45 M.

is obtained when one-half of the cobalt is replaced by manganese.[37]

2. There is less loss of acetic acid due to decarboxylation. This has been reported for *p*-xylene oxidations using tracer studies where 5-10% of the cobalt is replaced by manganese.[37]
3. The catalyst cost is decreased. The cost of manganese is usually around a factor of 10 less than cobalt; hence there is an obvious economic incentive to reduce the cobalt concentration.
4. The catalyst concentration can be decreased. This is obvious from the enhanced rate of oxidation when some cobalt is replaced by manganese.

The decrease in catalyst concentration allows for easier processing steps during the isolation and purification of the aromatic acids. Most aromatic acids are substantially insoluble in acetic acid under 50°C. Hence, the usual problem quoted for homogeneous catalytic processes—separation of the product from the catalyst—is not a significant problem for most MC-type oxidations. Normally, but not always, 85-95% of the catalyst stays in solution upon cooling. However, the remaining 5-15% of the catalyst with the product solids always needs to be removed for most commercial applications of the aromatic acid. Less catalyst concentration in the reactor will normally mean less catalyst on the product solids; hence, a purer product and less processing loss of the catalyst components.

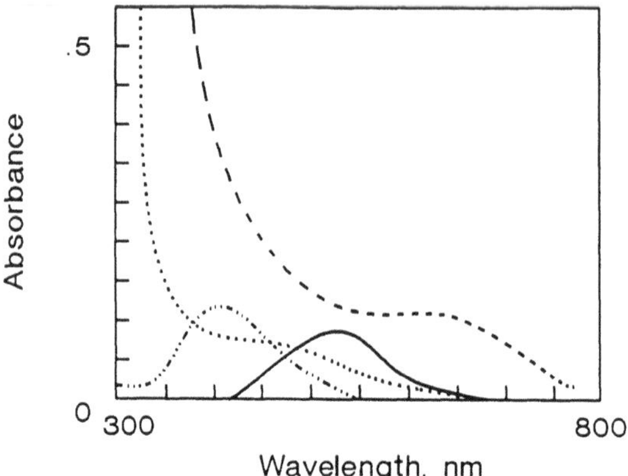

Fig. 6 The spectra of Co(III) acetate (0.0005 M, - - -), Co(II) acetate (0.0050 M, ——), Mn(III) acetate (0.0010 M, •••••), dibromine (0.0005 M, -•-•-). Mn(II) is transparent at these concentrations.

B. Rates of Some Selected Reactions of Co(III), Mn(II), Mn(III), and Br^-

These reactions are easily monitored via UV-VIS since the spectra of these species are very different (see Fig. 6). As previously discussed, Co(III) acetate reacts with itself to produce Co(II) acetate, carbon dioxide, and the methyl radical. Figure 7 illustrates this reaction as cobalt(III) has a half-life of 14 min at 100°C and the solution changes color from green to pink.

If one mixes equal molar amounts of cobalt(III) acetate with Mn(II) acetate in acetic acid at 100°C, there is an instantaneous color change from green to brown as the Mn(II) is oxidized by the Co(III) (see Fig. 7).[38]

The half-life of Mn(III) acetate in acetic acid is 790 min at 100°C.[39] This is easily observed since Mn(III) is an intense brown color and Mn(II) is essentially colorless. So what happens to the rate of acetic acid decomposition when Mn(III) is added to a solution of cobalt(II)? Cobalt(III) will be instantly reduced and the rate of acetic acid decomposition will decrease by a factor of 790/14 = 56.

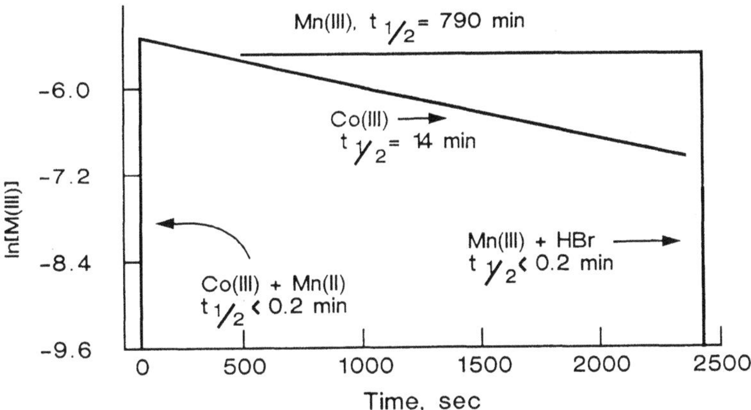

Fig. 7 Selected rates of species at 100°C under nitrogen. Initial concentrations: Co(III) acetate = 0.0045 M, Mn(II) acetate = 0.0089, HBr = 0.00113 M.

Finally, we illustrate in Fig. 7 that Mn(III) is instantaneously reduced by hydrobromic acid, just as cobalt(III) is. Thus, we have shown that the production of dibromine (and indirectly the bromine atom or metal-bromine atom complex) occurs rapidly with both Co(III) and Mn(III) in acetic acid but that the solvent decomposition will be less by a factor of 56 when the two metals are mixed together.

C. Some Reasons for the Increase in Selectivity and Activity

A catalytic cycle illustrating some of the functions of the Co/Mn/Br catalyst is given in Fig. 8.

1. The increase in selectivity is obvious since addition of manganese lowers the solvent and aromatic acid decomposition rates. Many of the byproducts (e.g., methyl acetate, methane, and carbon dioxide; see Sec. III.B) are caused by the solvent and aromatic acid decomposition.
2. The increase in activity occurs since less of the peroxide used to form cobalt(III) is "wasted" in solvent decomposition. More of it goes to generate the radical of *p*-toluic acid.

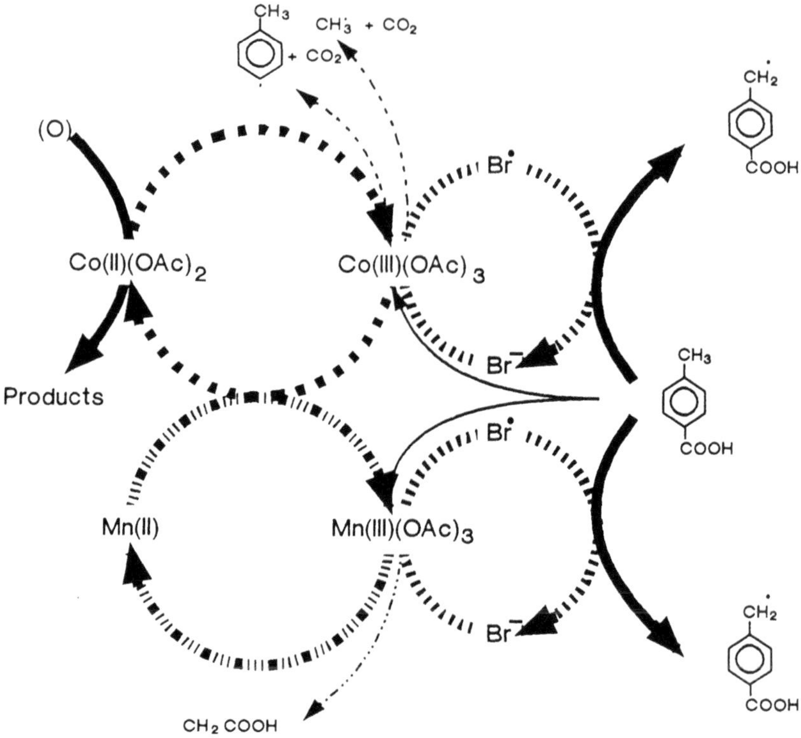

Fig. 8 Catalytic cycles for a Co/Mn/Br catalyst. The relative broadness of the lines indicates their relative importance.

VII. Summary

The summary is illustrated by a series of barriers given in Fig. 9. We started by giving thermodynamic and kinetic arguments why radicals can't form under 300°C. We then showed that there is a reaction of cobalt (and only cobalt) with acetic acid which can generate radicals of alkyl aromatics (*p*-xylene). When the first methyl group on *p*-xylene is oxidized, the rate of reaction diminishes greatly due to electronic deactivation of the ring. Various deactivation mechanisms result in termination of the oxygenation. Attempts to increase the temperature to increase the rate results in another stumbling block—the decarboxylation of acetic acid by cobalt. This problem is over-

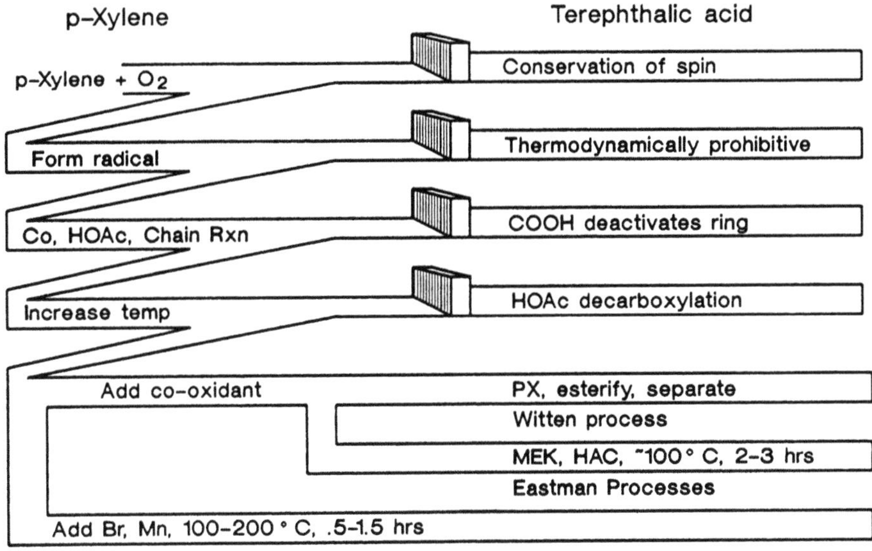

Fig. 9 An illustration describing the problems in autooxidation and how they have been overcome.

come by using cooxidants—*p*-xylene in the Witten process and acetaldehyde in the Eastman process. Finally, bromide is added to cobalt, which gets around the problem of acetic acid and aromatic acid decarboxylation by rapid electron transfer from cobalt to bromide. Manganese operates much the same way as bromide since another rapid electron transfer reaction occurs between cobalt and manganese and thus decreases decarboxylation reactions further.

NOTES AND REFERENCES

1. CEH Marketing Research Report by Carolyn S. Hughes, SRI International Chemical Economics Handbook, 695.4021 A, February, 1984.

2. "Autooxidation" is a generic term for the interaction of dioxygen with organic molecules. The primary product from this interaction is the peroxide, which then usually reacts further to give alcohols, aldehydes, and carboxylic acids.

3. K. Weissermel and H.-J. Arpe, *Industrial Organic Chemistry*, Verlag-Chemie, Weinheim, p. 342.

4. U.S. Patent 2,833,816.

5. Mid-Century is often abbreviated "MC"; hence, the process is often referred to as Amoco's MC process.

6. S. W. Benson and P. S. Nangia, *Acc. Chem. Res., 12,* 224 (1979).

7. Roger A. Sheldon and Jay K. Kochi, *Metal-Catalyzed Oxidations of Organic Compounds,* Academic Press, New York, 1981.

8. F. A. Cotton and G. Wilkinson, *Advanced Inorganic Chemistry,* 4th ed., Wiley-Interscience, New York, pp. 490-491.

9. K. Sawada and M. Tanaka, *J. Inorg. Nucl. Chem., 39,* 339 (1977). We have confirmed this observation using a bromide-selective electrode. The equilibrium constant does drastically decrease if 5% water is present in the acetic acid. This was also reported in Ref. 29.

10. S. W. Benson, *Thermochemical Kinetics, Methods for the Estimation of Thermochemical Data and Rate Parameters,* 2nd ed., Wiley-Interscience, New York, 1976, p. 204.

11. Reference 3, p. 56. The thermal dissociation of hydrocarbons to radicals is of great importance since it is the major industrial method for the preparation of olefins.

12. Ref. 7, Chap. 2.

13. S. W. Benson and P. S. Nangia, *Acc. Chem. Res., 12,* 224 (1979).

14. See, for example, K. Sakota, Y. Kamiya, and N. Ohta, *Can. J. Chem., 47,* 387 (1969).

15. A. Nakmura and M. Tsutsui, *Principles and Applications of Homogeneous Catalysis,* Wiley-Interscience, New York, 1980, p. 57. Another possible route is to form the $CH_3O_2\cdot$ radical which can subsequently react with the *p*-xylene. I would guess this is also a highly unselective species because of its relatively high free-energy content.

16. N. M. Emanuel, E. T. Denisov, and Z. K. Maizus, *Liquid-Phase Oxidation of Hydrocarbons,* Plenum Press, New York, 1967, p. 71.

17. See Ref. 8, Sec. 4-24.

18. P. Roffia, P. Callini, and L. Motta, *Ind. Eng. Chem. Prod. Res. Dev., 23,* 629 (1984). This reference describes the oxidation of *p*-xylene using a Co/Mn/Br catalyst, but references quoted therein indicate that similar byproducts form in cobalt-catalyzed autooxidations. We have also found this to be true in cobalt-catalyzed autooxidations promoted with methyl ethyl ketone.

19. J. Dermietzel, C. Wienhold, H. Grundmann, A. Staschok, J. Koch, and E. Bordes, *Chem. Tech., 35*(1), 29 (1983).

20. E. I. Heiba, R. M. Dessau, and W. J. Koehl, Jr., *J. Am. Chem. Soc., 91,* 24 (1969).

21. W. H. Starnes, Jr., *J. Org. Chem.*, *31*, 1436 (1966); Sheldon S. Lande and Jay K. Kochi, *J. Am. Chem. Soc.*, *90*, 19 (1968).

22. See Ref. 7, p. 29. We have confirmed this in our laboratory by spiking autooxidations with phenols.

23. W. Partenheimer, U.S. Patent 4,719,311.

24. The usual equation quoted (Ref. 7, p. 23) gives a 1/2 power dependence of the initiation rate on the rate of hydrocarbon disappearance. However, cooxidations, which are always present in autooxidations, give different expressions. See C. Walling, *J. Am. Chem. Soc.*, *91*, 27 (1969).

25. A variety of di- and tricarboxylic acids of diphenyl are found in autooxidations of *p*-xylene. See Ref. 18.

26. The Amoco MC process produces terephthalic acid rather than dimethyl terephthalate. The ratio of terephthalic acid to dimethyl terephthalate being produced commercially has been steadily increasing over the years. By 1988, the ratio should be about equal (see Ref. 1).

27. K. Nakaoka, Y. Miyama, S. Matsuhisa, and S. Wakamatsu, *Ind. Eng. Chem. Prod. Res. Develop.*, *12*(2), 150 (1973).

28. W. F. Brill, *Ind. Eng. Chem.*, *52*(10), 837 (1960).

29. M. G. Roelofs, E. Wasserman, J. H. Jensen, A. E. Nader, *J. Am. Chem. Soc.*, *198*, 4207 (1987).

30. An aliquot is removed from the reactor, diluted with acetic acid, and the UV-VIS spectrum immediately measured on a Hewlett-Packard 8451A diode array spectrometer. Using the accompanying HP software a curve fit is performed using solutions of known amounts of cobalt(II) and cobalt(III) acetate (the latter prepared via oxone).

31. Values of 40-80% are often quoted; see Refs. 27 and 28.

32. The color is caused by a few percent of cobalt-bromide complexes existing in the tetrahedral form as evidenced by the characteristic UV-VIS absorptions around 700 nm.

33. Experimental conditions were 1.9 M *p*-xylene, 0.080 M cobalt(II) acetate catalyst at 95°C and 0.42 M in 2-butanone. After 2.5 hr an equal molar amount of sodium bromide to cobalt is added to the flask.

34. The increased activity could be also because bromine radicals may be involved in the propagation step. Decomposition of acetaldehyde in the gas phase to give CO + CH_4 is strongly catalyzed by HBr in a typical chain fashion. It increases the rate by a factor of 300. W. D. Bardsley, R. L. Falles, R. Hunter, and V. R. Stimson, *J. Chem. Ed.*, *61*(8), 657 (1984).

35. See Ref. 3, p. 344.

36. D. A. S. Ravens, *J. Chem. Soc., 55,* 1768 (1959).

37. T. P. Kenigsberg, N. G. Ariko, N. I. Mitskevich, and V. F. Nazimok. Translated from *Kinetika i Kataliz, 26*(6), 1485 (1985).

38. These observations are consistent with the redox potential in acidic aqueous solution for Cobalt(III) of 1.8 V and manganese(III) of 1.50 V both from the M(II) state.

39. The decarboxylation of acetic acid with Mn(III) apparently produces a carbon-centered radical, $\cdot CH_2COOH$. See, for example, W. E. Fristad and J. R. Peterson, *J. Org. Chem., 50,* 12 (1985).

40. In 2 M H_2SO_4 aqueous media, the redox potential of Co(III) is 1.8 V. Using the redox potential of $Br^- \rightarrow Br_2$ of -1.08 V, the dissociation energy of dibromine of 46.3 kcal/mol, and ignoring solvation energies, we obtain a value of 1.9 V for the process $Br^- \rightarrow Br\cdot$. The formation of bromine radicals (atom) is hence not favored in water or, we assume, acetic acid.

Index

Milton Keynes UK
Ingram Content Group UK Ltd.
UKHW020016071024
449327UK00032B/2811